NELSON

VICscience

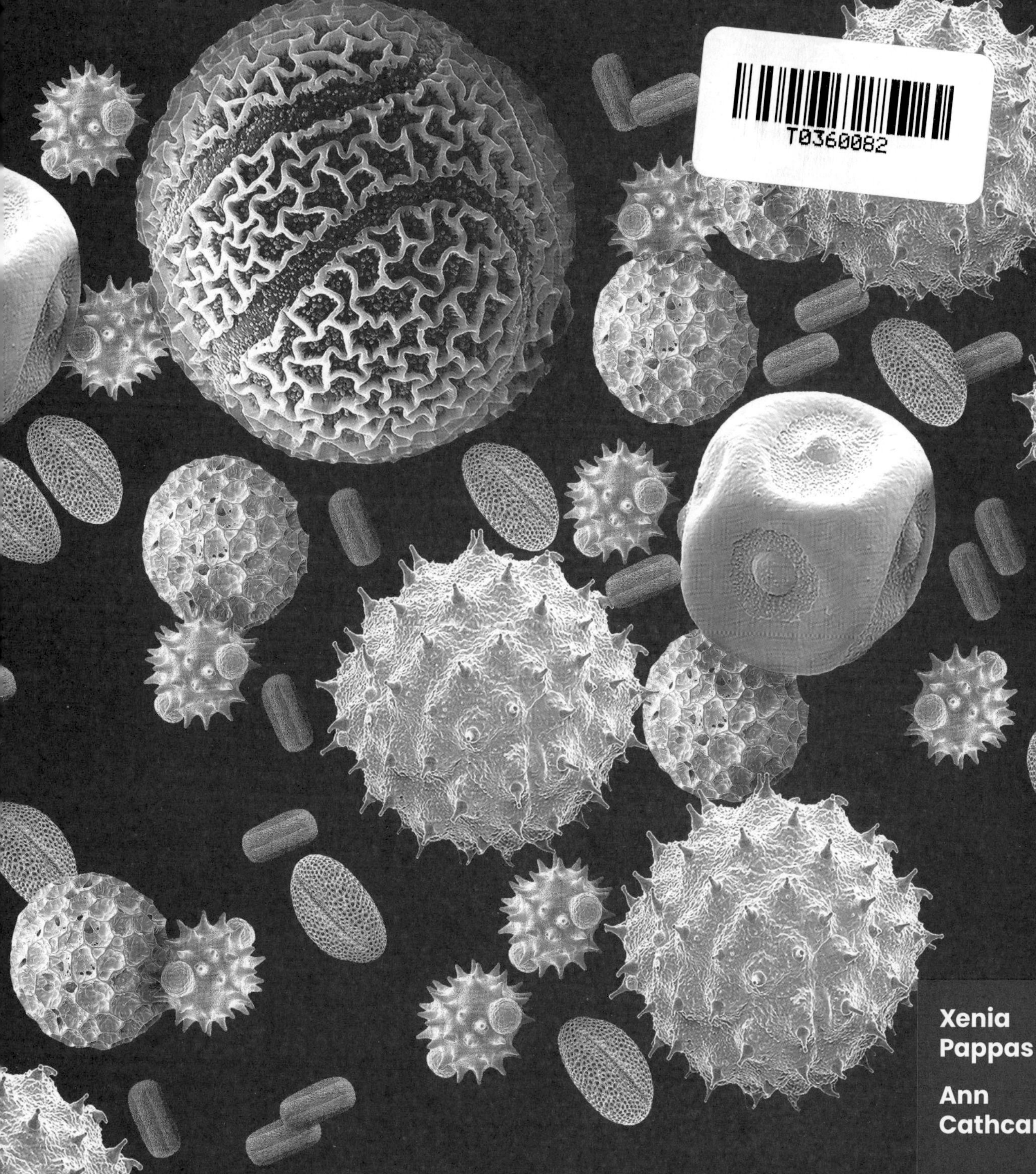

Xenia Pappas

Ann Cathcart

SKILLS WORKBOOK

biology

VCE UNITS 1 + 2

Nelson VICscience Biology Units 1 & 2 Skills Workbook
1st Edition
Xenia Pappas
Ann Cathcart
Julie Gould
ISBN 9780170452632

Publisher: Eleanor Gregory
Editor: Felicity Clissold
Cover design: Leigh Ashforth (Watershed Art & Design)
Text design: Alba Design
Project designer: James Steer
Permissions researcher: Wendy Duncan
Typeset by: SPi Global
Production controller: Renee Tome

ISBN 978 0 17 045263 2

Cengage Learning Australia
Level 7, 80 Dorcas Street
South Melbourne, Victoria Australia 3205

Cengage Learning New Zealand
Unit 4B Rosedale Office Park
331 Rosedale Road, Albany, North Shore 0632, NZ

For learning solutions, visit **cengage.com.au**

Printed in China by 1010 Printing International Limited.
3 4 5 6 7 25 24 23

Contents

4 Regulation of systems 86

5 Scientific investigations 125

6 Chromosomes to genomes 148

Introduction

Biology is the study of living organisms, their structure, function, growth, evolution, where they live and how they live. Like the study of any other science there is key knowledge and terminology that you need to know and understand and be able to use appropriately. However, no study of Biology would be complete without also addressing the key science skills. The study of Biology is not just about learning content, it is also about developing, using and demonstrating the skills that enable you to fully understand, experience and engage with the subject. It is about learning to think and work like a scientist.

Some of the key science skills can only be used and demonstrated in a laboratory situation but many can be developed, used and demonstrated away from the laboratory, and it is these skills that are the focus of this workbook.

Seven key science skills have been mandated by the Victorian Curriculum Assessment Authority (VCAA) across all VCE science subjects. These key science skills are transferable across subjects as well as being examinable in the VCE exam. Developing these key science skills means that you will be able to do the following.

- Develop aims and questions, formulate hypotheses and make predictions
- Plan and conduct investigations
- Comply with safety and ethical guidelines
- Generate, collate and record data
- Analyse and evaluate data and investigation methods
- Construct evidence-based arguments and draw conclusions
- Analyse, evaluate and communicate scientific ideas

(*VCAA VCE Biology Study Design 2022–2026* pages 7–9)

Each of the key science skills listed above is broken up into multiple sub-skills. The mapping provided on pages vii–x of this workbook allows you to see how these skills and sub-skills have been addressed in this workbook.

This workbook follows the structure of the *VCE Biology Study Design 2022–2026*. It is full of activities that have been carefully crafted to enable you to consolidate your knowledge on a topic and to develop, use and demonstrate key science skills. Developing any skill takes time and practice; the key science skills in this book have been introduced in a graduated way starting with **practising** skills that have been introduced to in previous years of science study. As you gain proficiency and confidence, you will go on to **reinforce** newer and more complex skills. Then there are the new skills requiring an increased level of proficiency and thinking that you will **develop** during the course.

Activities often require a mixture of skill levels for completion. Each activity has been signposted with an icon to indicate the highest level of skill required to complete the activity.

Shows you activities that require previously introduced skills and will require practice as you work through the activities.

Shows activities that will build on previously introduced skills.

Shows activities that introduce a new skill or skills that require development and challenge you at a high level of proficiency.

This workbook can be used with any VCE Biology textbook that covers the *VCAA VCE Biology Study Design 2022–2026*. It has been mapped to *VICscience Biology Units 3 & 4* using icons in both the workbook and the student textbook. The icons have been placed in the workbook notifying you of the pages in the student textbook where corresponding content occurs, and conversely icons have been placed in the student textbook to indicate the best place to undertake each activity.

The major headings in the workbook match the major headings in *VICscience Biology Units 3 & 4*. Applicable key knowledge is listed under each of the major headings. Skill activities show the corresponding key skills listed that students will be using or demonstrating.

Acknowledgements

Many people have been involved in the production of this workbook, and special thanks go to:

Aunty Zeta Thompson is a respected Elder and descendant of the Yarra Yarra Clan of the Wurundjeri people on her paternal side and a descendant of the Ulupna Clan of the Yorta Yorta people on her maternal side. Aunty Zeta provided invaluable information and anecdotes about Aboriginal culture and life that formed the basis of several activities in this workbook.

Rebecca Farmlonga is a proud Wadawurrung woman and Traditional Owner. Rebecca has taught and led in secondary schools for more than 20 years and is passionate about Aboriginal and Torres Strait Islander Education. Rebecca reviewed the activities relating to Aboriginal and Torres Strait Islander peoples.

Julie Gould teaches VCE Biology at Brunswick Secondary College. Julie reviewed all the activities in the workbook and wrote the answers that appear in the back of this book.

Key science skills grid

Key science skill	VCE Biology Units 1 & 2	Chapters									
		1	2	3	4	5	6	7	8	9	10
Develop aims and questions, formulate hypotheses and make predictions Practise	• identify, research and construct aims and questions for investigation	1.1.1	2.1.2	3.1.2		5.1.1			8.1.1	9.1.2 9.3.1 9.3.2	
	• identify independent, dependent and controlled variables in controlled experiments		2.1.2	3.1.2	4.3.2	5.1.1			8.1.1 8.1.2		
	• formulate hypotheses to focus investigation	1.1.1	2.1.2	3.1.2	4.3.2	5.1.1			8.1.1 8.1.2	9.3.2	
	• predict possible outcomes	1.5.1 1.5.2	2.1.2	3.1.2		5.1.1			8.1.1	9.3.1 9.3.3 9.4	
Plan and conduct investigations Practise	• determine appropriate investigation methodology: case study; classification and identification; controlled experiment; correlational study; fieldwork; literature review; modelling; product, process or system development; simulation		2.1.3			5.2.1			8.1.1	9.1.2	
	• design and conduct investigations; select and use methods appropriate to the investigation, including consideration of sampling technique and size, equipment and procedures, taking into account potential sources of error and uncertainty; determine the type and amount of qualitative and/or quantitative data to be generated or collated		2.1.3		4.3.2	5.2.2			8.1.1		
	• work independently and collaboratively as appropriate and within identified research constraints, adapting or extending processes as required and recording such modifications										
Comply with safety and ethical guidelines Reinforce	• demonstrate safe laboratory practices when planning and conducting investigations by using risk assessments that are informed by safety data sheets (SDS), and accounting for risks		2.1.3		4.3.2	5.3.1			8.1.1	9.3.2	
	• apply relevant occupational health and safety guidelines while undertaking practical investigations										
	• demonstrate ethical conduct when undertaking and reporting investigations										

Key science skill	VCE Biology Units 1 & 2	Chapters									
		1	2	3	4	5	6	7	8	9	10
Generate, collate and record data Reinforce	• systematically generate and record primary data, and collate secondary data, appropriate to the investigation, including use of databases and reputable online data sources			3.1.4		5.3.1					
	• record and summarise both qualitative and quantitative data, including use of a logbook as an authentication of generated or collated data			3.1.2	4.3.2	5.3.1					
	• organise and present data in useful and meaningful ways, including schematic diagrams, flow charts, tables, bar charts and line graphs	1.1.2 1.3.2 1.6.3	2.4.2	3.1.3 3.1.4 3.1.5 3.2.1 3.2.2 3.3.1 3.4.1	4.2.3 4.2.5 4.3.1 4.3.2 4.5.1		6.4.1		8.2.2	9.2.1 9.2.2 9.3.1 9.3.2 9.3.3	10.2.3
	• plot graphs involving two variables that show linear and non-linear relationships		2.1.4		4.2.5 4.3.2		6.4.1		8.1.2		10.2.3
Analyse and evaluate data and investigation methods Develop	• process quantitative data using appropriate mathematical relationships and units, including calculations of ratios, percentages, percentage change and mean	1.1.2		3.1.4	4.2.2 4.3.2	5.3.2 5.5.1	6.4.1	7.1.1	8.1.2		10.2.3
	• identify and analyse experimental data qualitatively, handling where appropriate, concepts of: accuracy, precision, repeatability, reproducibility and validity of measurements; errors (random and systematic); and certainty in data, including effects of sample size in obtaining reliable data			3.1.4	4.2.2 4.3.3	5.3.2	6.1.4			9.3.2	10.2.3
	• identify outliers, and contradictory or provisional data						6.3.1		8.1.2		10.2.3
	• repeat experiments to ensure findings are robust										
	• evaluate investigation methods and possible sources of personal errors/mistakes or bias, and suggest improvements to increase accuracy and precision, and to reduce the likelihood of errors		2.1.4								10.2.3

Key science skill	VCE Biology Units 1 & 2	Chapters									
		1	2	3	4	5	6	7	8	9	10
Construct evidence-based arguments and draw conclusions Develop	• distinguish between opinion, anecdote and evidence, and scientific and non-scientific ideas									9.1.2	10.2.1 10.2.2
	• evaluate data to determine the degree to which the evidence supports the aim of the investigation, and make recommendations, as appropriate, for modifying or extending the investigation		2.1.4					7.2.1	8.1.2	9.3.2	
	• evaluate data to determine the degree to which the evidence supports or refutes the initial prediction or hypothesis	1.1.1									
	• use reasoning to construct scientific arguments, and to draw and justify conclusions consistent with the evidence and relevant to the question under investigation		2.2.1 2.2.2 2.2.3	3.2.3 3.4.2	4.2.2		6.2.4		8.1.2		
	• identify, describe and explain the limitations of conclusions, including identification of further evidence required		2.4.1								
	• discuss the implications of research findings and proposals										
Analyse, evaluate and communicate scientific ideas Develop	• use appropriate biological terminology, representations and conventions, including standard abbreviations, graphing conventions and units of measurement	1.1.1 1.5.1 1.5.2		3.2.4			6.2.1 6.2.3	7.1.1 7.1.2 7.1.3 7.3.1 7.3.2 7.4.1 7.5.1 7.5.2			
	• discuss relevant biological information, ideas, concepts theories and models and the connections between them	1.3.1			4.3.3 4.4.2		6.4.2		8.2.1		
	• analyse and explain how models and theories are used to organise and understand observed phenomena and concepts related to biology, identifying limitations of selected models/theories	1.1.1 1.1.2 1.3.2 1.4.1	2.2.2		4.2.5 4.3.3 4.5.3		6.1.1 6.4.1				
	• critically evaluate and interpret a range of scientific and media texts (including journal articles, mass media communications and opinions in the public domain), processes, claims and conclusions related to biology by considering the quality of available evidence				4.6.1		6.1.3				10.4

Key science skill	VCE Biology Units 1 & 2	Chapters									
		1	2	3	4	5	6	7	8	9	10
	• analyse and evaluate bioethical issues using relevant approaches to bioethics and ethical concepts, including the influence of social, economic, legal and political factors relevant to the selected issue		2.3.1 2.5.1			5.4.1	6.1.4 6.1.5		8.3.1		10.1.1 10.3 10.4
	• use clear, coherent and concise expression to communicate to specific audiences and for specific purposes in appropriate scientific genres, including scientific reports and posters	1.6.1 1.6.2	2.1.1 2.4.1		4.1.1 4.3.3	5.6.1	6.1.1			9.5	10.4
	• acknowledge sources of information and assistance, and use standard scientific referencing conventions			3.2.4							10.1.2 10.4

1 Cellular structure and function

Getty Images/DE AGOSTINI PICTURE LIBRARY/Contributor

Remember

In your study of science in Years 7–10, you have already been introduced to content that you will now build upon in Unit 1 Biology. Take some time to refresh your knowledge of this content before you enter this chapter. Try to answer the following questions from memory. If you cannot do this, then use a reference to assist you.

PAGE 6

1 What are cells?

2 What are organelles?

3 Name as many organelles that are found inside cells as you can. If you know what each organelle does, then write this down as well.

4 What is an organism?

5 Distinguish between unicellular organisms and multicellular organisms.

6 Name the six kingdoms used to classify living organisms.

9780170452632

1.1 Cells are the basic structural unit of life

Key knowledge

Cellular structure and function

- cells as the basic structural feature of life on Earth, including the distinction between prokaryotic and eukaryotic cells

1.1.1 The cell theory

Key science skills

Develop

Develop aims and questions, formulate hypotheses and make predictions

- identify, research and construct aims and questions for investigation
- formulate hypotheses to focus investigation

Construct evidence-based arguments and draw conclusions

- evaluate data to determine the degree to which the evidence supports or refutes the initial prediction or hypothesis

Analyse, evaluate and communicate scientific ideas

- use appropriate biological terminology, representations and conventions, including standard abbreviations, graphing conventions and units of measurement
- analyse and explain how models and theories are used to organise and understand observed phenomena and concepts related to biology, identifying limitations of selected models/theories

Biology is the pursuit of knowledge and understanding of the natural world. This knowledge follows a systematic investigation based on the scientific method and supported by evidence. During your studies in biology you will be presented with many opportunities to undertake scientific investigations using the scientific method. As part of these investigations, you are going to meet the terms 'research question' and 'hypothesis, and 'theory' and 'law'. These terms are not interchangeable; they have their own specific meaning and need to be used in the correct way. Read the definitions below, and then use this knowledge to answer the sets of questions that follow.

Research question

The research question is the question that you want to answer by undertaking scientific investigation. A good research question:

- is specific
- guides the design of the investigation
- identifies the independent and dependent variables that will be investigated
- provides criteria for judging whether your results have answered the research question
- is feasible; that is, answerable with the time and equipment available.

Hypothesis

A hypothesis is written as a statement and is a tentative explanation that is based on an existing theory or model about what you think the outcome of the investigation is going to be. A hypothesis is written before the investigation is undertaken. A good hypothesis:

- is framed as a prediction
- is falsifiable (it can be refuted)
- cannot be proven; rather it is supported
- can be written as: If ... *the independent variable is changed* ... then ... *something will happen to the dependent variable.*

1 You have found a green gelatinous-like substance in a puddle in your backyard. You are not sure what it is but think it may be a living organism. Write a research question that you could use to investigate this substance further.

2 You have discovered that this green gelatinous-like substance is a living thing but you cannot decide if it is a plant or an animal. As it is green, you guess that it might be a plant. Write a hypothesis that you could investigate to find out if this blob is a plant or an animal.

Theory

Nothing is 100 per cent certain. A theory is an explanation of an observation that is overwhelmingly supported by tested hypotheses. Remember, you cannot prove a hypothesis, so you also cannot prove a theory. You have probably heard of the theory of evolution by natural selection. This was proposed by Charles Darwin in 1859 and was based on a large amount of evidence that was collected by Darwin over many years of work. So far, no scientific evidence has been presented that disproves Darwin's theory; however, his theory *has* been adjusted over the years in light of new evidence resulting from new technologies such as the electron microscope and gene sequencing.

The theory of spontaneous generation was first proposed by Aristotle in 384–322 BCE and it stood the test of time for the next 1900 years. The theory proposed that living organisms developed from non-living matter. This theory was supported by evidence such as when you left out pieces of cheese and bread wrapped in rags, mice appeared. If you left meat out on the bench, maggots appeared. The theory was disproved by Redi in 1668 with one experiment (Figure 1.1).

Figure 1.1 **Redi's experiment disproved the theory of spontaneous generation.**

3 Predict the results of the experiment shown in Figure 1.1 if the theory of spontaneous generation was supported.

4 Redi found that maggots appeared in Jar 1 and on the cloth that was used to seal Jar 3. Explain why Jar 1 contained maggots and Jars 2 and 3 did not.

5 Explain why these results refuted the theory of spontaneous generation.

One of the underlying theories of biological science is the cell theory. The cell theory states that:

- the cell is the basic structural and functional unit of an organism (Schwann and Schleiden, 1839)
- all new cells are produced from pre-existing cells (Virchow, 1849).

6 Consult other sources, such as your Biology textbook, and list the people and the evidence they discovered that contributed to the formulation of the cell theory.

Law

If you drop a pencil, you would hypothesise that it will fall to the ground. This is because over millennia, when people dropped things, they fell to the ground. Isaac Newton put this evidence together to formulate the law of universal gravitation. This law is supported overwhelmingly by the results of many investigations, so it has been accepted that this is how the natural world behaves.

A law is a detailed description of how some aspect of the natural world behaves. It usually involves mathematics. Laws are common in physics and chemistry, but biology has laws as well. You will meet the law of independent assortment and law of segregation when you study heredity later in this course.

7 What is the difference between a theory and a law?

1.1.2 Prokaryotic and eukaryotic cells

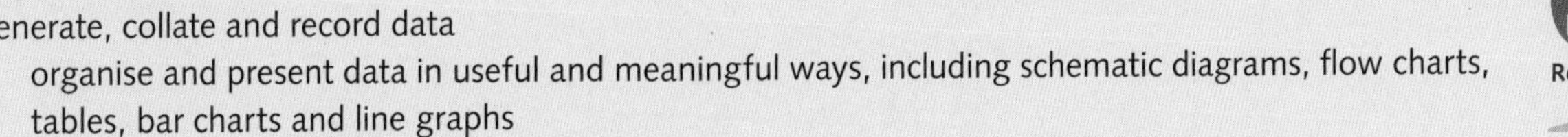

Key science skills

Generate, collate and record data

- organise and present data in useful and meaningful ways, including schematic diagrams, flow charts, tables, bar charts and line graphs

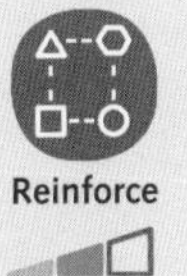

PAGE 9

The two types of cell structure in living things are prokaryotic and eukaryotic. Living things are grouped into kingdoms based on the structure of their cells. Prokaryotic cells are found in kingdom Eubacteria and kingdom Archaebacteria, and eukaryotic cells are found in the kingdoms called Animalia, Plantae, Fungi and Protista. Prokaryotic cells and eukaryotic cells have unique characteristics but also have some characteristics in common.

You can show these characteristics using a Venn diagram (Figure 1.2). On the left-hand side of the Venn diagram, list the characteristics that only prokaryotic cells possess. On the right-hand side of the Venn diagram, list the characteristics that only eukaryotic cells possess. Where the two circles of the Venn diagram intersect, list the characteristics that both prokaryotic and eukaryotic cells have. When you have completed these lists, answer the question below.

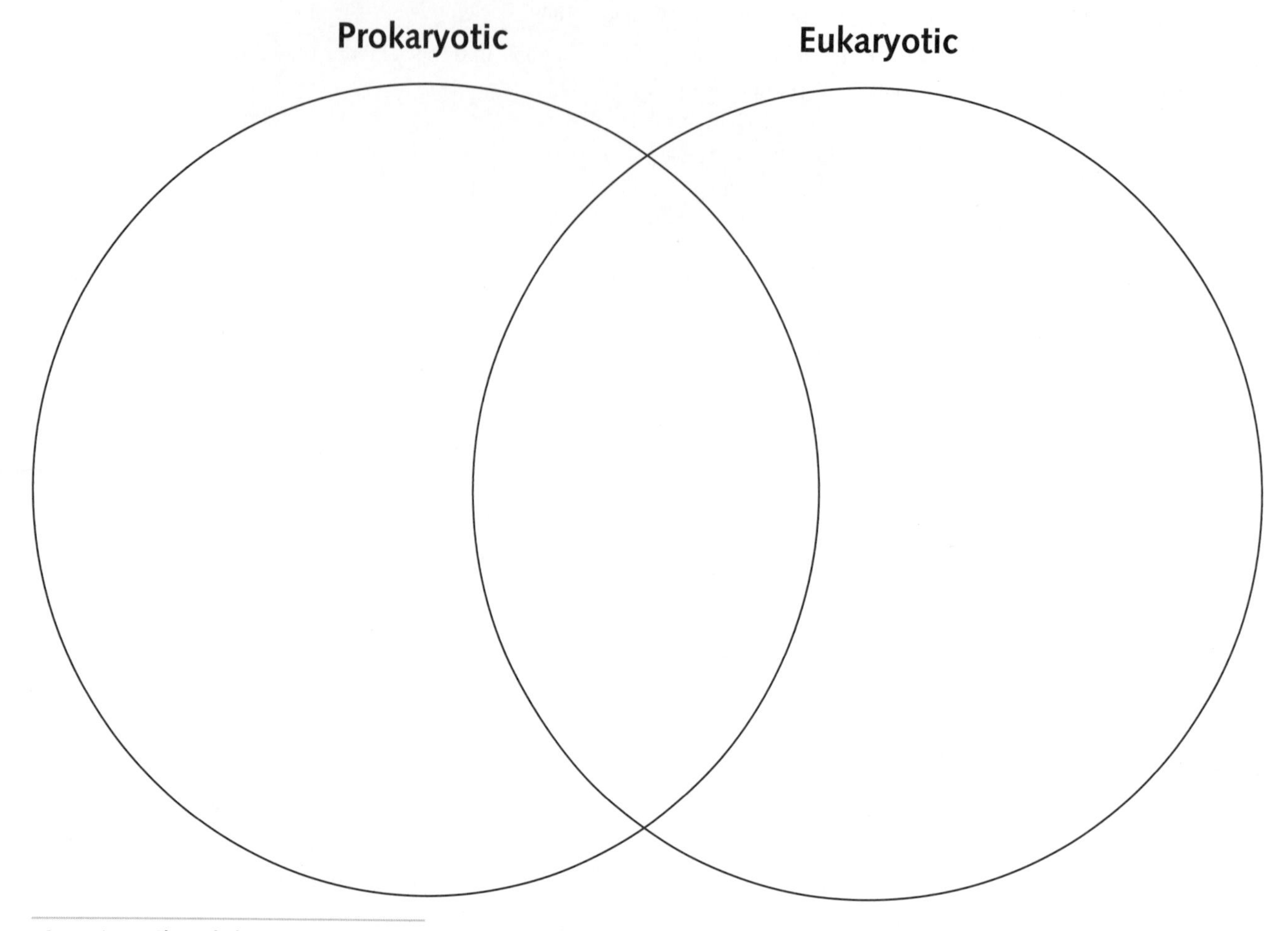

Figure 1.2 Shared characteristics of prokaryotic and eukaryotic cells

Question: Compare prokaryotic cells and eukaryotic cells.

Hint: When you compare, you describe how two things are the same and different. If you use your Venn diagram above to describe how the two types of cells are the same, you need to use the list you compiled in the overlapping area. To describe how the two types of cells are different, you need to use the lists on the left-hand side and the right-hand side.

9780170452632

1.2 Size and shape of cells

Key knowledge

Cellular structure and function

- surface area to volume ratio as an important factor in the limitations of cell size and the need for internal compartments (organelles) with specific cellular functions

1.2.1 Surface area to volume ratio

Key science skills

Analyse and evaluate data and investigation methods

- process quantitative data using appropriate mathematical relationships and units, including calculations of ratios, percentages, percentage change and mean

Analyse, evaluate and communicate scientific ideas

- analyse and explain how models and theories are used to organise and understand observed phenomena and concepts related to biology, identifying limitations of selected models/theories

TB PAGE 11

Cells are very small. You need a microscope to see them and you need an electron microscope to see them really well. They can be a variety of shapes depending on their function. Cells obtain all their requirements such as oxygen and glucose from outside. These requirements pass across the plasma membrane to get into the cell. The requirements are used by the organelles that are located within the cytosol of the cell.

As the organelles process the requirements, waste products are produced. These waste products cannot build up within the cell or they will get to a toxic level. They must be removed from the cell across its plasma membrane to outside the cell.

The plasma membrane is the surface area of the cell; the cytoplasm is the volume of the cell. For a cell to be able to function efficiently, it must have an optimal relationship between its surface area and volume. This is known as the surface area to volume ratio or SA:V.

Activity: modelling cells

Sometimes in science, the ideas or objects that are being studied are too big (the universe), too small (cells), too fragile (bubbles) or too dangerous (radioactive isotopes) to be able to study them effectively. In these cases, models are used to represent the object, system or idea. Models can be physical, virtual or mathematical; they can be manipulated and can be used to make predictions about how the idea or object would behave under certain conditions. In this activity, you will use physical models to represent cells. You will calculate their surface area to volume ratios and determine how these are affected by the shape of the cell.

It is important to remember that these models are not the objects; they are simplified representations of a much more complex object. As such, they are limited in what they can represent. You will investigate these limitations as part of this activity.

To complete this activity, you will need scissors and sticky tape or a glue stick, and access to a calculator (optional).

What to do

Step 1: Go to pages 9 and 11 to locate Figure 1.3a & b. Carefully cut out each shape.

Step 2: For each shape, calculate its surface area either by counting the number of square centimetres in each part of the object, or by using mathematics (SA = 2(length × width + width × height + length × height)). Enter your result for each shape into the second column of Table 1.1. Enter the correct units in the column header.

Step 3: Fold each shape along its fold lines and use sticky tape or glue to form the three-dimensional shape.

Step 4: For each shape, calculate its volume by using mathematics ($V = L \times W \times H$). Enter your result for each shape into the third column of Table 1.1. Enter the correct units in the column header.

Step 5: Show your results as SA:V in the fourth column of Table 1.1. Remember to show this ratio by using the smallest whole numbers possible; for example, 24:6 = 4:1.

Step 6: Answer the questions below.

Table 1.1 Results

Shape of cell	Surface area	Volume	SA:V
Cube			
Rectangle			

1 What effect did shape have on the SA:V of your cells?

2 Consider both of your cells. Which one would be able to gain nutrients and remove wastes more efficiently? Explain your answer.

3 List two ways that could enable cells to increase their SA:V to improve their ability to gain requirements and remove waste.

4 Explain one way that this model has helped you understand SA:V as it relates to cells.

5 Remember, models are representations of the objects, not the objects themselves. As such, they are not perfect in representing the actual object. State two limitations of this model.

a

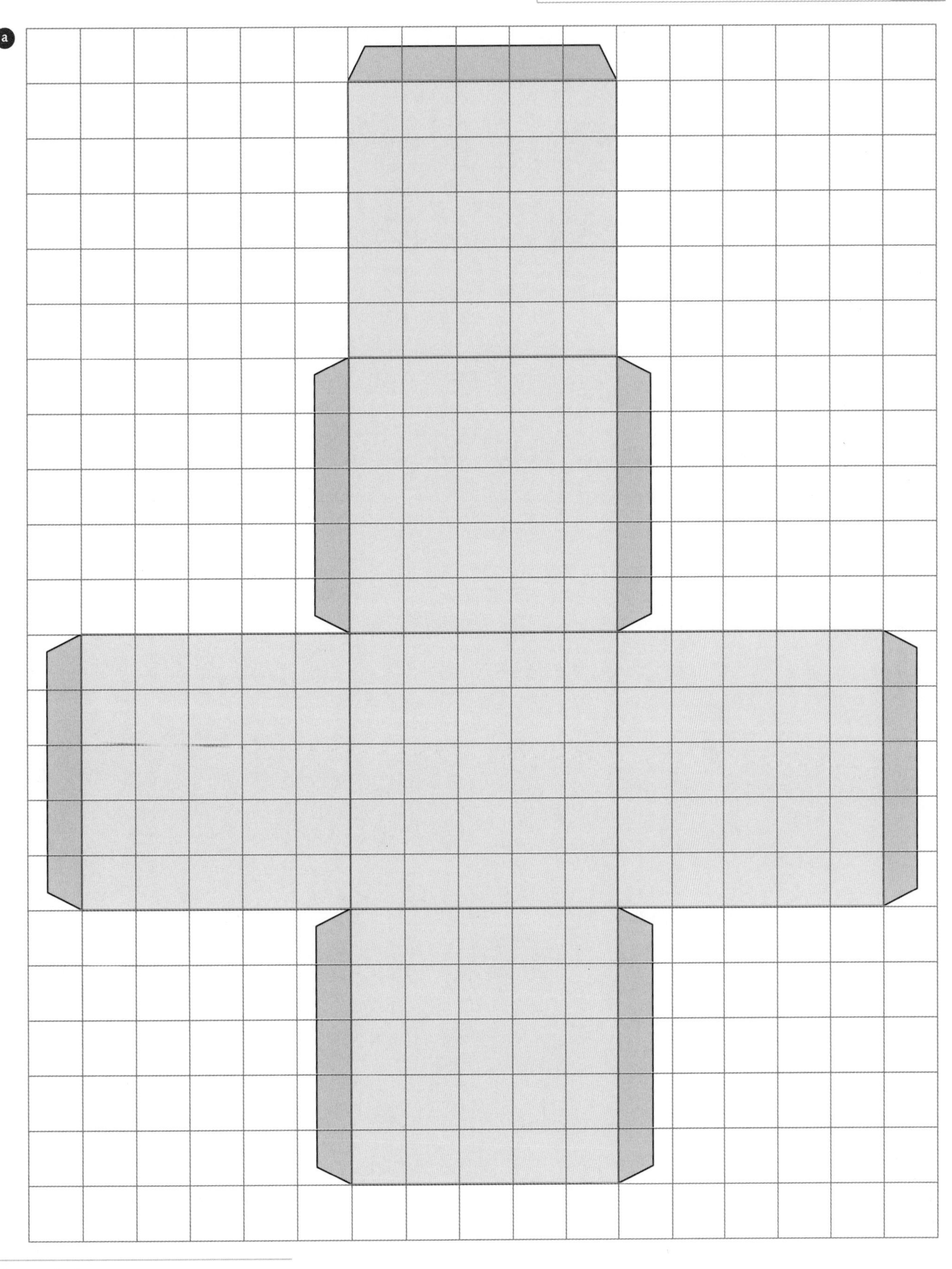

Figure 1.3a

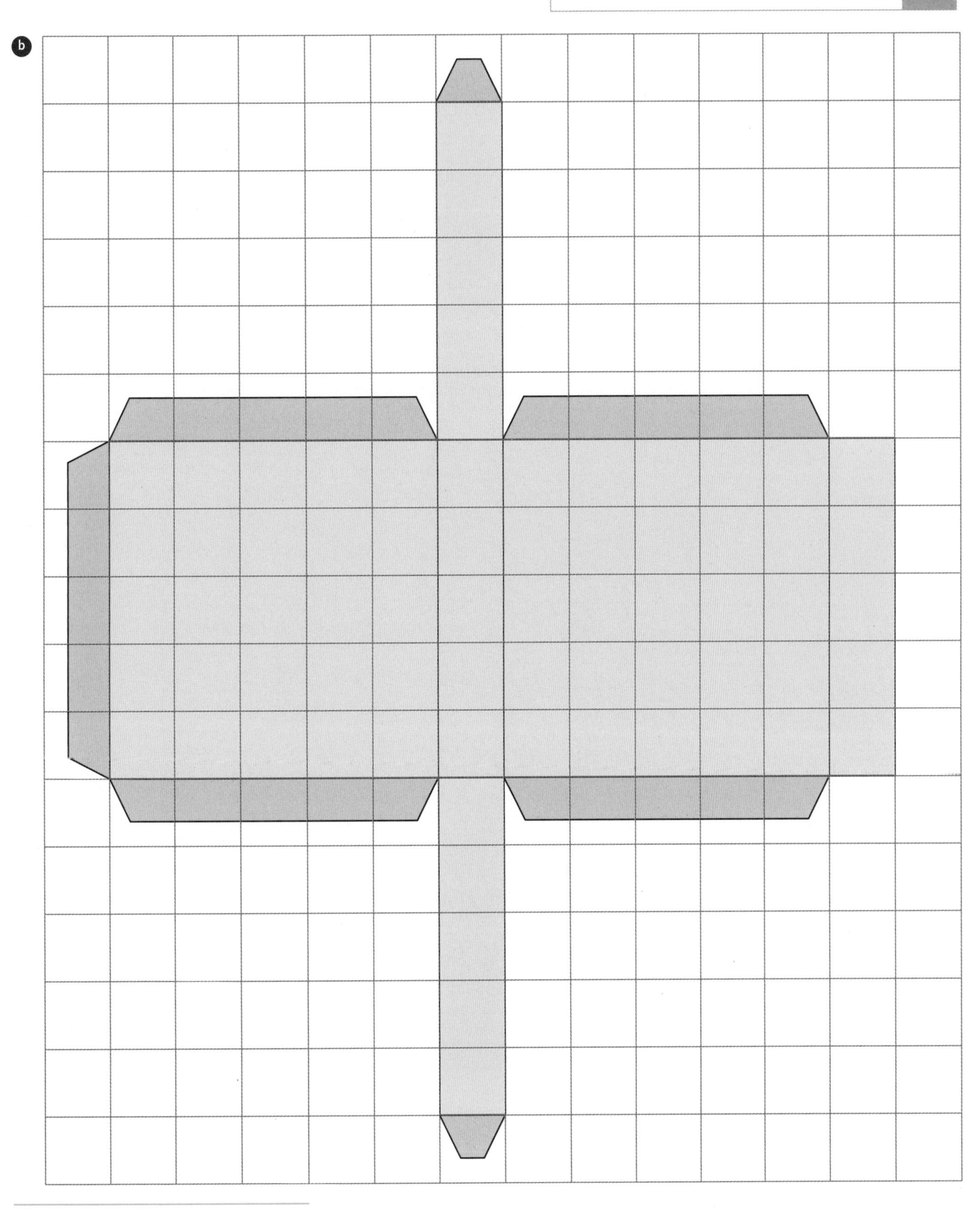

Figure 1.3b

1.3 What is inside a cell?

Key knowledge

Cellular structure and function

- the structure and specialisation of plant and animal cell organelles for distinct functions, including chloroplasts and mitochondria

1.3.1 Compartments in cells

Key science skills

Analyse, evaluate and communicate scientific ideas

- discuss relevant biological information, ideas, concepts, theories and models and the connections between them

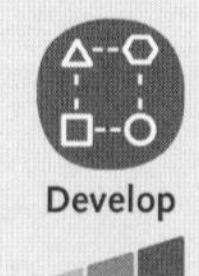

If you look at a plant leaf cell under an electron microscope, you will be able to see a prominent nucleus, a large vacuole, and organelles such as chloroplasts and mitochondria (Figure 1.4a). If you were to further magnify the chloroplasts and mitochondria, you would see their internal structure (Figures 1.4b & c). Carefully study Figure 1.4 and answer the questions that follow.

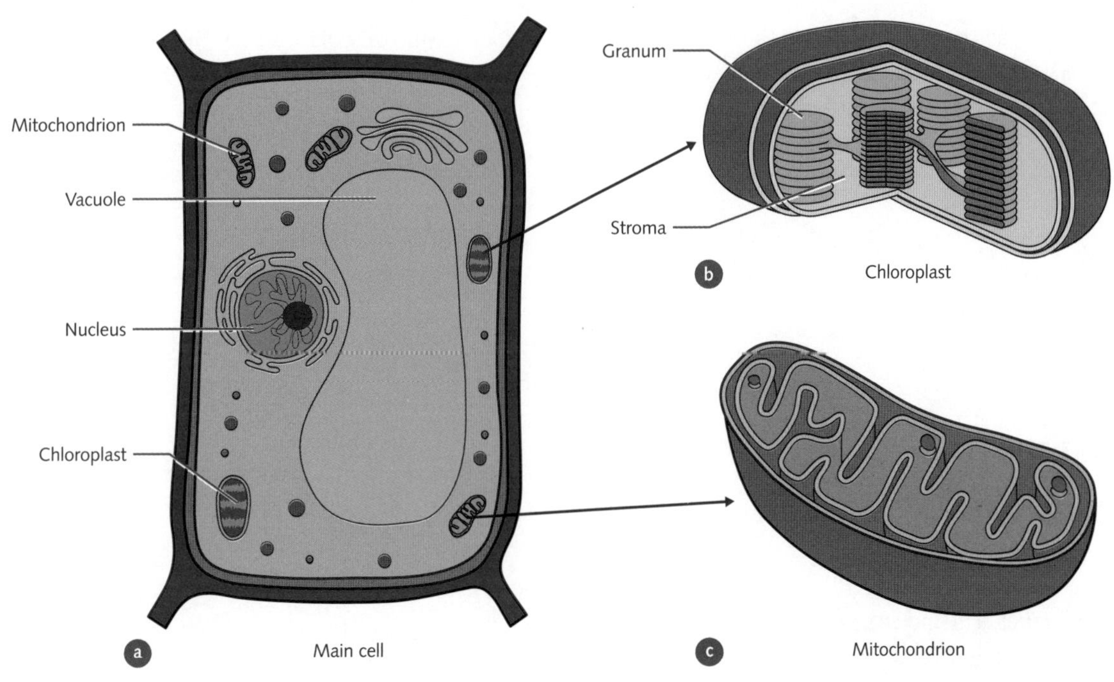

Figure 1.4

1 What function does the chloroplast perform in the plant cell?

2 What function does the mitochondrion perform in the plant cell?

3 List the functions that the membrane surrounding the chloroplast and mitochondrion perform.

4 Look at the structure inside the chloroplast (Figure 1.4b). Explain why the inside of the chloroplast is made up of many small flattened discs (called grana) rather than one large disc.

5 Look at the structure inside the mitochondrion (Figure 1.4c). Explain why the inside of the mitochondrion is made up of a folded membrane system (called cristae) rather than just a simple circle of membrane.

6 During the day, a plant cell can perform many functions simultaneously. It can photosynthesise, and at the same time release energy from glucose (respiration), produce proteins and transport protein products, to name just a few. Explain how this is possible in the small space inside a cell.

7 Prokaryotic cells have a diameter in the range of 0.1–5.0 micrometres (μm). Eukaryotic cells have a diameter in the range of 10–100 μm. Explain the difference between the size of these cells. In your explanation include discussion of SA:V and the presence/absence of organelles.

1.3.2 Organelles

PAGE 21

Key science skills

Generate, collate and record data

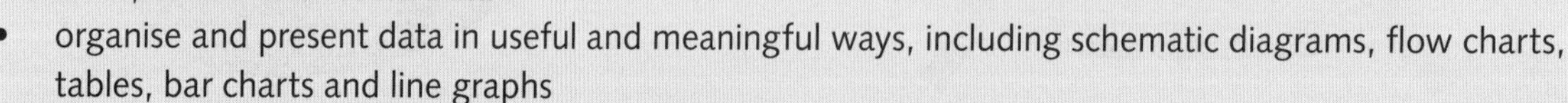

- organise and present data in useful and meaningful ways, including schematic diagrams, flow charts, tables, bar charts and line graphs

Analyse, evaluate and communicate scientific ideas

- analyse and explain how models and theories are used to organise and understand observed phenomena and concepts related to biology, identifying limitations of selected models/theories

Eukaryotic cells are able to take on a range of functions due to the compartmentalisation provided by organelles within the cell. Each organelle carries out its own specific function and is separated by membranes from other parts of the cell. This arrangement enables the cell to carry out many functions at the same time, ensuring that the needs of the cell, and the organism are met.

Activity: modelling a eukaryotic cell

This activity is a modelling activity where you will create your own specialised eukaryotic cells. To complete this activity, you will need scissors and a glue stick.

What to do

Step 1: Figures 1.5 and 1.6 on pages 16 and 17 present you with the outlines of an animal and a plant cell respectively.

Step 2: Figure 1.7 on page 19 shows a variety of organelles. Cut out these organelles.

Step 3: Use these figures to make the following specialised cells:

a a pancreatic cell that produces large amounts of insulin, the hormone that regulates blood sugar levels
b a leaf cell of plant that is actively photosynthesising.

Step 4: When you have arranged the organelles in each cell to meet the above requirements, glue the organelles in place.

Step 5: Use a pencil to draw any smaller organelles as required.

Step 6: Label your cells to show the names of each of the different types of organelles. Review and follow the rules for labelling scientific diagrams.

- Use a ruler to draw the lines.
- Do not use arrows for label lines.
- Ensure the label line points to the centre of the structure being labelled.
- Print all labels horizontally.
- Place all labels on the right-hand side of the diagram where possible.
- Do not cross label lines.

Using the cells you have made, answer the following questions.

1 Justify your choice of organelles in the pancreatic cell you have produced in terms of its specialised function.

2 Justify your choice of organelles in the plant cell you have produced in terms of its specialised function.

3 On page 18, create a Venn diagram to compare the structure and function of animal cells and plant cells. Remember, when you compare, you look at both the similarities and differences.

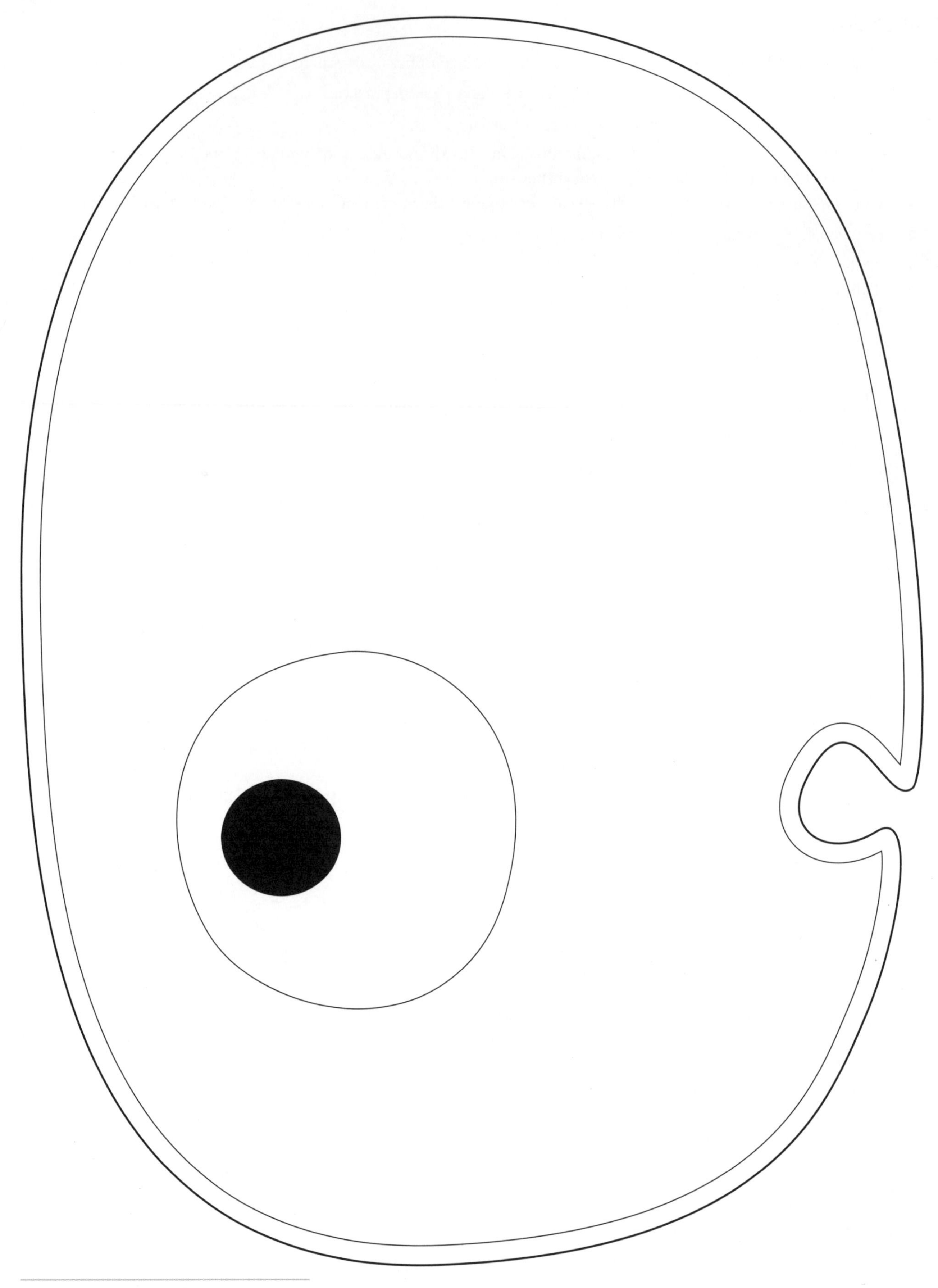

Figure 1.5 Animal cell

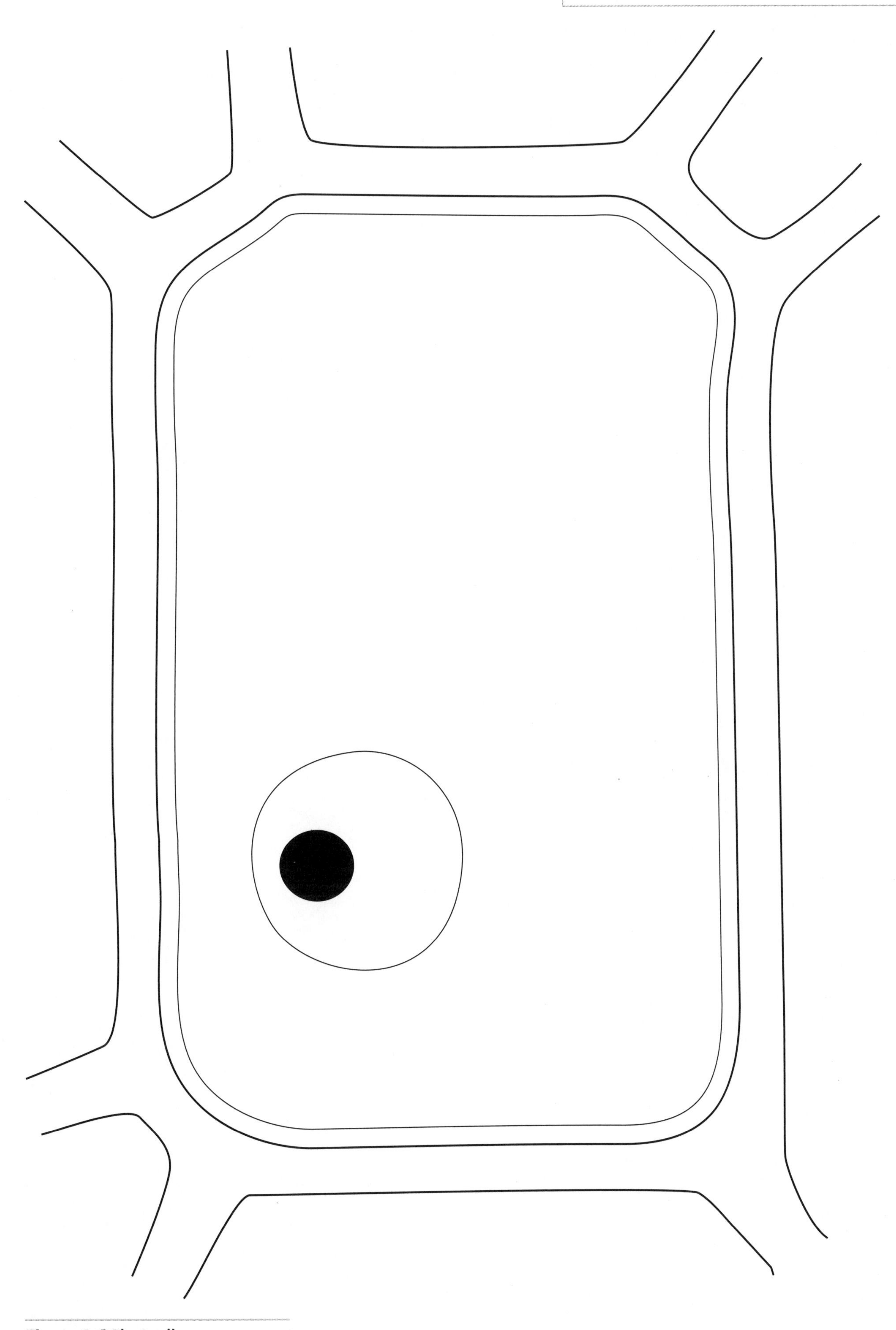

Figure 1.6 Plant cell

Activity: modelling a eukaryotic cell

Use this space to create your Venn diagram for question 3 on page 15: compare animal cells and plant cells.

9780170452632

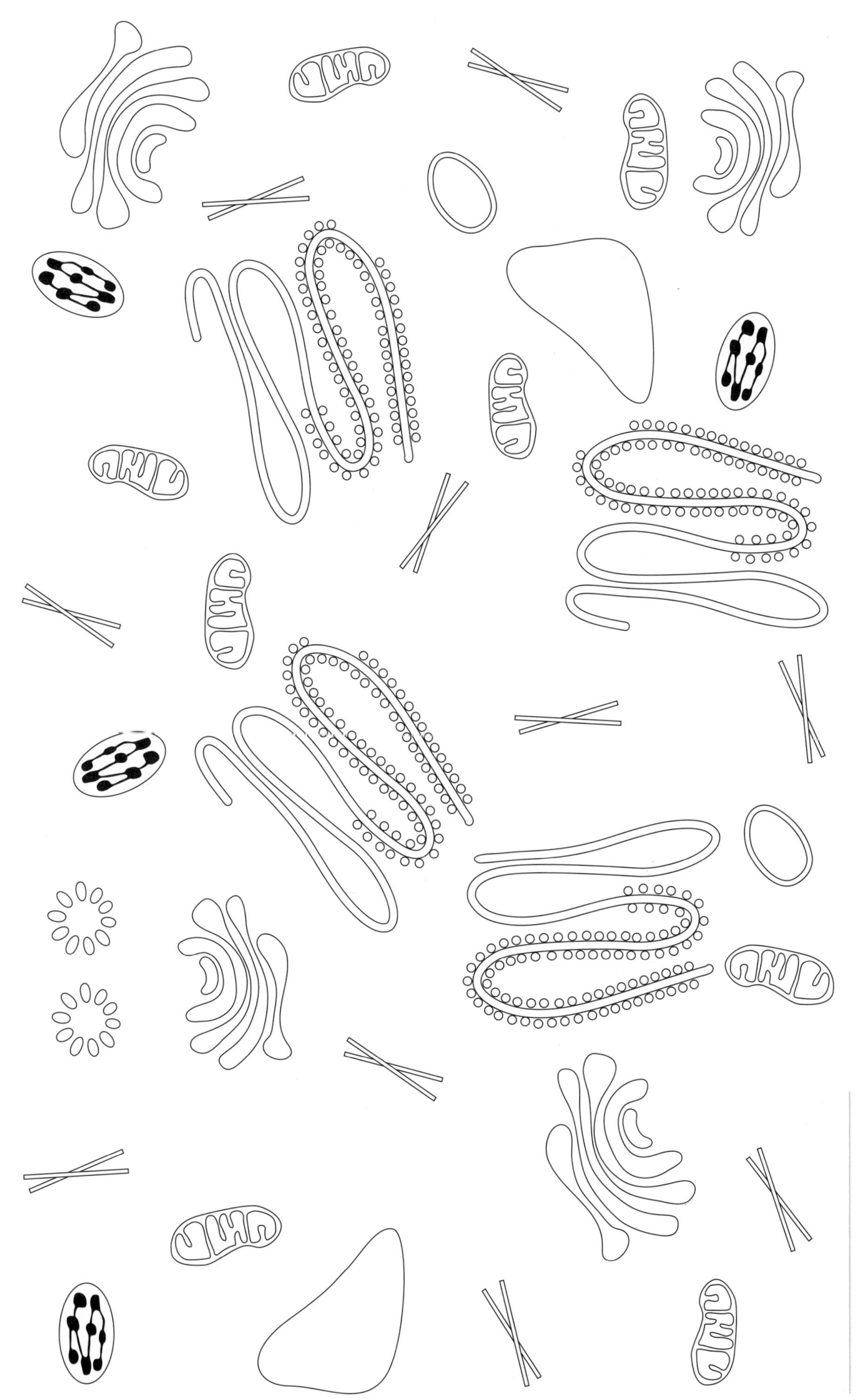

Figure 1.7 Organelles – cut these out and use them to create the specialised animal and plant cells on pages 16 and 17.

1.4 Plasma membrane

Key knowledge

Cellular structure and function

- the structure and function of the plasma membrane in the passage of water, hydrophilic and hydrophobic substances via osmosis, facilitated diffusion and active transport

1.4.1 Structure of plasma membrane

Key science skills

Analyse, evaluate and communicate scientific ideas

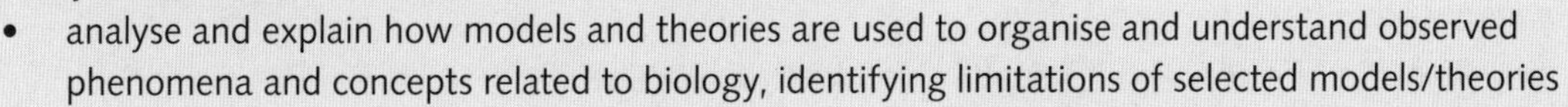

- analyse and explain how models and theories are used to organise and understand observed phenomena and concepts related to biology, identifying limitations of selected models/theories

TB PAGE 28

The plasma membrane forms the boundary between the internal environment of the cell (the cytoplasm) and its external environment. Figure 1.8 below represents the structure of the plasma membrane that surrounds cells. The structure of the plasma membrane is known as a fluid mosaic model because the embedded proteins make it look like a mosaic, and the movement of the phospholipids, proteins and cholesterol within the membrane cause it to flex and bend much like a fluid.

1 Consider Figure 1.8.

  a Label the hydrophobic and hydrophilic ends of the phospholipid molecules.

  b Label the phospholipid bilayer.

  c Draw the following proteins in the gaps in the membrane below: transport protein, receptor protein, recognition protein, adhesion protein. Label these and state each of their functions.

  d Add cholesterol to the phospholipid bilayer. State its function.

External environment of cell

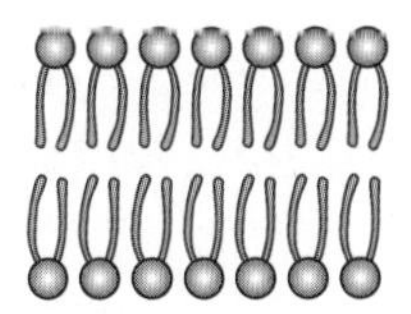

Internal environment of cell

Figure 1.8

2 The plasma membrane is selectively permeable. How does the structure of the plasma membrane give it this ability?

3 The hydrophobic ends of the phospholipid molecules face inwards to the middle of the plasma membrane. Explain why.

4 State two reasons why the plasma membrane needs to be flexible.

5 Do you think the fluid mosaic model is a good description of the plasma membrane? Explain your answer.

____

____

____

1.5 Passive movement across membranes

Key knowledge
Cellular structure and function
- the structure and function of the plasma membrane in the passage of water, hydrophilic and hydrophobic substances via osmosis, facilitated diffusion and active transport

1.5.1 Diffusion

TB PAGE 32

Key science skills
Develop aims and questions, formulate hypotheses and make predictions
- predict possible outcomes

Analyse, evaluate and communicate scientific ideas
- use appropriate biological terminology, representations and conventions, including standard abbreviations, graphing conventions and units of measurement

Develop

1 Complete the following sentences using words from the word list below. Some words can be used more than once.

concentration	gas	low
energy	gradient	solute
equilibrium	high	solvent

a Diffusion is the net movement of ______ or ______ particles from a region of ______ concentration to a region of ______ concentration without the addition of ______ until ______ is reached.

b Diffusion takes place when there is a ______ ______ between two areas in relation to the substance being diffused.

c Once ______ has been reached, the ______ or ______ particles will move randomly at equal rates in all directions.

d A solution is a mixture of a ______ and a ______.

2 Figure 1.9 shows two different types of gas particles in a confined space. Oxygen is represented by the pink circles and carbon dioxide is represented by the grey circles. Draw the location of the gas particles in the box on the right-hand side to show what it would look like when equilibrium is reached.

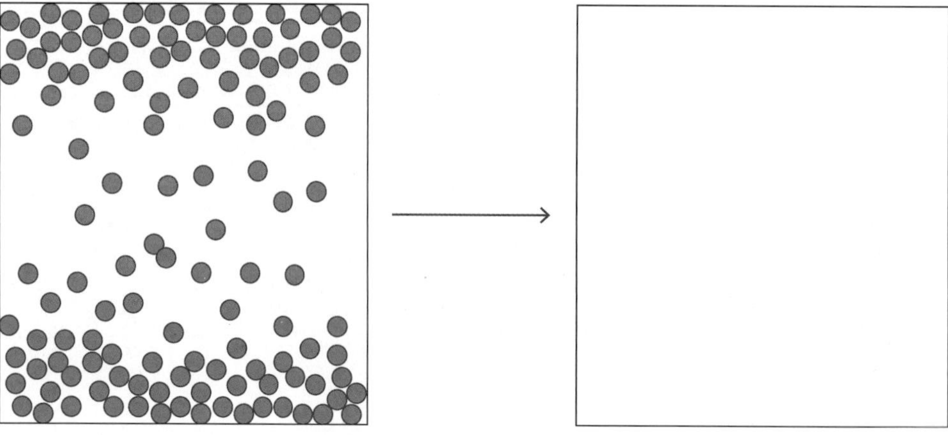

Figure 1.9

3 Figure 1.10 shows a spoonful of sugar about to be added to Cup A of water. The sugar (solute) dissolves in the water (solvent) to become a solution. Draw the locations of the solute particles in Cup B once the sugar has been added and when equilibrium is reached in Cup C.

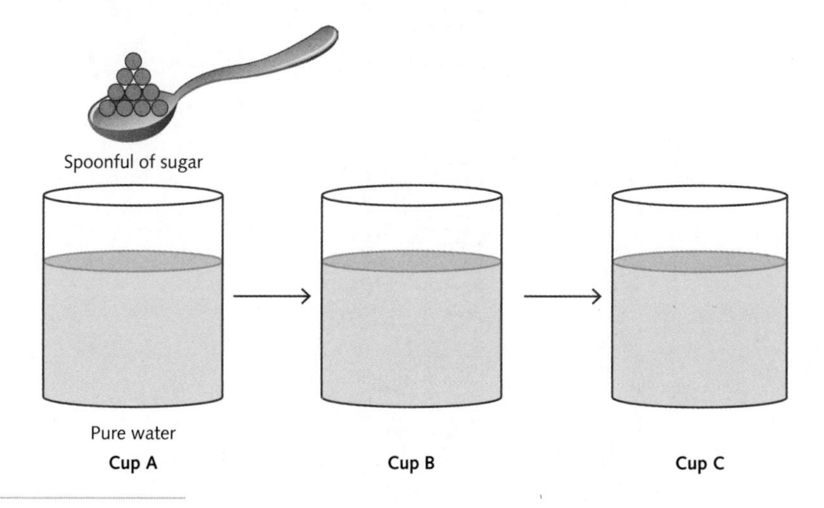

Figure 1.10

4 Diffusion also occurs through a plasma membrane. Oxygen moves into the cell and carbon dioxide moves out of the cell by the process of diffusion. Figure 1.11 represents part of the plasma membrane of a muscle cell, which is working hard and using large amounts of oxygen and producing large amounts carbon dioxide. Draw the process of diffusion that would be occurring in this cell.

External environment of the cell 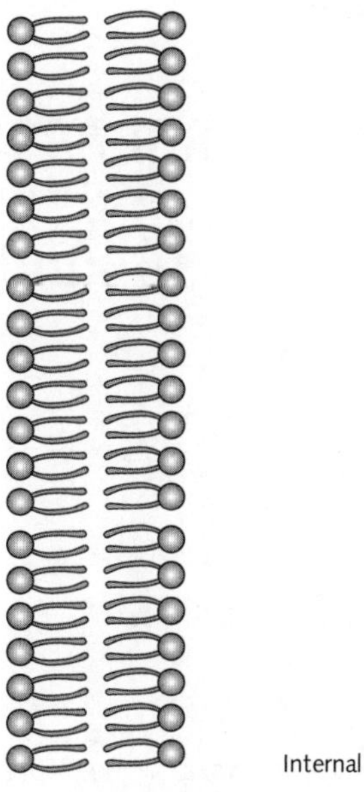 Internal environment of the cell

Figure 1.11

5 Figure 1.12 represents the plasma membrane of a leaf cell of a plant. It is daylight and the cell is actively photosynthesising. Draw the process of diffusion that would be occurring in this cell.

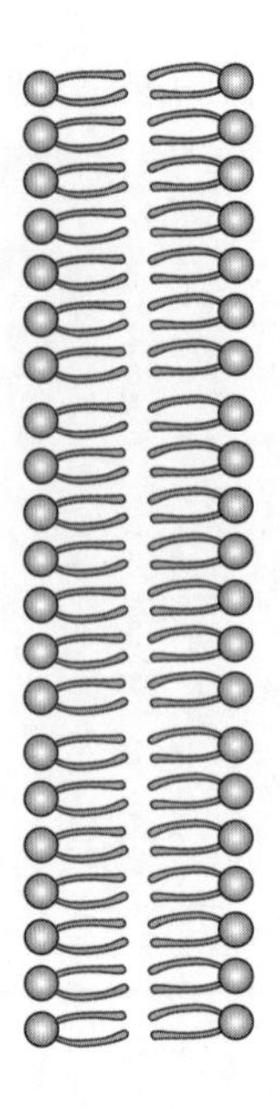

Figure 1.12

6 Figure 1.13 represents the plasma membrane of a photosynthetic plant cell at night. It is actively respiring, but not photosynthesising. Glucose is a large partially hydrophilic molecule; therefore, it cannot cross the plasma membrane by simple diffusion. Draw the process of facilitated diffusion of glucose that would be occurring in this cell as glucose moves to other actively respiring cells.

External environment of the cell

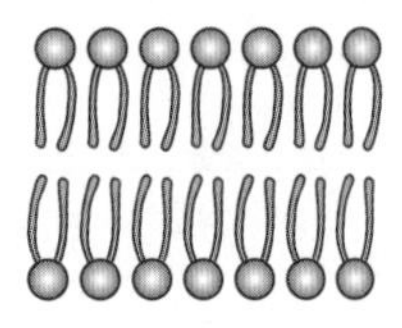
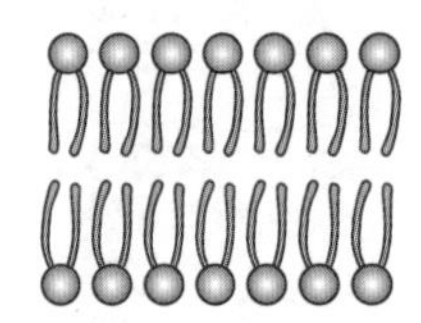
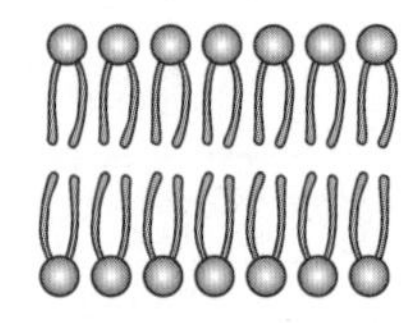

Internal environment of the cell

Figure 1.13

1.5.2 Osmosis: a special type of diffusion

PAGE 35

Key science skills

Develop aims and questions, formulate hypotheses and make predictions
- predict possible outcomes

Analyse, evaluate and communicate scientific ideas
- use appropriate biological terminology, representations and conventions, including standard abbreviations, graphing conventions and units of measurement

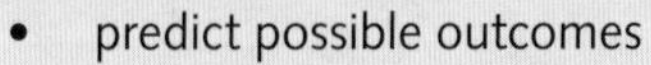

1 Complete the following sentences using words from the word list below. Some words can be used more than once.

energy	permeable	solvent
gradient	selectively	water

a ________________ is the universal ________________.

b Osmosis is the movement of a ________________ across a ________________ ________________ membrane, moving down a concentration ________________. Like diffusion, osmosis requires no input of ________________.

2 Figure 1.14 shows a container that is divided into two halves by a membrane that is selectively permeable to water but not glucose. There are different concentrations of sugar solution on either side of the membrane. In the container on the right-hand side, draw what you predict will happen after two hours.

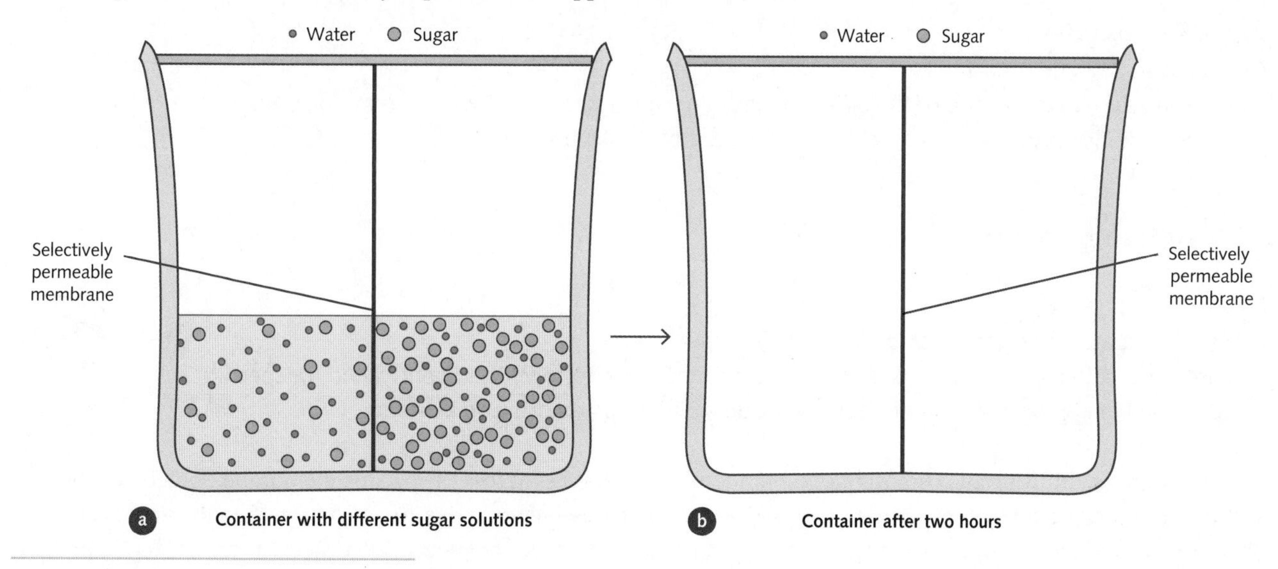

Figure 1.14

3 Predict and draw what will happen when the following cells are placed in:

a salt water

b distilled water.

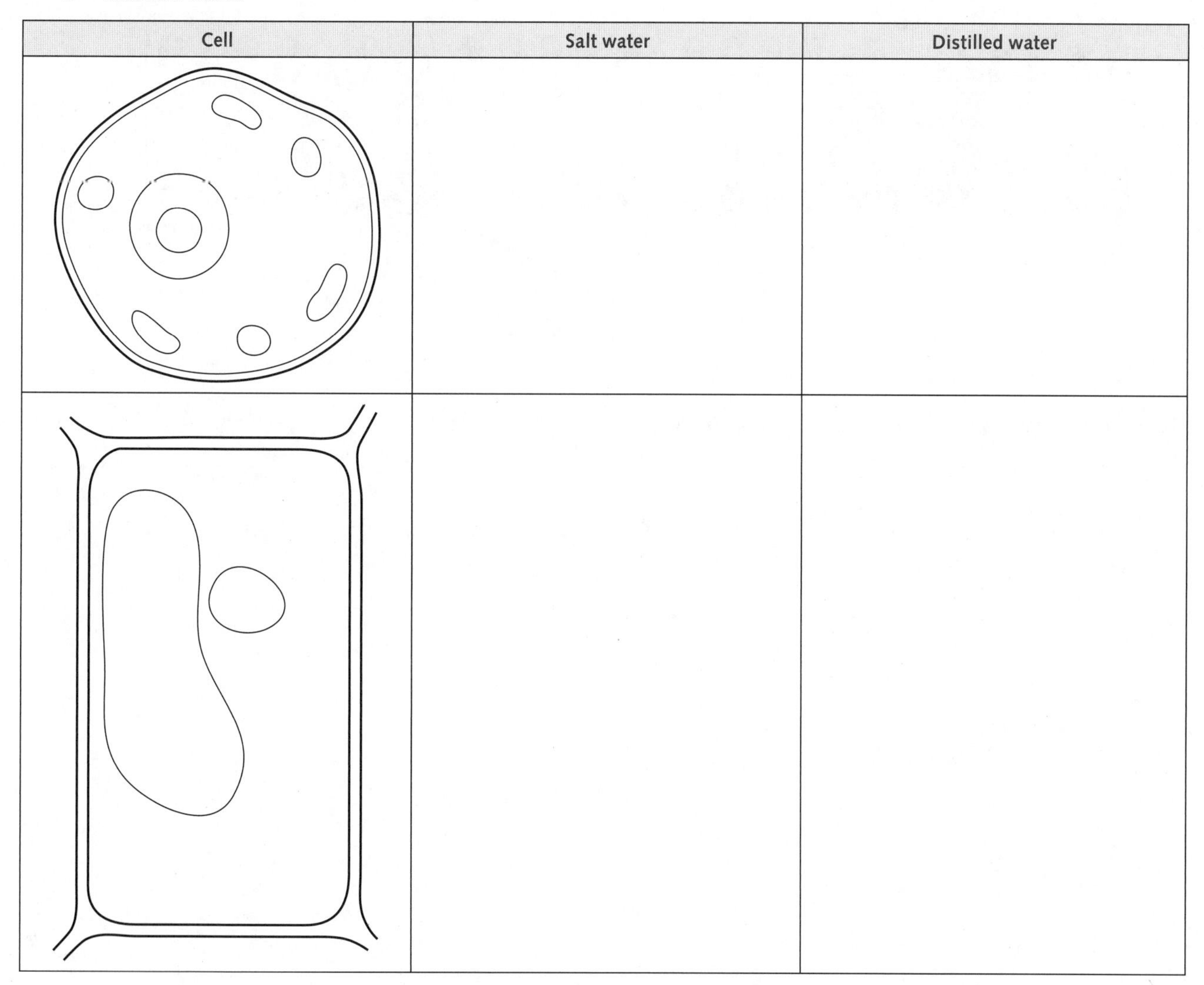

Cell	Salt water	Distilled water

1.6 Movement across membranes using energy

Key knowledge

Cellular structure and function

- the structure and function of the plasma membrane in the passage of water, hydrophilic and hydrophobic substances via osmosis, facilitated diffusion and active transport

1.6.1 Active transport

PAGE 39

Key science skills

Analyse, evaluate and communicate scientific ideas

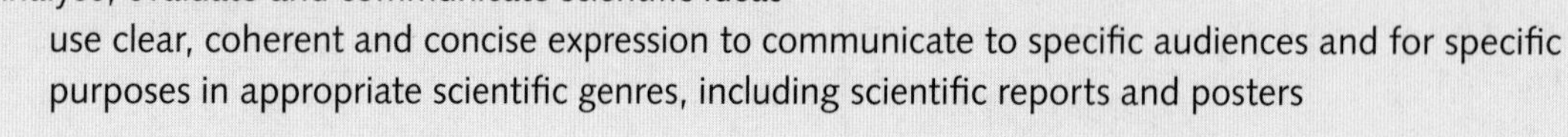

- use clear, coherent and concise expression to communicate to specific audiences and for specific purposes in appropriate scientific genres, including scientific reports and posters

In some instances, diffusion and osmosis alone cannot move substances across the plasma membrane. Sometimes the substance is moving against a concentration gradient. In these cases, energy is required to pump the substances across the plasma membrane by using active transport. Figure 1.15 shows active transport occurring via a carrier protein in the plasma membrane of a cell.

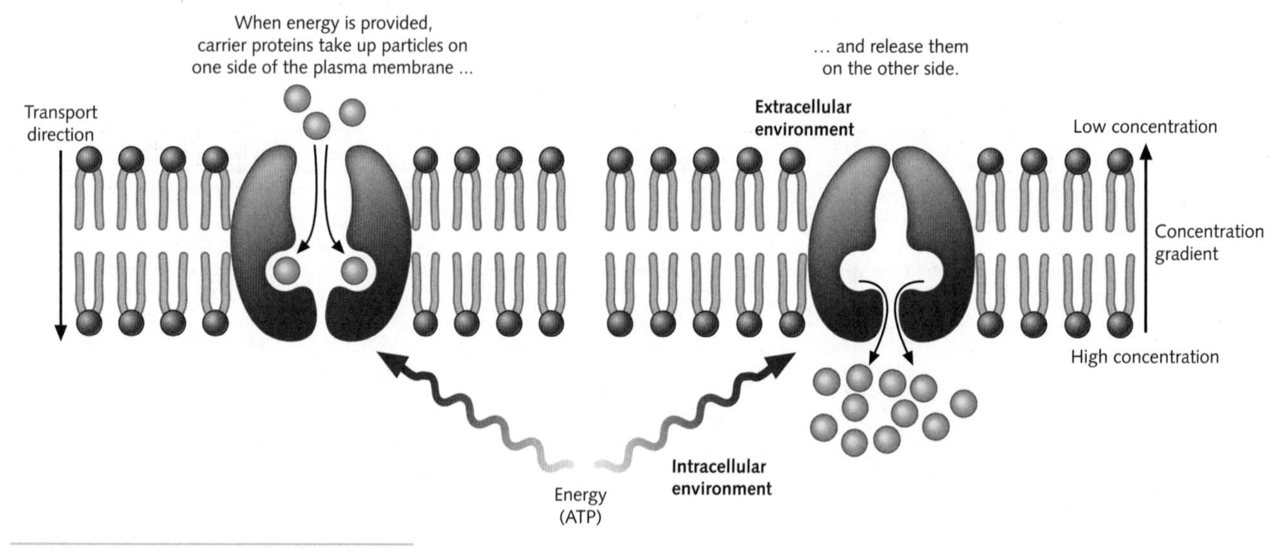

Figure 1.15 Active transport via a carrier protein in the plasma membrane of a cell

Write 500 words to explain what is happening in Figure 1.15. Your language, terminology and explanations need to be able to be understood by a Year 7 student.

Include in your explanation:

- what the diagram is representing
- reasons why substances would move across a plasma membrane
- the concept of concentration gradient
- why simple diffusion is not able to move these substances across the plasma membrane
- why energy is required to move some substances across the plasma membrane
- where the energy comes from for active transport
- the structures involved and how substances are moved across a plasma membrane using active transport.

1.6.2 Bulk transport: movement of large molecules across membranes

TB PAGE 40

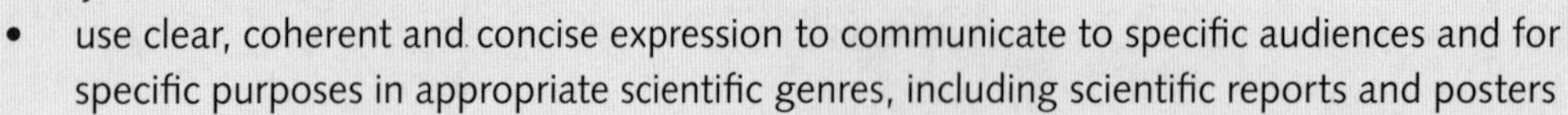

Key science skills

Analyse, evaluate and communicate scientific ideas

- use clear, coherent and concise expression to communicate to specific audiences and for specific purposes in appropriate scientific genres, including scientific reports and posters

Below are two scenarios relating to moving substances into or out of cells. Read these scenarios and draw a picture in the boxes provided to show how each cell enables movement of the substance across the plasma membrane.

Scenario 1

The hormone insulin regulates the level of sugar in the blood. Beta cells in the pancreas contain many ribosomes and endoplasmic reticulum that produce and process insulin. From there, the hormone travels to the Golgi apparatus where it is further processed and packaged into large secretory vesicles. The hormone leaves the cell by exocytosis and travels in the bloodstream to body tissues where it acts to lower blood sugar level.

Scenario 2

White blood cells travel around in the blood. They form part of the immune system and spring into action when bacteria are detected in the body. The white blood cell extends around a bacterial cell so that it is completely enclosed in a large vesicle. Digestive enzymes are secreted into the vesicle to break down the bacterial cell into large particles. These large particles are moved across the plasma membrane by phagocytosis.

 9780170452632

1.6.3 Comparing and contrasting diffusion, osmosis and active transport

Key science skills

Generate, collate and record data

- organise and present data in useful and meaningful ways, including schematic diagrams, flow charts, tables, bar charts and line graphs

TB PAGE 41

Use the Venn diagram below to compare diffusion, osmosis and active transport.

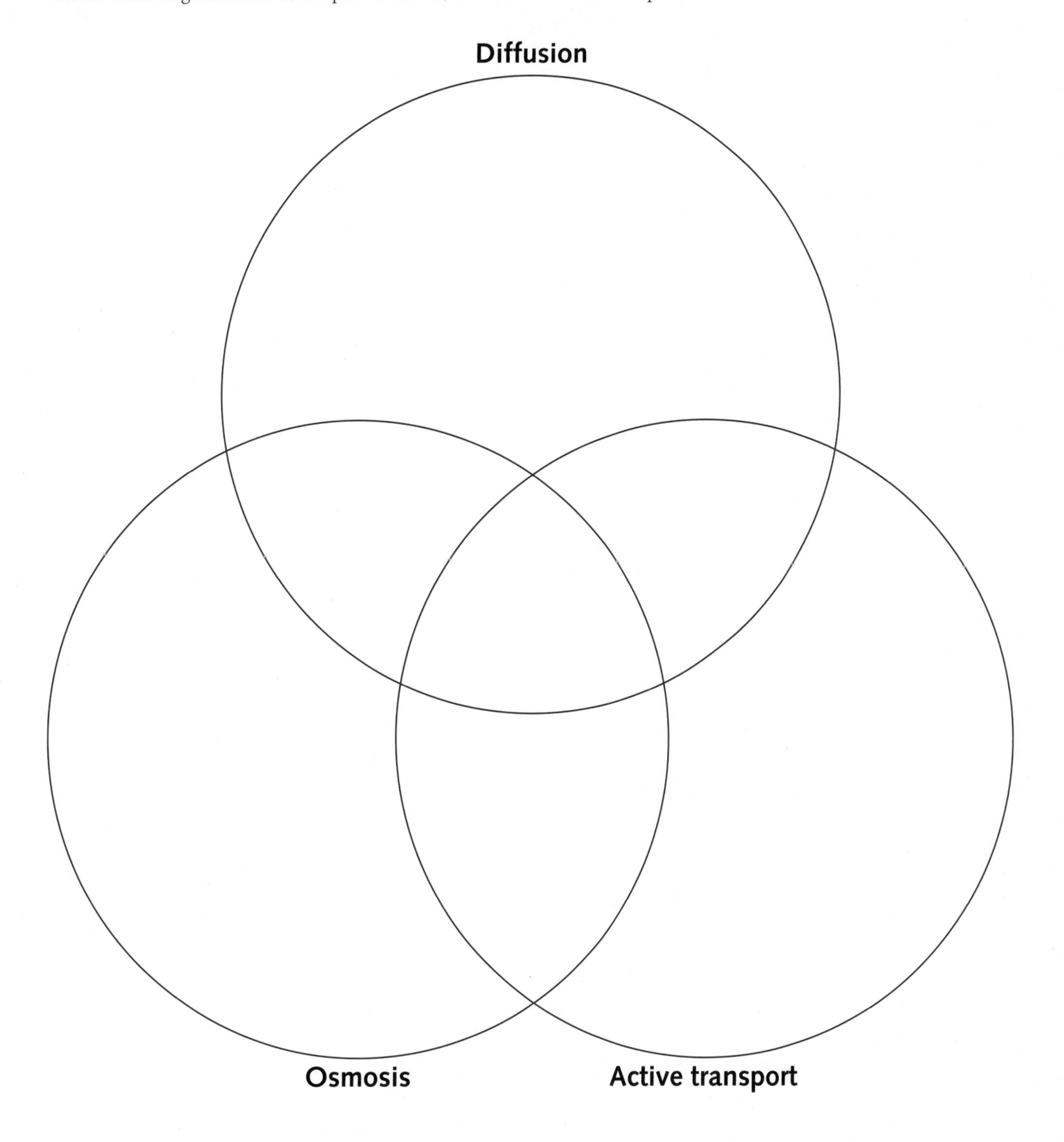

1.7 Chapter review

PAGE 47

1.7.1 Key terms

You will need scissors and a glue stick to complete this activity.

Cut out each of the dominoes on page 31. Starting with the Beginning domino, match the question and answer, organising them so they fit in the space below, until you finally finish with the End domino. Stick the dominoes in place.

Unicellular	What is the term for 'water fearing'?
Metabolism	What is a cell that has a membrane-bound nucleus and organelles called?
Solute	What is the relationship between the amount of plasma membrane and cytoplasm in a cell?
Fluid mosaic model	What is a single-celled organism called?
Mitochondria	The movement of water through a selectively permeable membrane is called _____.
SA:V	A membrane that allows substances to pass through is said to be _____.
Solution	What is the name of the model used to describe the structure of the plasma membrane?
Concentration gradient	What is an organism made up of more than one cell called?
Permeable	What is it called when molecules move equally in all directions in a solution?
Hydrophobic	What is the name of the organelle in a cell where photosynthesis occurs?
Diffusion	Where in a eukaryotic cell does respiration occur?
Passive transport	What is a cell with no membrane-bound organelles called?
Solvent	What is the name given to all the activities that occur within a cell?
Chloroplast	What is the liquid part of a solution called?
Equilibrium	END
Eukaryotic	What is the movement of a gas or a solute from an area of high concentration to an area of low concentration called?
Multicellular	Movement of a substance along a concentration gradient is called _____.
Selectively permeable	What is the substance that is dissolved in a solvent to make a solution called?
Osmosis	What is the name given to the difference in concentration of a substance from one area to another?
Active transport	Which type of membrane allows some substances to pass through but not others?
Prokaryotic	What form of diffusion uses carrier molecules embedded in the plasma membrane?
BEGINNING	What do you get when you mix a solute with a solvent?
Facilitated diffusion	What type of transport moves molecules from an area of low concentration to an area of high concentration?

Chapter review continued

PAGE 49

1.7.2 Practice test questions

Multiple-choice questions

1 ©VCAA 2002 EXAM 1 Q1 (adapted) A typical eukaryotic cell has

A a nucleoid.
B leucoplasts.
C plasmids.
D membrane-bound organelles.

2 ©VCAA 2002 EXAM 1 Q20 (adapted) A researcher examined four different and appropriately stained cells with an electron microscope. These were:

- a red blood cell
- a human skin cell
- a photosynthesising cell and
- an onion epidermal cell.

It would be reasonable to expect the researcher to be able to see which of the following organelles in all cells

A chloroplasts.
B ribosomes.
C cell wall.
D a nucleus.

The following information applies to Questions 3 and 4.

In a series of experiments, animal cells and plant cells were placed in solutions of different concentrations.

Solution X: distilled water
Solution Y: concentration the same as the cytosol of the cells at the start of the experiment
Solution Z: higher concentration of solute than the cytosol of the cells at the start of the experiment

The appearance of the cells at the start of the experiments was as follows.

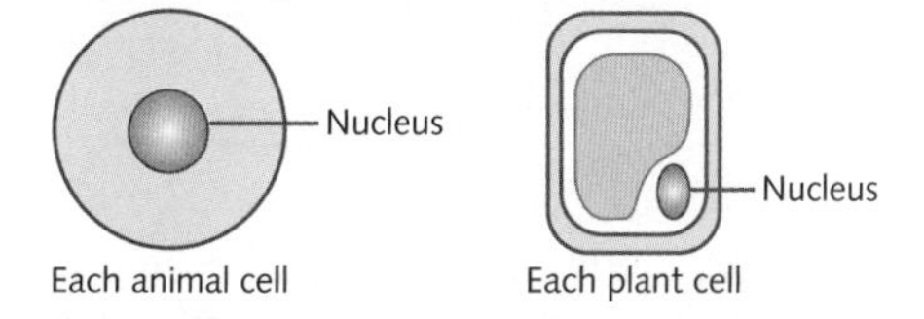

After several minutes in the solutions, the cells appeared as shown below.

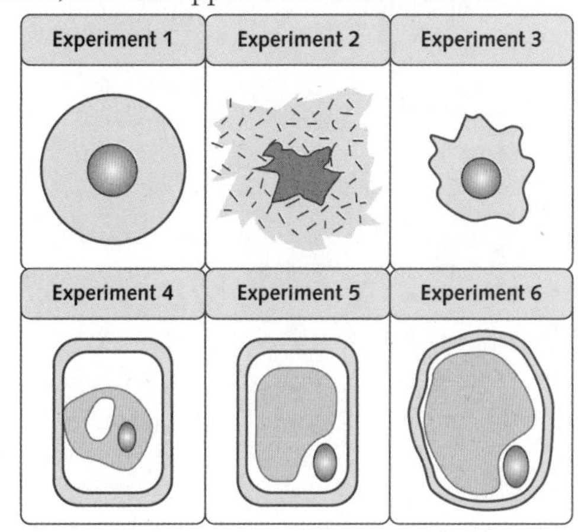

3 ©VCAA 2003 EXAM 1 Q5 (adapted) Analysis of the data reveals that

A experiment 5 represents a plant cell in solution X.
B experiment 6 represents a plant cell in solution Z.
C experiment 2 represents an animal cell in solution X.
D experiment 3 represents an animal cell in solution Y.

4 Which of the following statements is correct in relation to the data?

A If the cell in experiment 6 was placed in solution Z, it would return to how it appeared at the start of the experiment.

B If the cell in experiment 3 was placed in solution Z, it would return to how it appeared at the start of the experiment.

C If the cell in experiment 4 was placed in solution Z, it would return to how it appeared at the start of the experiment.

D If the cell in experiment 2 was placed in solution Z, it would return to how it appeared at the start of the experiment.

5 Which cell is best able to obtain sufficient requirements and remove wastes to stay alive?

A large round cell

B small round cell

C large square cell

D small flat cell

Short-answer questions

6 ©VCAA 2002 EXAM 1 SECTION B Q1 (adapted) The diagram below represents a cross-section of part of a plasma membrane.

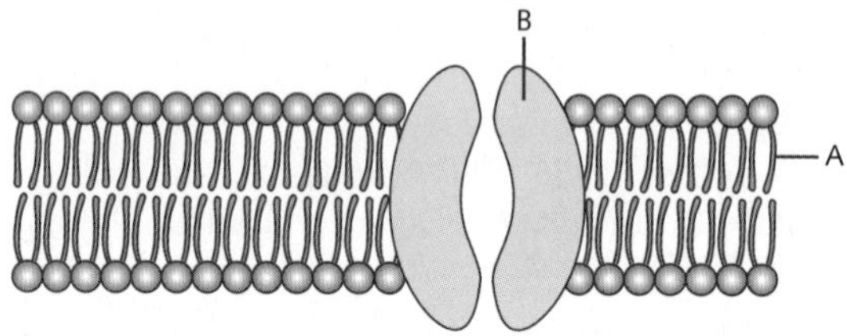

a Name the structures labelled A and B.

Structure A: ____________________

Structure B: ____________________

2 marks

The concentration of sodium ions, Na^+, in human blood plasma is approximately 150 mmol L^{-1}. In the cytosol of red blood cells, the concentration of these ions is between 25 and 30 mmol L^{-1}.

b What is this difference called and which way would you expect sodium ions to flow?

2 marks

When a plant cell is placed into saline (salty) water, water moves out of the cell.

d What term is used to describe this movement of water out of a cell?

1 mark

e Explain why water moves out of the plant cell in this situation.

1 mark

The cell cycle 2

Remember

To gain full advantage of the activities in Chapter 2, there is some content from Chapter 1 and from your previous studies in science, that you will need to understand. Take some time to refresh your knowledge of this content before you enter this chapter. Try to answer the following questions from memory. If you cannot do this, then use a reference to assist you.

PAGE 56

1 What tiny unit makes up all living things?

2 All living things are classified into kingdoms according to their cell type. What are these two main types of cells?

3 Distinguish between these two main types of cells.

4 What are centrioles and where are they found?

5 What are chromosomes and what are they made of?

6 What are genes? What do they code for?

2.1 Binary fission

Key knowledge

The cell cycle and cell growth, death and differentiation

- binary fission in prokaryotic cells

2.1.1 Binary fission

Key science skills

Analyse, evaluate and communicate scientific ideas

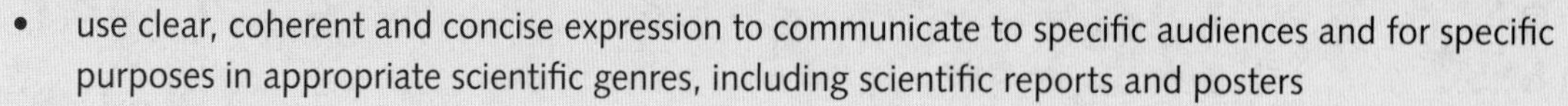

- use clear, coherent and concise expression to communicate to specific audiences and for specific purposes in appropriate scientific genres, including scientific reports and posters

Add to the diagram in Figure 2.1 to show what is occurring in the DNA of a prokaryotic cell undergoing binary fission. In the lined space below, explain what is happening at each stage of binary fission. Your explanation needs to be able to be understood by a Year 7 student.

Figure 2.1

2.1.2 Effect of temperature on binary fission – part A

Key science skills

Develop aims and questions, formulate hypotheses and make predictions

- identify, research and construct aims and questions for investigation

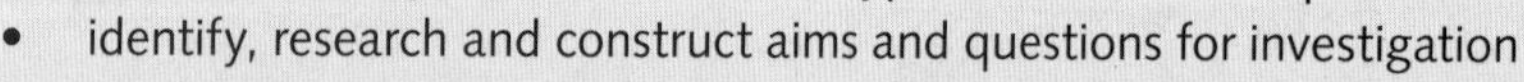

- identify independent, dependent and controlled variables in controlled experiments
- formulate hypotheses to focus investigation
- predict possible outcomes

Developing a hypothesis

A research scientist wants to find out if external factors such as temperature affect the rate at which binary fission occurs in bacteria. She decides to test the reproduction rate of the bacteria *Escherichia coli* at two different temperatures: 4°C and 35°C.

1 The aim of an investigation is the reason why an investigation is undertaken. Write an aim for this investigation.

2 An effective research question is a question that is specific and can be answered by performing experimental work or investigation with the resources and equipment that are available. It details what will be varied and what will be measured. It also gives a criterion for judging whether the obtained results have answered the research question. Write a research question for this investigation.

3 Determine the variables for this investigation.

a The independent variable is the factor that is being manipulated or changed. State the independent variable.

b The dependent variable is the factor being measured. State the independent variable.

c Controlled variables are any other variables that are kept constant during the investigation in order to determine the relationship between the independent and dependent variable. State any controlled variables.

4 A hypothesis is a tentative prediction, or a tentative explanation of an observation, based on an existing model or theory. A hypothesis should provide a prediction that can be tested by performing an investigation. This means it should at least be falsifiable; that is, it should be able to be refuted. Write a hypothesis for this investigation in the following format: If ... *the independent variable is changed* ... then ... *something will happen to the dependent variable.*

5 Predict what you think will be the outcome of this investigation.

2.1.3 Effect of temperature on binary fission – part B

Key science skills

Plan and conduct investigations

- determine appropriate investigation methodology: case study; classification and identification; controlled experiment; correlational study; fieldwork; literature review; modelling; product, process or system development; simulation
- design and conduct investigations; select and use methods appropriate to the investigation, including consideration of sampling technique and size, equipment and procedures, taking into account potential sources of error and uncertainty; determine the type and amount of qualitative and/or quantitative data to be generated or collated

Comply with safety and ethical guidelines

- demonstrate safe laboratory practices when planning and conducting investigations by using risk assessments that are informed by safety data sheets (SDS), and accounting for risks

PAGE 57

Planning the investigation

1 Your research question would be best tested by a controlled experiment methodology. What is a controlled experiment methodology?

2 The research scientist will grow bacteria on a suitable growth medium on agar plates. She wants to test the ability of the bacteria to grow at temperatures of 4°C and 35°C. Using this information, write a method that the research scientist could follow to test the hypothesis. The method is the step-wise process that the research scientist will undertake during the investigation.

3 Risk assessment is the process of evaluating potential risks involved in an investigation. *Escherichia coli* causes severe stomach upsets if ingested. Develop the risk assessment for this investigation using the risk assessment table below.

What are the risks in doing this investigation?	How can you manage these risks to stay safe?

4 Will the data that will be collected be qualitative or quantitative, or a combination of both? Describe the type of data that will be collected.

5 What steps have you incorporated into the method to ensure that the data collected is:

a accurate?

b precise?

c repeatable?

d valid?

2.1.4 Effect of temperature on binary fission – part C

TB PAGE 57

Key science skills

Generate, collate and record data

- plot graphs involving two variables that show linear and non-linear relationships

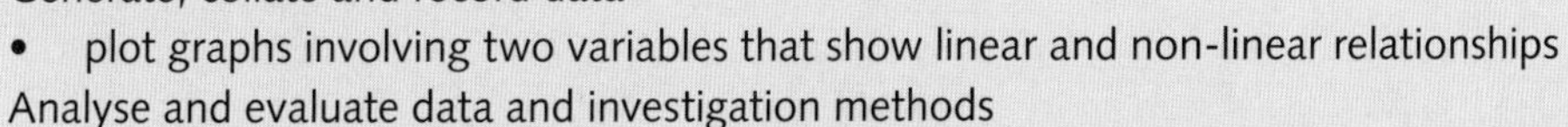

Analyse and evaluate data and investigation methods

- evaluate investigation methods and possible sources of personal errors/mistakes or bias, and suggest improvements to increase accuracy and precision, and to reduce the likelihood of errors

Construct evidence-based arguments and draw conclusions

- evaluate data to determine the degree to which the evidence supports or refutes the initial prediction or hypothesis

Analysing and evaluating the data

A research scientist carried out the investigation into the effect of temperature on the reproduction of bacteria. She inoculated two agar plates with one colony of *Escherichia coli* bacteria. She incubated one of the plates at 4°C for 8 hours and the other plate at 35°C for 8 hours. She counted the number of bacterial colonies on each plate every hour. Her results appear in Table 2.1.

Table 2.1 Results table

Time (hours)	Number of bacterial colonies at 4°C	Number of bacterial colonies at 35°C
0	1	1
1	1	2
2	2	8
3	2	32
4	4	43
5	3	67
6	8	72
7	8	74
8	16	81

1 Graph the results in Table 2.1 using the graph paper below.

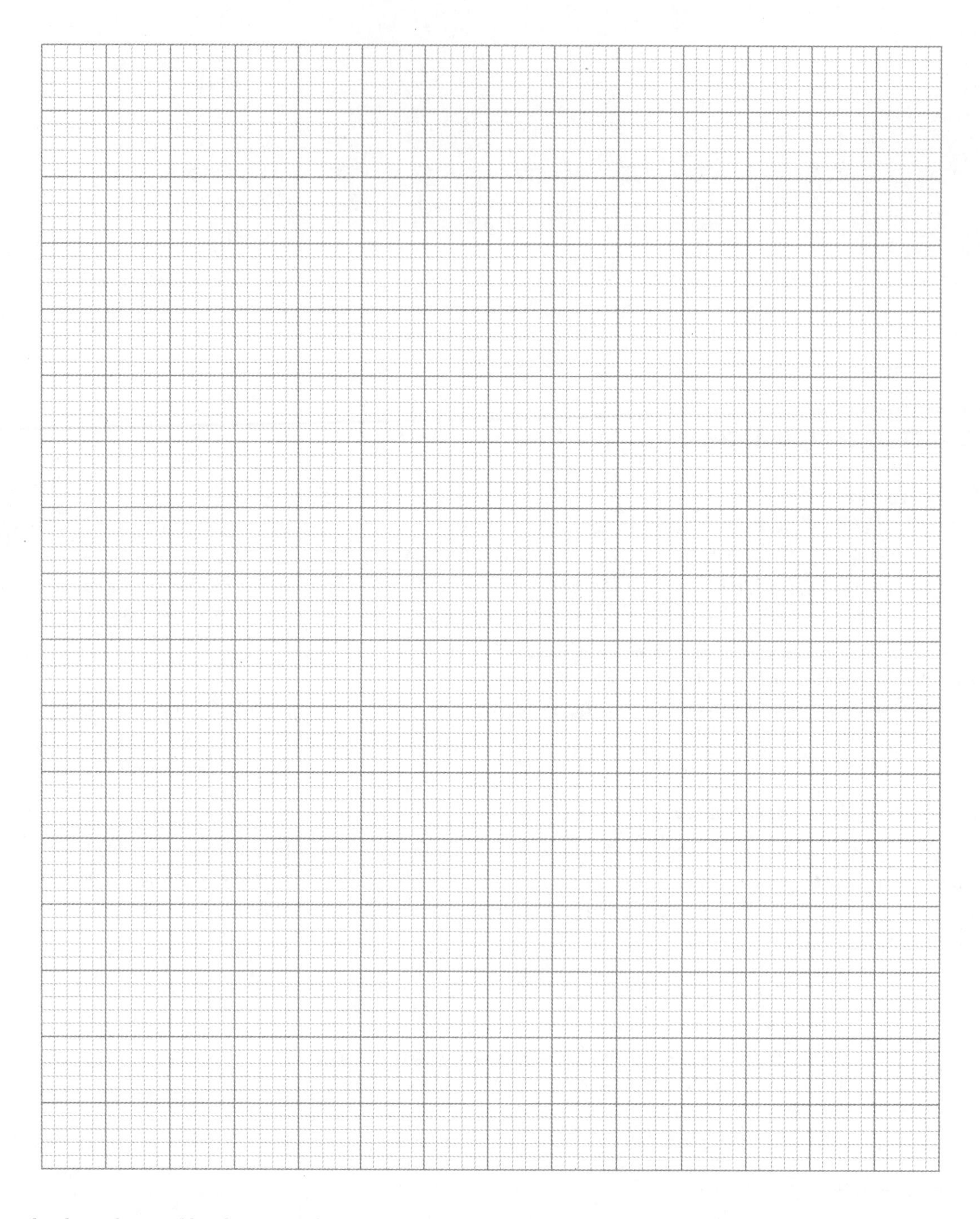

2 Does the data obtained by the research scientist support or refute the hypothesis? (Refer to Activity 2.1.2) Explain.

3 Identify at least one improvement to the method that could remove the possibility of error, or increase the accuracy or precision of the data.

4 What does the researcher's results mean for the storage of easily spoilt food such as soft cheese?

2.2 Eukaryotic cell cycle

Key knowledge

The cell cycle and cell growth, death and differentiation

- the eukaryotic cell cycle, including the characteristics of each sub-phases of mitosis and cytokinesis in plant and animal cells

2.2.1 The cell cycle

Key science skills

Construct evidence-based arguments and draw conclusions

- use reasoning to construct scientific arguments, and to draw and justify conclusions consistent with the evidence and relevant to the question under investigation

PAGE 59

Figure 2.2 shows a graph of cell volume against time (diagram **a**) and the amount of DNA during a cell cycle (diagram **b**) in a cell. Study these graphs carefully so you understand what they are telling you and use the information provided in these graphs to answer the questions below.

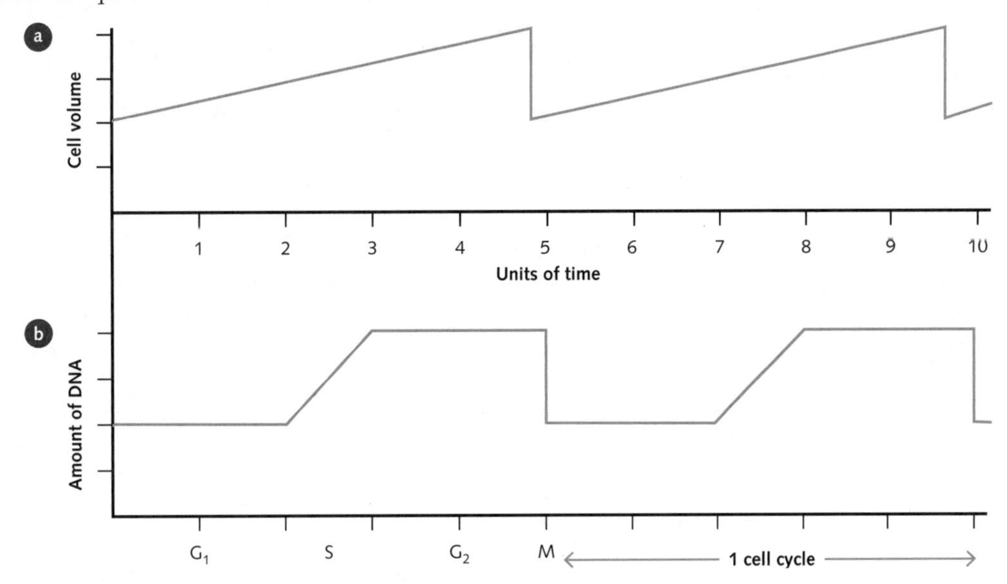

Figure 2.2 a The cell volume and b amount of nuclear DNA change during a cell cycle, reflecting the different phases of the cycle

1 Do these graphs represent an actively dividing cell? Provide evidence for your answer.

2 During what time is the cell in the G_1 phase of the cell cycle? Provide evidence for your answer.

3 During what time is the cell in the G_2 phase of the cell cycle? Provide evidence for your answer.

4 At what time does each chromosome appear as two chromatids joined at the centromere?

5 At what time does the cell divide into two daughter cells? Provide evidence for your answer.

6 What is the length of time required for this cell to complete one cell cycle?

2.2.2 The phases of mitosis in eukaryotic cells

TB PAGE 60

Key science skills

Analyse, evaluate and communicate scientific ideas

- discuss relevant biological information, ideas, concepts, theories and models and the connections between them

Develop

Actively dividing cells are undergoing mitosis. An onion root tip grows quickly in order to establish the plant in the soil and start absorbing water for the cell's needs. Onion root tip cells are mitotically active. These cells can be frozen in time by putting them into ethanoic alcohol. The chromosomes are stained and a 'squash' of the root tips is prepared for viewing under the microscope. A squash is prepared by putting a small piece of root tip onto a microscope slide, placing a coverslip over it, and then using the thumb to squash the root material. Figure 2.3 on page 45 shows onion root tip cells that have been prepared in this way.

What to do

You will need scissors and a glue stick to complete this activity.

Step 1: Locate Figure 2.3 on page 45, which shows mitosis in the cells of an onion root tip.

Step 2: Cut out each stage of mitosis from Figure 2.3.

Step 3: Place one example of a cell at each stage of mitosis under the correct heading below and glue in place.

Step 4: Describe precisely what is happening at each stage of mitosis in the space on the right.

Step 5: When you have completed the activity, answer the questions on page 44.

Interphase

Prophase

Metaphase

Anaphase

Telophase

1 The root squash from Figure 2.3 shows cells at various stages of the cell cycle. Explain why the cells are all at different stages.

2 State three reasons why a cell would undergo mitosis.

3 What occurs in the cell once mitosis is complete?

Use Figure 2.3 to complete the activity on pages 42–43.

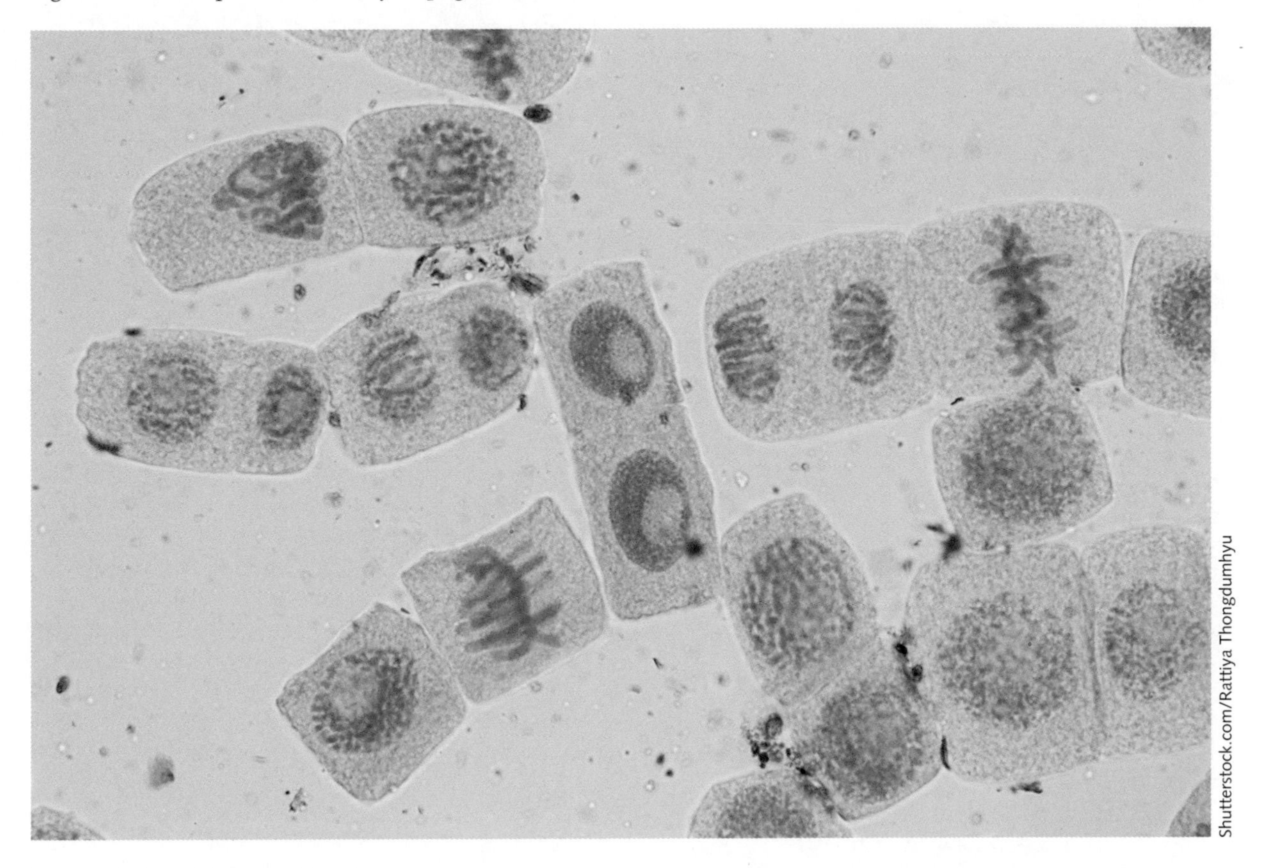

Shutterstock.com/Rattiya Thongdumhyu

Figure 2.3 **Cells undergoing mitosis in an onion root tip. The chromosomes have been stained to make them easier to see.**

2.2.3 Cell cycle checkpoints

Key science skills

Construct evidence-based arguments and draw conclusions

- use reasoning to construct scientific arguments, and to draw and justify conclusions consistent with the evidence and relevant to the question under investigation

Mitosis is a complex process occurring in the confined space of a cell. There is a high chance that mistakes (mutations) will be made either in the DNA replication or in the chromosome division phases of mitosis. Mutations can be detrimental to the cell, and so checkpoints are in place in the cell cycle to detect mutations. Mutations can sometimes be repaired or, if not, the cell cycle is stopped and the cell will be destroyed by a process called apoptosis (programmed cell death).

A group of molecules called cyclins form a complex with cycline-dependent kinases (CDKs) control the progression of the cell cycle through each checkpoint. Figure 2.4 shows the concentration of various cyclins and CDKs within a cell during the various phases of the cell cycle.

From your knowledge of the cell cycle and using information from Figure 2.4, answer the questions below.

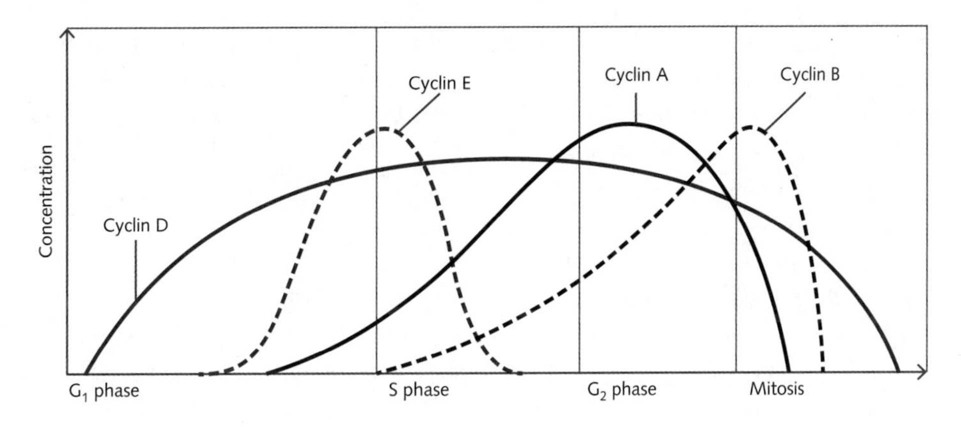

Figure 2.4 Control of the cell cycle by various cyclins and associated CDKs

1 Which cyclin prepares the cell for DNA replication? Provide the reasoning behind your choice.

2 Which cyclin promotes the assembly of the mitotic spindle? Provide the reasoning behind your choice.

3 Which cyclin triggers the cell to move between G_0, G_1 and S phases? Provide the reasoning behind your choice.

2.3 Apoptosis

Key knowledge

The cell cycle and cell growth, death and differentiation

- apoptosis as a regulated process of programmed cell death

2.3.1 When apoptosis fails: HeLa cells

PAGE 66

Key science skills

Analyse, evaluate and communicate scientific ideas

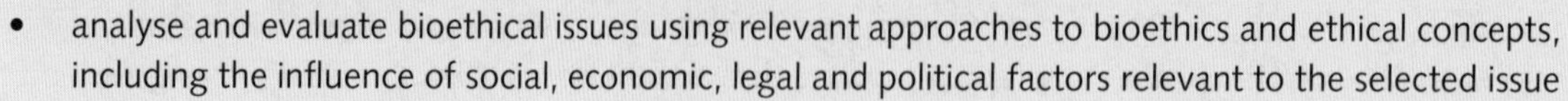

- analyse and evaluate bioethical issues using relevant approaches to bioethics and ethical concepts, including the influence of social, economic, legal and political factors relevant to the selected issue

The importance of HeLa cells

Cancer is a disease where abnormal cells in the body divide and grow uncontrollably, often caused by mutations in DNA and chromosomes. Mutations could trigger the acceleration of cell division rates, inhibiting the normal control of systems, including stopping the cell cycle or apoptosis. In cancer, old cells do not die; rather, they grow out of control and form new, abnormal cells. When these abnormal cells accumulate, they form a mass of tissue, known as a tumour. Cancer cells are frequently called immortal. Normal cells divide about 50 times and then die, cancer cells can go on dividing indefinitely if supplied with their requirements.

HeLa cells were derived from a young African American woman named Henrietta Lacks, who died of an aggressive cervical cancer in 1951. She attended the gynaecology department of Johns Hopkins Hospital in Baltimore, Maryland, [USA], where doctors found a tumour on her cervix. When a biopsy (sample) was taken from the tumour, one specimen was given to George Gey and his colleagues at the Tissue Culture Laboratory at Johns Hopkins, who had ambitions to isolate and maintain normal or diseased tissue as 'temporary derived cell strains'.

What made Henrietta Lacks' tumour different from other tumours that had been brought to the laboratory was that the cells grew vigorously. Before this, it was possible to grow cells from cancer specimens; however, these usually did not last and died before studies could be completed.

In contrast, the cells that grew from Henrietta Lacks' tumour kept on dividing as long as the appropriate growth conditions were met.

Over the past several decades, this cell line has contributed to many medical breakthroughs, from research on the effects of zero gravity in outer space and the development of the polio vaccine, to the study of leukaemia, the AIDS virus and cancer worldwide. The National Institute of Health in the US analysed and evaluated the scientific literature involving HeLa cells and found over 110 000 publications that cited the use of HeLa cells between 1953 and 2018.

Although many other cell lines are in use today, HeLa cells have supported advances in most fields of medical research in the years since HeLa cells were isolated.

Adapted from https://www.news-medical.net/news/20200116/Scientists-discover-how-B-cells-in-tumor-cells-promote-a-favorable-immunotherapy-response.aspx; https://www.news-medical.net/life-sciences/What-are-HeLa-Cells.aspx; https://www.hopkinsmedicine.org/henriettalacks/importance-of-hela-cells.html

The HeLa story provides many bioethical dilemmas for researchers. The HeLa cell line was established in 1951 without the consent of Henrietta Lacks or her family. The cell line has now contributed to the studies and cures of many diseases, yet the Lacks' family has not received any compensation for their involvement. In 2013, the genome of the HeLa cells was published in a public database to aid all researchers using the cell line, but this disclosed genetic traits of the surviving Lacks' family members. The genome had to be removed from the public databases.

There are three major approaches to resolving bioethical issues. (See box on the next page.)

9780170452632

Ethical approaches

1. **A consequences-based approach** places central importance on the consideration of the consequences of an action (the ends), with the aim of maximising positive outcomes and minimising negative effects.
2. **A duty- and/or rule-based approach** is concerned with how people act (the means) and places central importance on the idea that people have a duty to act in a particular way, and/or that certain ethical rules must be followed, regardless of the consequences.
3. **A virtues-based approach** is person rather than action based. Consideration is given to the virtue or moral character of the person carrying out the action, providing guidance about the characteristics and behaviours a good person would seek to achieve to then be able to act in the right way.

1 Tumour cells from Henrietta Lacks were initially cultured and grown without her permission or the permission of her family. Discuss the bioethical approach(es) (above) that could be considered in this situation. Provide the reasons for your choice of approach.

2 Although the HeLa cells were the first cells that could be easily shared and multiplied in a laboratory setting, Johns Hopkins has never sold or profited from the discovery or distribution of HeLa cells and does not own the rights to the HeLa cell line. Rather, Johns Hopkins offered HeLa cells freely and widely for scientific research (https://web.archive.org/web/20210307160442/www.hopkinsmedicine.org/henriettalacks/importance-of-hela-cells.html).

Does this information change your view of the use of HeLa cells? Which ethical approach would now be best to consider in this situation? Provide reasons for your choice of approach.

2.4 Disruption to the regulation of the cell cycle

Key knowledge

The cell cycle and cell growth, death and differentiation

- disruption to the regulation of the cell cycle and malfunctions in apoptosis that may result in deviant cell behaviour: cancer and characteristics of cancer cells

2.4.1 Proto-oncogenes and tumour suppressor genes

Key science skills

Construct evidence-based arguments and draw conclusions

- use reasoning to construct scientific arguments, and to draw and justify conclusions consistent with the evidence and relevant to the question under investigation

Analyse, evaluate and communicate scientific ideas

- use clear, coherent and concise expression to communicate to specific audiences and for specific purposes in appropriate scientific genres, including scientific reports and posters

An important skill in science is being able to communicate in a clear, accurate and precise fashion. You need to choose your words carefully so they convey exactly what you want to communicate.

Below is a topic that you will discuss in no more than 150 words. Use the scaffolding provided to structure your response.

Topic: Proto-oncogenes act in the opposite way from tumour suppressor genes.

Explain the context.

Explain the function of proto-oncogenes.

Explain the function of tumour suppressor genes.

Comment on whether these functions could be considered 'opposite'.

2.4.2 Action of mutagens on the cell cycle

Key science skills

Generate, collate and record data

- organise and present data in useful and meaningful ways, including schematic diagrams, flow charts, tables, bar charts and line graphs

Mutagens are agents that either induce or increase the frequency of mutation in DNA. There are several types of mutagens.

Graphic organisers, such as concept maps, spider maps, mind maps and lotus diagrams, are used to organise information so that it is easy to understand and remember. Use the space below to devise a graphic organiser to show the three main types of mutagens, examples of each and their effect on cells.

2.5 Stem cells

Key knowledge

The cell cycle and cell growth, death and differentiation

- properties of stem cells that allow for differentiation, specialisation and renewal of cells and tissues, including the concepts of pluripotency and totipotency

2.5.1 Bioethical issues: embryonic stem cells

Key science skills

Analyse, evaluate and communicate scientific ideas

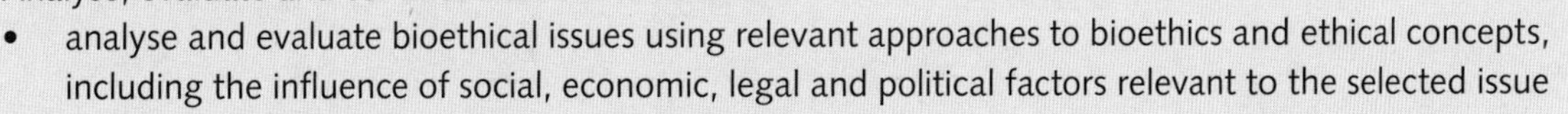

- analyse and evaluate bioethical issues using relevant approaches to bioethics and ethical concepts, including the influence of social, economic, legal and political factors relevant to the selected issue

Carefully read the following article and answer the questions below.

Japanese newborn gets new liver from stem cells

Doctors in Japan have successfully transplanted liver cells derived from embryonic stem cells (ES cells) into a newborn baby, in a world first that could provide new treatment options for infants. The newborn was suffering from urea cycle disorder, where the liver is not capable of breaking down toxic ammonia. But the six-day-old was too small to undergo a liver transplant, generally not considered safe until a child weighs around six kilograms at around three to five months old.

Embryonic stem cells are harvested from unwanted embryos and using them in research has raised ethical issues because embryos are subsequently destroyed. Use of these embryos has been approved by both the egg and the sperm donors after having already completed fertility treatment, according to the institute.

Doctors at the National Center for Child Health and Development, [Japan] decided to try a 'bridge treatment' until the baby was big enough, injecting 190 million liver

Continued on next page

cells derived from ES cells into the blood vessels of the baby's liver. Following the treatment, 'the patient did not see an increase in blood ammonia concentration and was able to successfully complete the next treatment', namely a liver transplant, the institute said in a press release.

The baby, whose sex has not been disclosed, received a liver transplant from its father and was discharged from the hospital six months after birth.

'The success of this trial demonstrates safety in the world's first clinical trial using human ES cells for patients with liver disease', the institute said. It noted that in Europe and the United States, liver cells are often available after being removed from brain-dead donors, but the supply in Japan is more limited. That has created difficulties in managing the health of small children as they wait to grow big enough for liver transplants.

Extract from https://medicalxpress.com/news/2020-05-japan-newborn-liver-stem-cells.html

1 What are stem cells?

2 Were the stem cells used for this treatment totipotent or pluripotent? What is the difference?

3 Identify the bioethical issue that this treatment raises.

4 Which bioethical approach do you think the doctors using this treatment would have argued? (Refer to p. 49.) Provide your reasoning.

Sometimes there are other factors that influence bioethical issues. These can be social (how does it affect people, families and society?), economic (how much does it cost? is it affordable?), legal (is it within the law? are there laws to deal with this?) or political (does the issue affect government policy or that of another country?). These factors need to be identified and included in any discussion of bioethical issues.

5 List any social or economic factors that you can think of that would be relevant to this bioethical issue.

9780170452632

2.6 Chapter review

TB PAGE 82

2.6.1 Key terms

1 Differentiate (explain how they are different) between each pair of terms shown below.

a daughter cell; parent cell

b multipotent stem cell; pluripotent stem cell

c adult stem cell; embryonic stem cell

d mutagen; mutation

e binary fission; mitosis

f chromatin; chromatid

g embryo; foetus

Chapter review continued

2.6.2 Practice test questions

PAGE 84

Multiple-choice questions

1 Which event would you expect to occur during the process of binary fission?

A Chromosomes line up along the equator of the cell.
B Replicated chromosomes separate to either end of the parent cell.
C Chromosomal material crosses over.
D Organelles replicate and divide equally between new daughter cells.

2 During the eukaryotic cell cycle

A G_1 phase is when cells are replicating chromosomal material.
B mitotic phase results in the production of four identical daughter cells.
C G_2 phase is when the cell prepares to undergo mitosis.
D G_0 phase is when the cells are replicating chromosomal material.

3 Which stage of mitosis is represented in Figure 2.5?

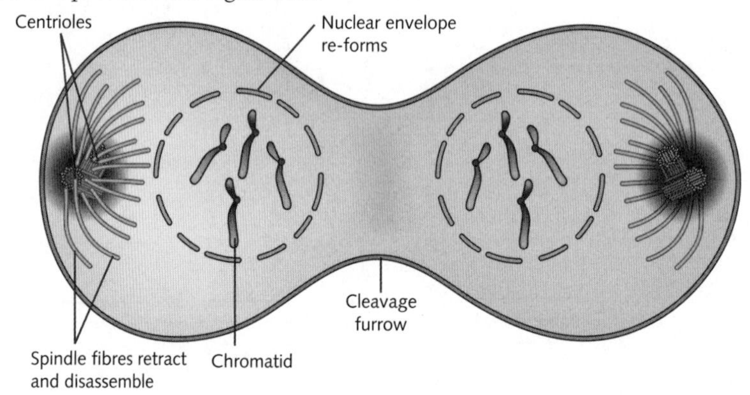

Figure 2.5

A Anaphase
B Metaphase
C Telophase
D Prophase

4 During the cell cycle there are checkpoints to check for, and repair, any mistakes that are made during the DNA replication. These checkpoints

A are controlled by cyclins, which stop the cell cycle if any mistakes are detected.
B are controlled by cyclins, which support the cell through the cell cycle.
C result in apoptosis if any mistake is detected.
D are evenly distributed throughout the cell cycle.

5 When a cell begins to divide uncontrollably, a cancerous tumour can be formed. In this situation you would expect

A proto-oncogenes to promote further cell division.
B differentiation of cells to occur.
C the p53 gene to be activated to promote further cell division.
D the BRCA1 gene to be activated to promote further cell division.

6 An example of a chemical mutagen is

A X-rays.
B viruses.
C sunlight.
D tobacco smoke.

7 A fertilised human egg that has the ability to develop into a complete embryo is called

A multipotent.
B pluripotent.
C impotent.
D totipotent.

Short-answer questions

8 Mitosis occurs in plant and animal cells, but it is slightly different in each.

a Contrast mitosis in plant and animal cells by stating two differences.

2 marks

b A parent plant cell contains 38 chromosomes at prophase. How many chromosomes will each daughter cell contain at telophase?

1 mark

9 Figure 2.6 shows a summary of potential cell fates. Using information from this figure, answer the questions below.

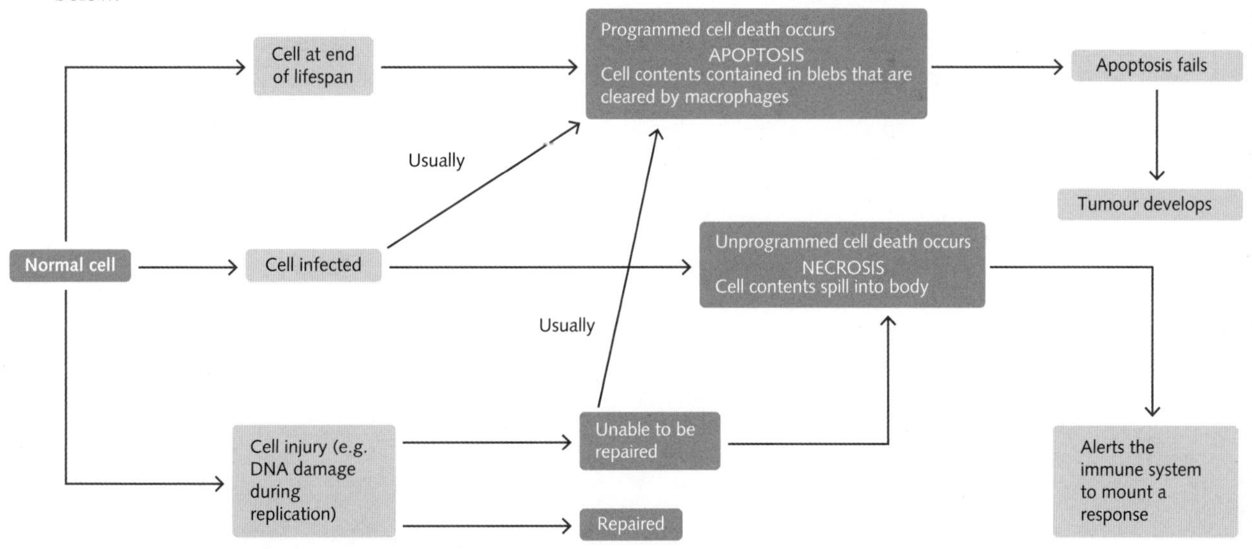

Figure 2.6 Summary of potential cell fates

a What is one fate of an infected cell?

1 mark

b What is one fate of a cell with DNA damage that cannot be repaired?

1 mark

c Under what circumstances would a tumour develop?

1 mark

3 Functioning systems

Shutterstock.com/Barbol

Remember

Chapter 3 Functioning systems will call on content that you have already met in your science studies from previous years or from earlier this year. Take some time to refresh your knowledge of this content before you enter this chapter. Try to answer the following questions from memory. If you cannot do this, then use a reference to assist you.

PAGE 92

1 Cells make up tissues, tissues make up organs, and organs make up systems. Use a flow chart to identify the cells, tissues and organs that make up the human circulatory system.

2 State the names of the two tube-like systems that conduct water and nutrients throughout vascular plants. Identify what each system carries.

3 An autotroph produces its own organic molecules using inorganic molecules, whereas a heterotroph ingests organic molecules from the environment.

a List the inorganic molecules that are inputs into the autotrophic process called photosynthesis.

b List the three types of complex molecules that humans take in as food.

4 The digestive system physically and chemically breaks down food into smaller pieces. Why is this an essential process?

5 The circulatory system carries nutrients and oxygen to cells and removes waste.

a List two examples of nutrients carried by the circulatory system.

b List two examples of waste carried by the circulatory system.

6 State the names of three processes that occur in cells to move substances either around the cell or across the plasma membrane.

3.1 Vascular plants

Key knowledge

Functioning systems

- specialisation and organisation of plant cells into tissues for specific functions in vascular plants, including intake, movement and loss of water

3.1.1 Water transport: xylem

Figure 3.1 shows the water transport system in a plant, from water uptake at the root surface, to xylem tissue extending from the roots to the leaves, and water loss from the leaves. Use your knowledge of the water transport system in a plant to complete the information in the boxes by following the instructions below.

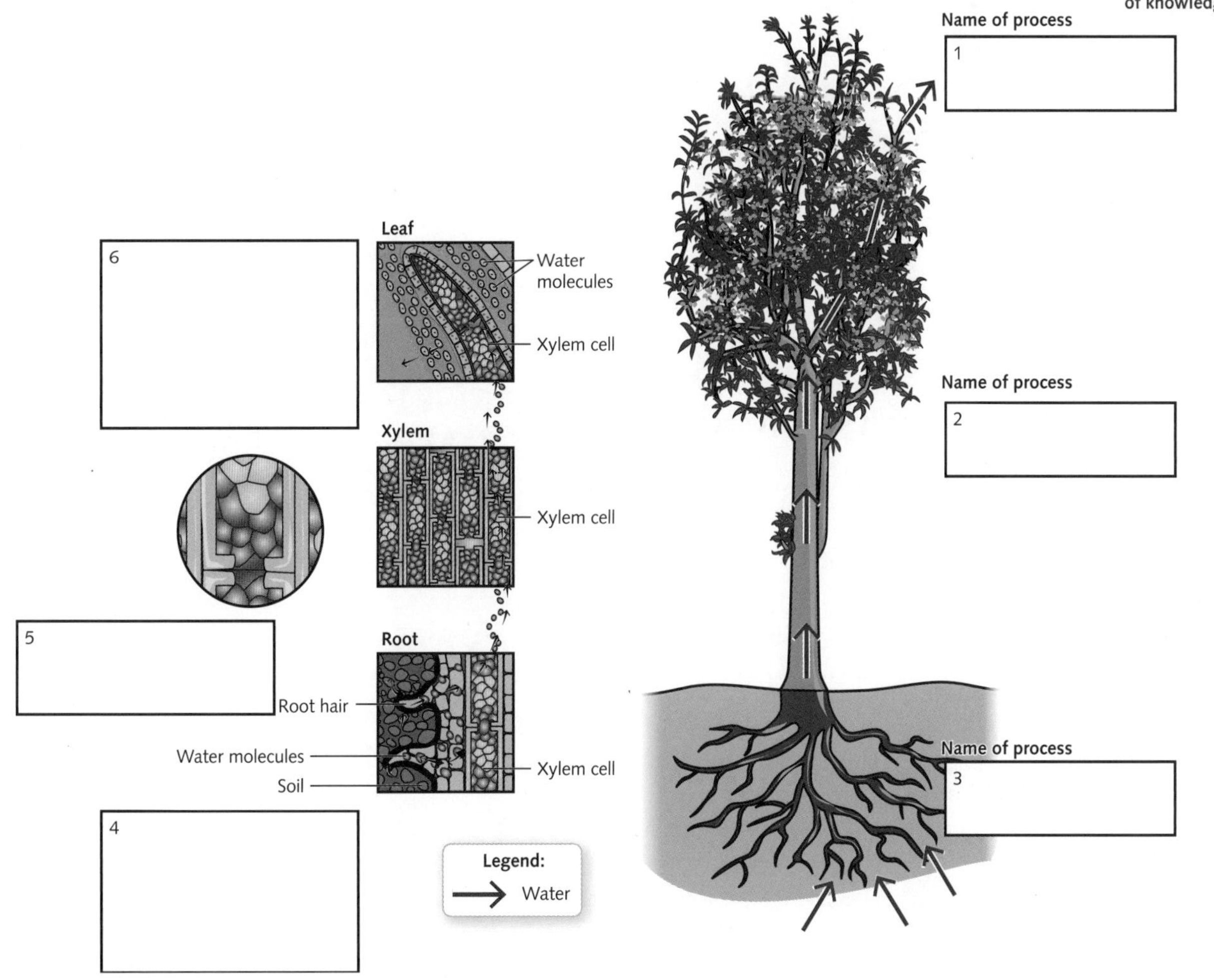

Figure 3.1 Water transport system in a plant

Using Figure 3.1 on page 57, complete the boxes as follows.

a Boxes 1–3: State the name of the process.

b Boxes 4–6: Describe in three lines what is happening to the water molecules.

3.1.2 Water loss from the shoot system

TB PAGE 97

Key science skills

Reinforce

Develop aims and questions, formulate hypotheses and make predictions

- identify, research and construct aims and questions for investigation
- identify independent, dependent and controlled variables in controlled experiments
- formulate hypotheses to focus investigation
- predict possible outcomes

Generate, collate and record data

- record and summarise both qualitative and quantitative data, including use of a logbook as an authentication of generated or collated data

A potometer is an instrument that measures the uptake of water by a plant, effectively measuring the rate of transpiration.

A research scientist wants to find out if external factors, such as air temperature, light intensity or wind speed over a leaf's surface, affect the rate of transpiration in a plant. He decides to test this by using a potometer, as shown in Figure 3.2.

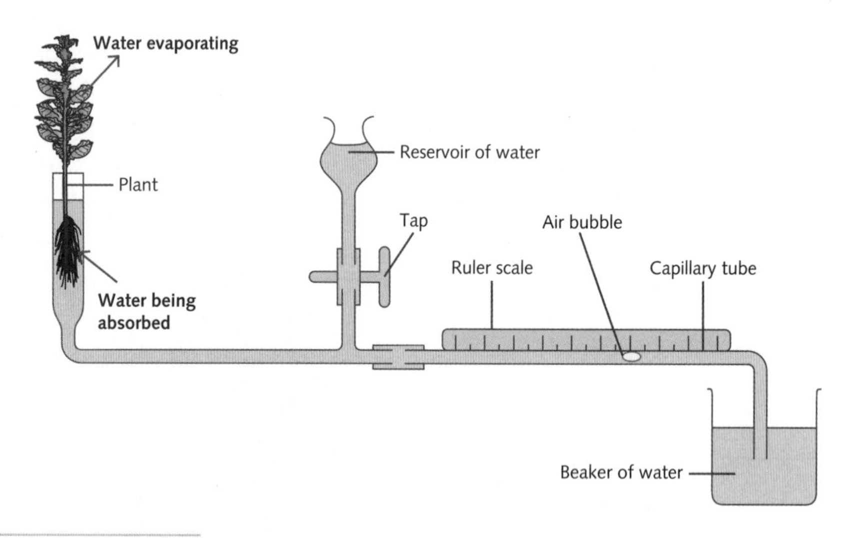

Figure 3.2 A potometer

Experiment A: varying intensity of light source

1 Write an aim for this investigation.

2 Write a research question for this investigation.

3 Determine the variables for this investigation.

a independent variable

b dependent variable

c controlled variables

4 Write a hypothesis for this investigation in the following format: If ... then ...

5 Predict what you think will be the outcome of this investigation.

Experiment B: effect of air movement at varying speeds

6 Write an aim for this investigation.

7 Write a research question for this investigation.

8 Determine the variables for this investigation.

a independent variable

b dependent variable

c controlled variables

9 Write a hypothesis for this investigation in the following format: If ... then ...

10 Predict what you think will be the outcome of this investigation.

Experiment C: effect of warm air

11 Write an aim for this investigation.

12 Write a research question for this investigation.

13 Determine the variables for this investigation.

a independent variable

b dependent variable

c controlled variables

14 Write a hypothesis for this investigation in the following format: If ... then ...

15 Predict what you think will be the outcome of this investigation.

3.1.3 Nutrient transport: phloem

PAGE 99

Key science skills

Generate, collate and record data

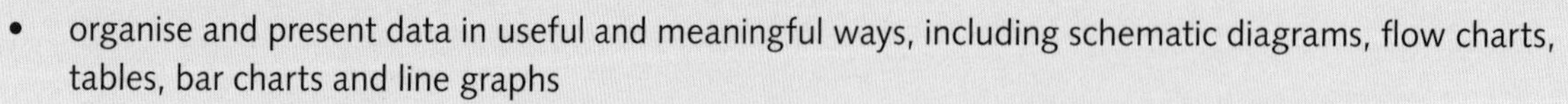

- organise and present data in useful and meaningful ways, including schematic diagrams, flow charts, tables, bar charts and line graphs

Figure 3.3 is a diagram of phloem tissue showing sieve tube cells with the associated companion cell. Also shown is the close connection between the phloem and xylem. Start at number 1 and study this figure carefully to make sure you fully understand what it is showing you.

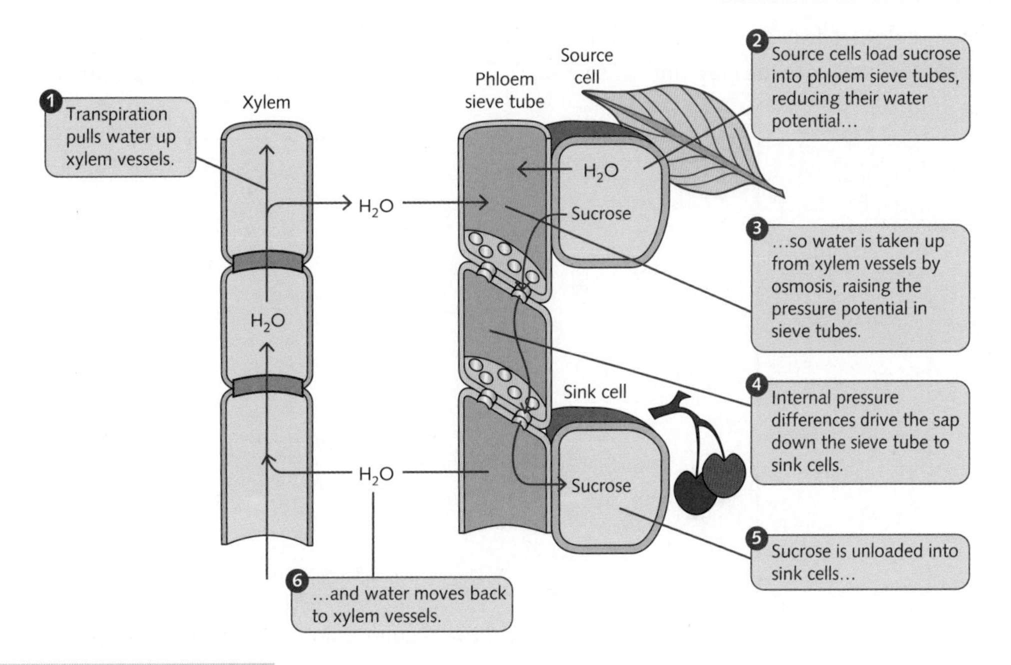

Figure 3.3 Nutrient movement in plants

Figure 3.4 a & b shows a longitudinal section of phloem cells and a cross-section of a sieve plate. Both diagrams were drawn by a scientist in their logbook.

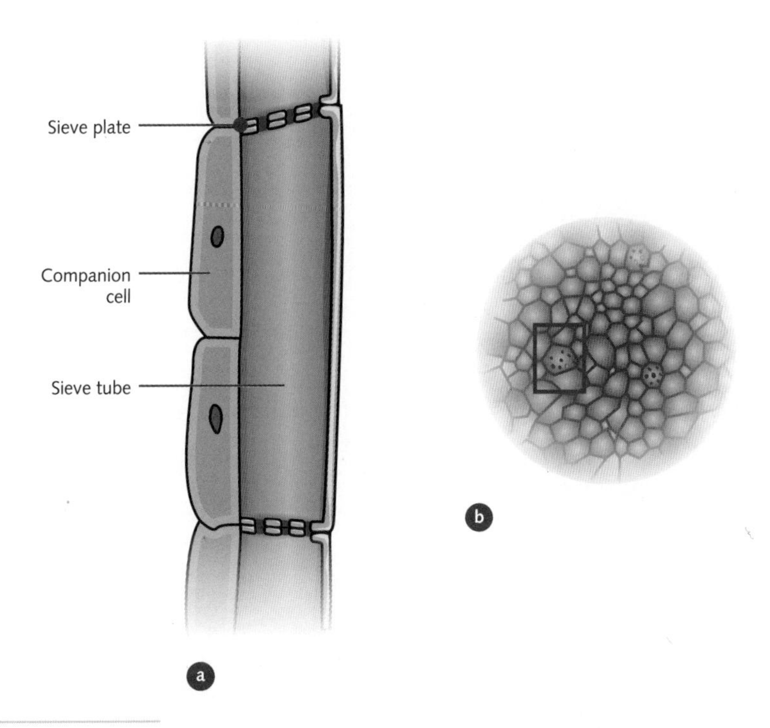

Figure 3.4 a Longitudinal section of phloem cells. b Cross-section of a sieve plate

1 In the space below, redraw the cells in the boxed part of the cross-section in Figure 3.4 on page 61, using scientific drawing conventions. Use the longitudinal section to help you label the three marked structures on the cross-section.

2 Describe how the structures of the sieve plate, sieve tube and companion cells make them well suited to their function.

3 Refer to Figure 3.3 on page 61. In plants, the cells that synthesise organic compounds such as glucose are called source cells. Identify the source cells in a plant.

4 Describe how the sucrose moves from the source cells to the sink cells (fruit or root cells) of the plant for storage as starch.

3.1.4 Translocation in plants

PAGE 99

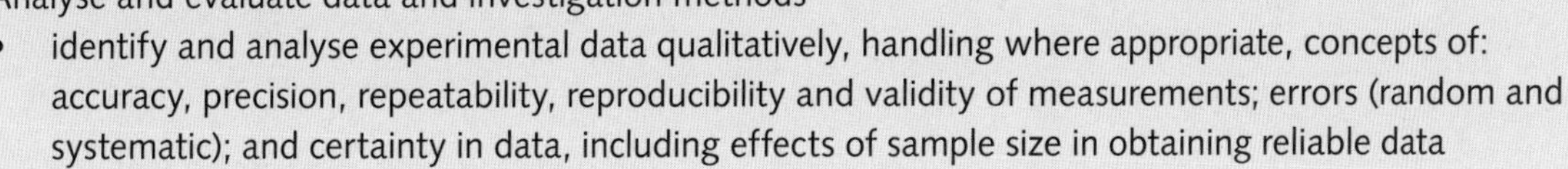

Key science skills

Analyse and evaluate data and investigation methods

- identify and analyse experimental data qualitatively, handling where appropriate, concepts of: accuracy, precision, repeatability, reproducibility and validity of measurements; errors (random and systematic); and certainty in data, including effects of sample size in obtaining reliable data

Data in biology is often presented in graphical form. Being able to analyse graphical data and work out what the graph is telling you is a key science skill. Study Figure 3.5 and read the axes labels carefully to work out what the graph is showing. Use this information to answer the questions below.

Translocation is the movement of nutrients, particularly sucrose, throughout the vascular plant. Sucrose is a product of photosynthesis in the green parts of plants, mainly the leaves. Sucrose moves from the photosynthetic cells by facilitated diffusion across the plasma membrane and into adjacent cells. The graph in Figure 3.5 shows the rate of translocation in a sugar beet plant. PCMBS is a substance that slows the rate of the uptake of sucrose by plant cells.

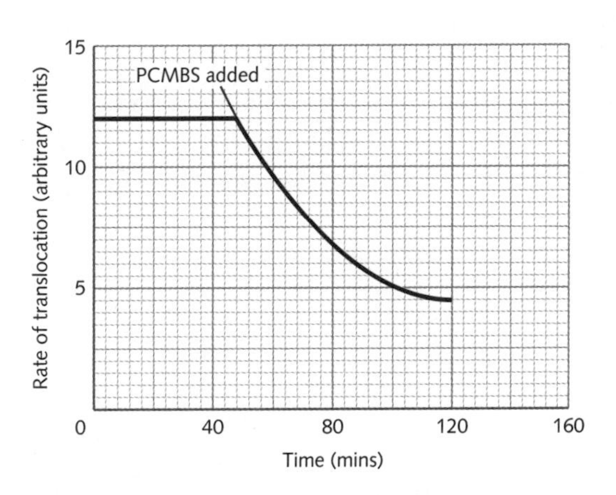

Figure 3.5 Rate of translocation of sugar in a beet plant

1 Describe what you see happening in the graph in Figure 3.5.

2 State two tissues that would be affected by the action of PCMBS. Provide an explanation for your choice of tissues.

3 This graph is known as second-hand data. What is second-hand data?

4 Extrapolate the data (extend the graph) to determine what the rate of transpiration would be after 160 minutes. Do you think that the plant would die due to the addition of PCMBS? Provide reasons for your answer.

3.1.5 Interaction of transpiration and translocation in plants

PAGE 99

Key science skills

Generate, collate and record data

- organise and present data in useful and meaningful ways, including schematic diagrams, flow charts, tables, bar charts and line graphs

Reinforce

Figure 3.6 shows a transverse section (or slice) through a vascular bundle. A vascular bundle contains both xylem and phloem tissue, which extend from the roots of the plant to the tips of the leaves. Vascular bundles also contain cambium tissue that separates the phloem from the xylem. Cambium tissue contains mitotically active cells. Use this information to answer the questions below.

1 a What do you think is the function of the vascular cambium?

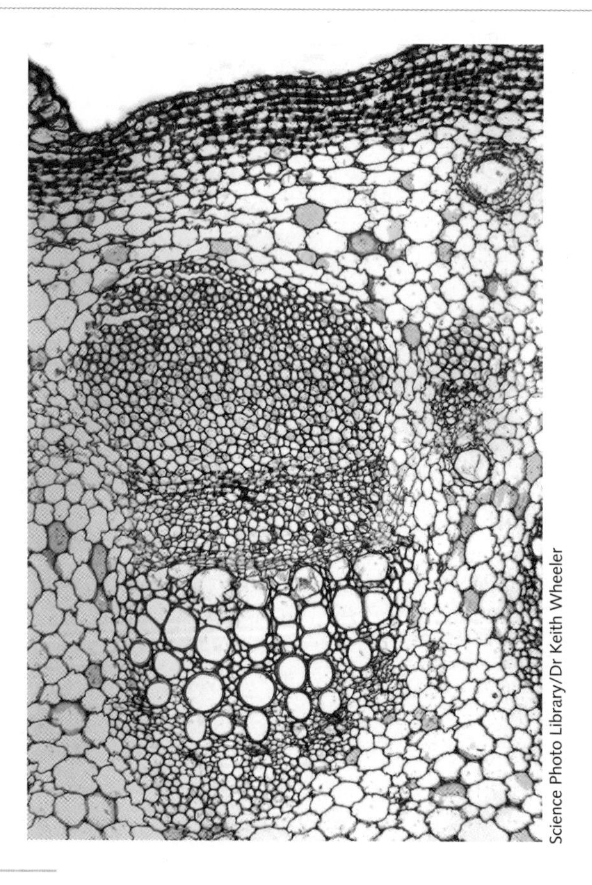

Figure 3.6 A vascular bundle

b Vascular bundles also contain protective and strengthening tissues. What do you think is the role of this tissue?

c Use the space below to draw a generalised diagram of Figure 3.6, showing the different areas in the cross-section. Do not show individual cells. Label each area.

Comparison of transport tissues in vascular plants

2 Figure 3.7 shows the outline of xylem and phloem cells. Use the information provided in this figure and your knowledge of xylem and phloem to complete Table 3.1.

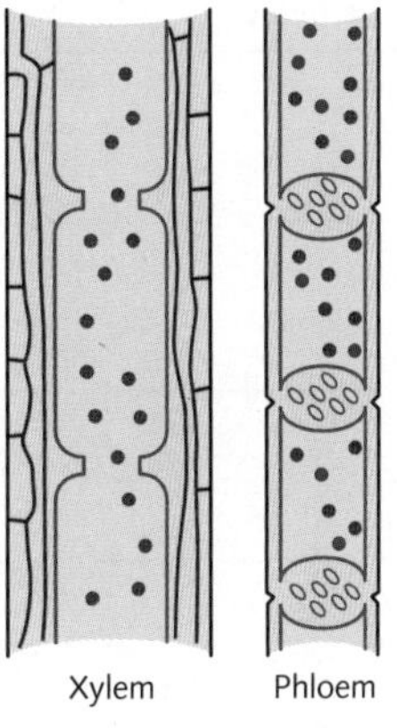

Figure 3.7 Comparison of xylem and phloem

Table 3.1

	Xylem	Phloem
Which way does the material flow?		
Are the cells living?		
Do the cells contain cytosol?		
Do the cells have a nucleus?		
What type of materials are transported?		
What is the end wall on each cell?		

3.2 Mammalian systems: digestive system

Key knowledge

Functioning systems

- specialisation and organisation of animal cells into tissues, organs and systems with specific functions: digestive, endocrine and excretory

3.2.1 Mechanical digestion

PAGE 100

Key science skills

Generate, collate and record data

- organise and present data in useful and meaningful ways, including schematic diagrams, flow charts, tables, bar charts and line graphs

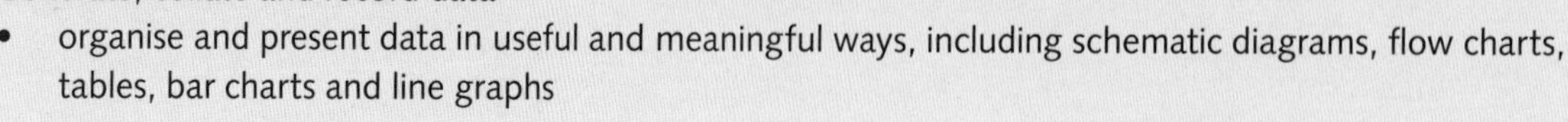

1 Mechanical digestion is also called physical digestion. Mechanical means that physical forces are used to break down large pieces of food into smaller pieces of food.

Complete the diagram of the human digestive system in Figure 3.8. Use different coloured pens or similar to annotate the diagram with the following.

a Name each structure.

b Using arrows and labels, show where mechanical digestion is occurring and name the type of mechanical digestion by summarising the functional nature of the process.

c Focusing specifically on peristalsis, use another colour as the code for peristalsis and identify areas of:

- peristaltic activity
- the type of muscle involved
- the contents of the gut being moved around.

2 Contents of the digestive system change as the contents proceed from the mouth to the anus. Using the diagram of the human digestive system as a reference, complete Table 3.2 below to show the mechanical or physical structures involved in mechanical digestion at each location and the contents of the digestive system at that point in the process.

Table 3.2

Location in digestive tract	Structures involved	Contents
Mouth		
Oesophagus		
Stomach		
Duodenum		
Jejunum		
Ileum		
Large intestine		
Anus		

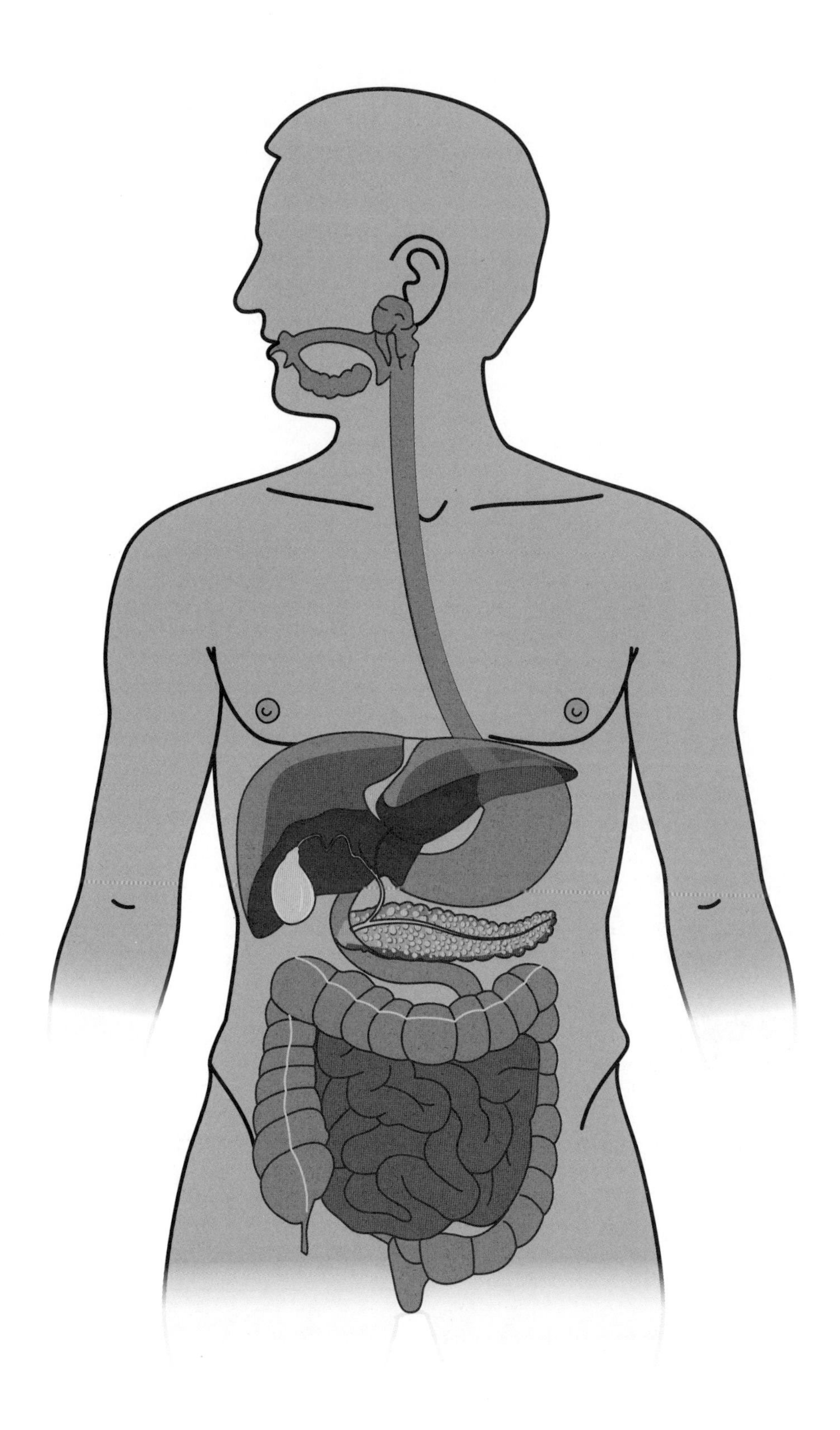

Figure 3.8 Human digestive system

3.2.2 Chemical digestion

PAGE 102

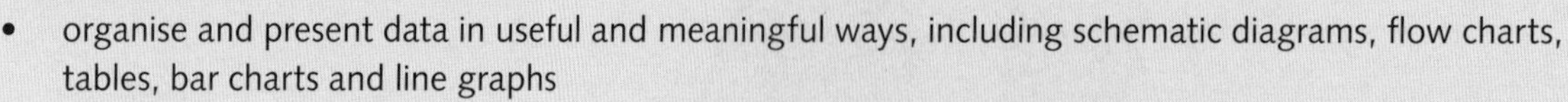

Key science skills

Generate, collate and record data

- organise and present data in useful and meaningful ways, including schematic diagrams, flow charts, tables, bar charts and line graphs

1 The three main food groups that humans ingest to be broken down by digestion are carbohydrates, fats and proteins. A number of substances are involved in their chemical digestion, including enzymes. Enzymes speed up chemical reactions that would otherwise take place very slowly. Digestive enzymes break down large molecules into smaller molecules. Fill in the boxes in Figure 3.9 with the names of the enzymes that are involved in the chemical digestion of each food group at the site specified.

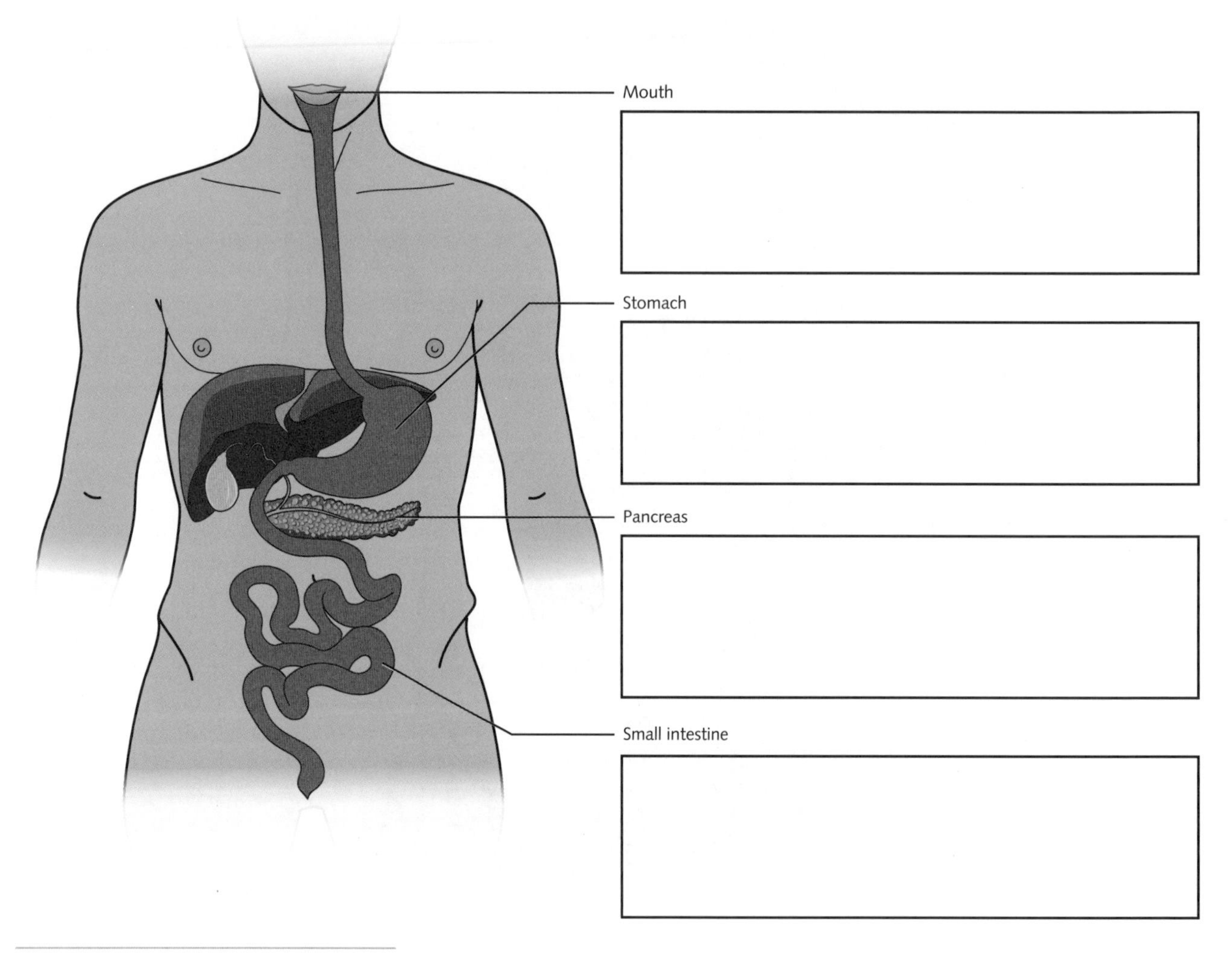

Figure 3.9 Enzymes involved in the chemical digestion of food

Structure of the digestive system

2 Source the missing words from the word list below and use them to fill in the blank spaces in Figure 3.10.

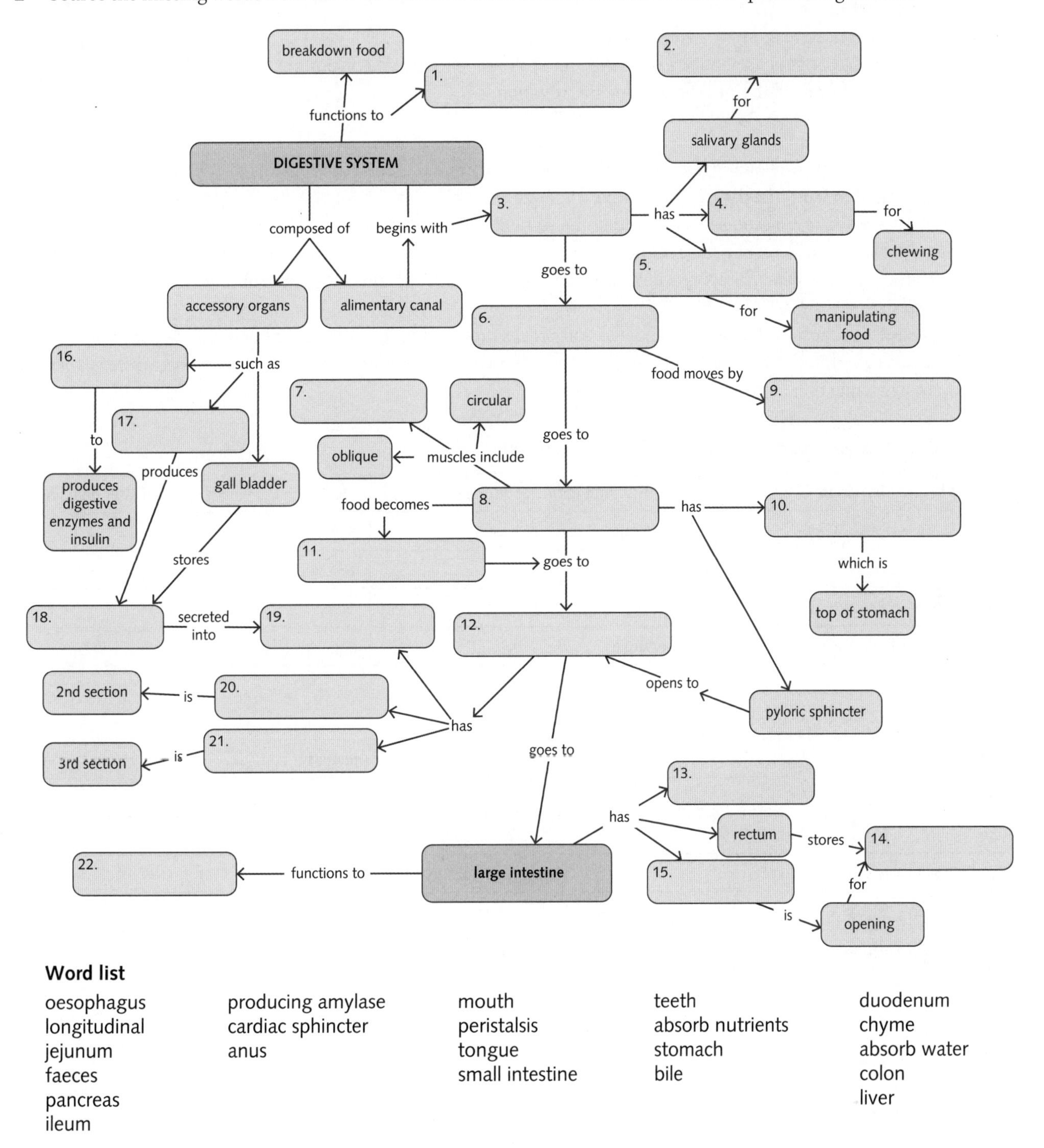

Word list

oesophagus
longitudinal
jejunum
faeces
pancreas
ileum
producing amylase
cardiac sphincter
anus
mouth
peristalsis
tongue
small intestine
teeth
absorb nutrients
stomach
bile
duodenum
chyme
absorb water
colon
liver

Figure 3.10 Mind map of human digestive system

3.2.3 Digestive enzymes

PAGE 102

Key science skills

Construct evidence-based arguments and draw conclusions

- use reasoning to construct scientific arguments, and to draw and justify conclusions consistent with the evidence and relevant to the question under investigation

Two students completed the following tests in the school laboratory and recorded their results in Tables 3.3 and 3.4 below. The students wanted to study the digestion of starch, a complex carbohydrate or polysaccharide. Starch molecules break down into smaller maltose molecules (disaccharide). The students used iodine to test for starch. Iodine turns dark purple in the presence of starch; it remains yellow in the absence of starch. They used Benedict's solution to test for the presence of maltose. In the presence of maltose, Benedict's solution turns yellow; it remains blue in the absence of maltose. The students' results are shown in Table 3.3.

Table 3.4 shows the action of pepsin, an enzyme that breaks down protein, under a number of different conditions. The protein used was egg white, which was incubated with the pepsin under the conditions shown. When egg white is digested by pepsin, the egg white breaks down into smaller pieces.

Table 3.3 Digestion of carbohydrate (starch) by salivary amylase

Contents before incubation	Test with iodine after one hour	Test with Benedict's solution after one hour
Tube 1: Starch + distilled water	Dark purple	Blue
Tube 2: Starch + saliva	Yellow	Yellow
Tube 3: Starch + saliva + HCl	Purple	Green
Tube 4: Starch + boiled saliva	Dark purple	Blue

Table 3.4 Digestion of protein (egg albumin) by pepsin

Incubation condition	Appearance of egg white after one hour
Tube 1: Protein + pepsin at 37°C	Some pieces of egg floating
Tube 2: Protein + pepsin + HCl at 37°C	Many pieces of egg floating
Tube 3: Protein + pepsin + HCl at 0°C	Egg not broken down
Tube 4: Protein + HCl at 37°C	Egg not broken down
Tube 5: Protein + pepsin + NaOH at 37°C	Few pieces of egg floating

1 What conclusions can be drawn from the data shown in:

a Table 3.3?

____

____

____

____

____

____

b Table 3.4?

____

____

9780170452632

2 You are investigating the activity of digestive enzymes. You gain a set of results for three enzymes, A, B and C, and plot the data on a graph, as shown in Figure 3.11.

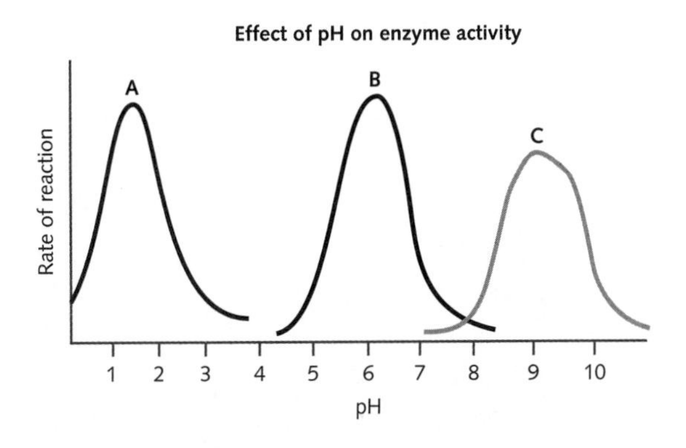

Figure 3.11 Results of activity of enzymes A, B and C

From your knowledge of digestive enzymes, complete the following.

a Probable name of enzyme A:

Location:

Optimum pH of enzyme A:

Substrate:

Product:

b Probable name of enzyme B:

Location:

Optimum pH of enzyme B:

Substrate:

Product:

c Probable name of enzyme C:

Location:

Optimum pH of enzyme C:

Substrate:

Product:

3.2.4 Absorption in the human gut

TB PAGE 107

Analyse, evaluate and communicate scientific ideas

- use appropriate biological terminology, representations and conventions, including standard abbreviations, graphing conventions and units of measurement
- acknowledge sources of information and assistance, and use standard scientific referencing conventions

In humans, nearly all nutrients are absorbed across the moist walls of the gut. These nutrients include:

- water and mineral ions
- carbohydrates, after digestion to monosaccharides
- proteins, after digestion to small peptides and amino acids
- neutral fat, after digestion to glycerol and free fatty acids.

Figure 3.12 shows a diagram of the absorption structures of the human ileum. Figure 3.13 shows a transmission electron micrograph (TEM) of the human small intestine and Figure 3.14 shows a cross-section of the wall of the human stomach.

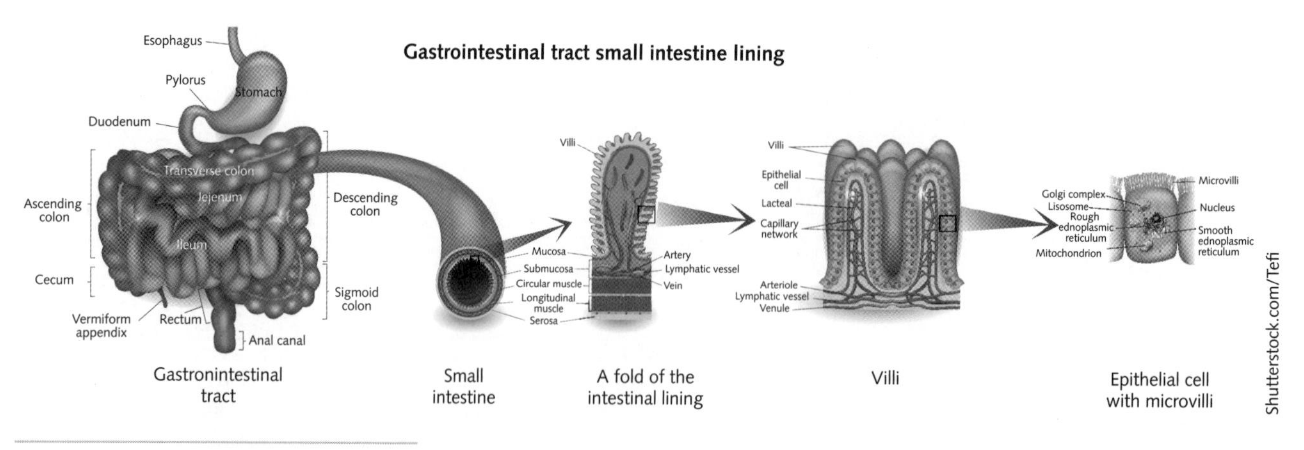

Figure 3.12 Absorption structures in the human ileum

9780170452632

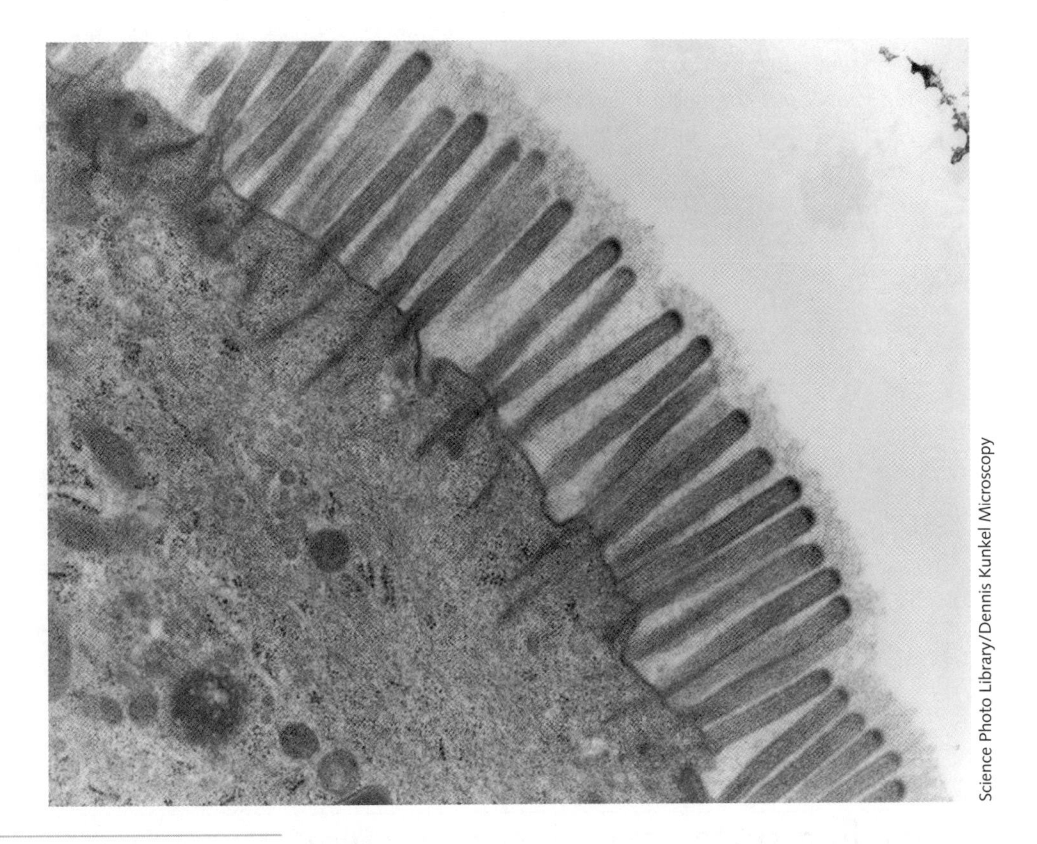

Figure 3.13 Microvilli of the small intestine (TEM)

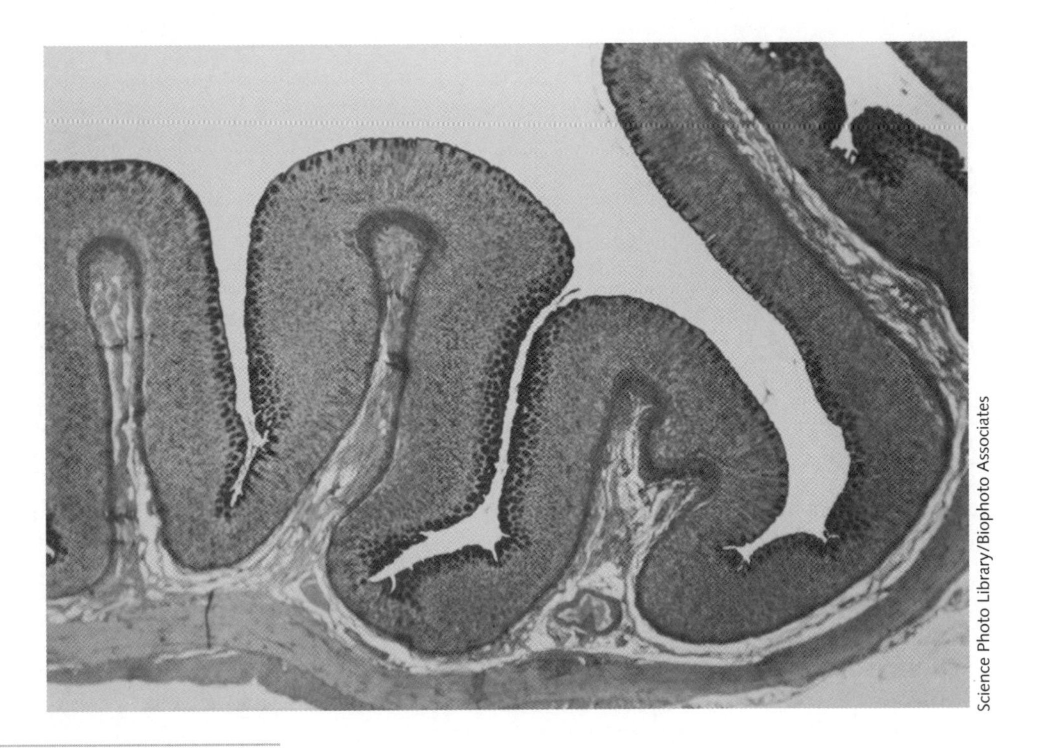

Figure 3.14 Histology slides showing the cross-section of the human stomach

7 Use Figures 3.12–3.14 as secondary source material to create a summary of the passage of the nutrients listed above from the lumen (inside) of the gut into the internal environment. Make sure that you use appropriate biological terminology and acknowledge the sources of your information. See the hints provided to help you structure your response.

Hint: talk about if any absorption occurs through the stomach walls

Hint: talk about the villi and the microvilli in the small intestine

Hint: talk about absorption into the lymphatic system

Hint: finish at absorption into the blood stream

3.3 Mammalian systems: endocrine system

Key knowledge

Functioning systems

- specialisation and organisation of animal cells into tissues, organs and systems with specific functions: digestive, endocrine and excretory

3.3.1 Endocrine system

PAGE 110

Key science skills

Generate, collate and record data

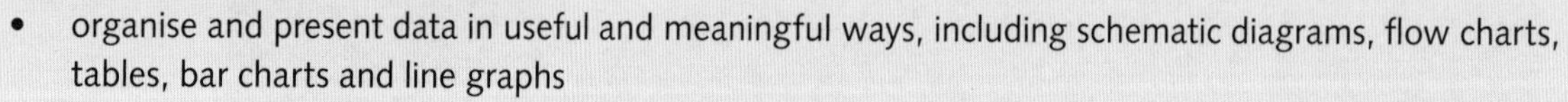

- organise and present data in useful and meaningful ways, including schematic diagrams, flow charts, tables, bar charts and line graphs

Hormones form part of the body's homeostatic control mechanisms to help you grow, repair tissues and stay alive and healthy. The endocrine system produces hormones and it has specific structures to carry out this function.

1 Complete the diagram in Figure 3.15 of the human body by drawing the main organs such as brain, heart, lungs, kidneys, stomach, intestines and oesophagus. Label each organ.

2 Use different colours to draw in and label the following main glands.

- hypothalamus
- pituitary and pineal glands
- pancreas
- thyroid and parathyroid
- adrenal gland
- liver
- testes and ovaries

3 Annotate the diagram, using boxes, with the following.

a name of hormones produced by each gland

b specific target tissue or cells of each hormones

c effect of each hormone on target tissue, and overall response

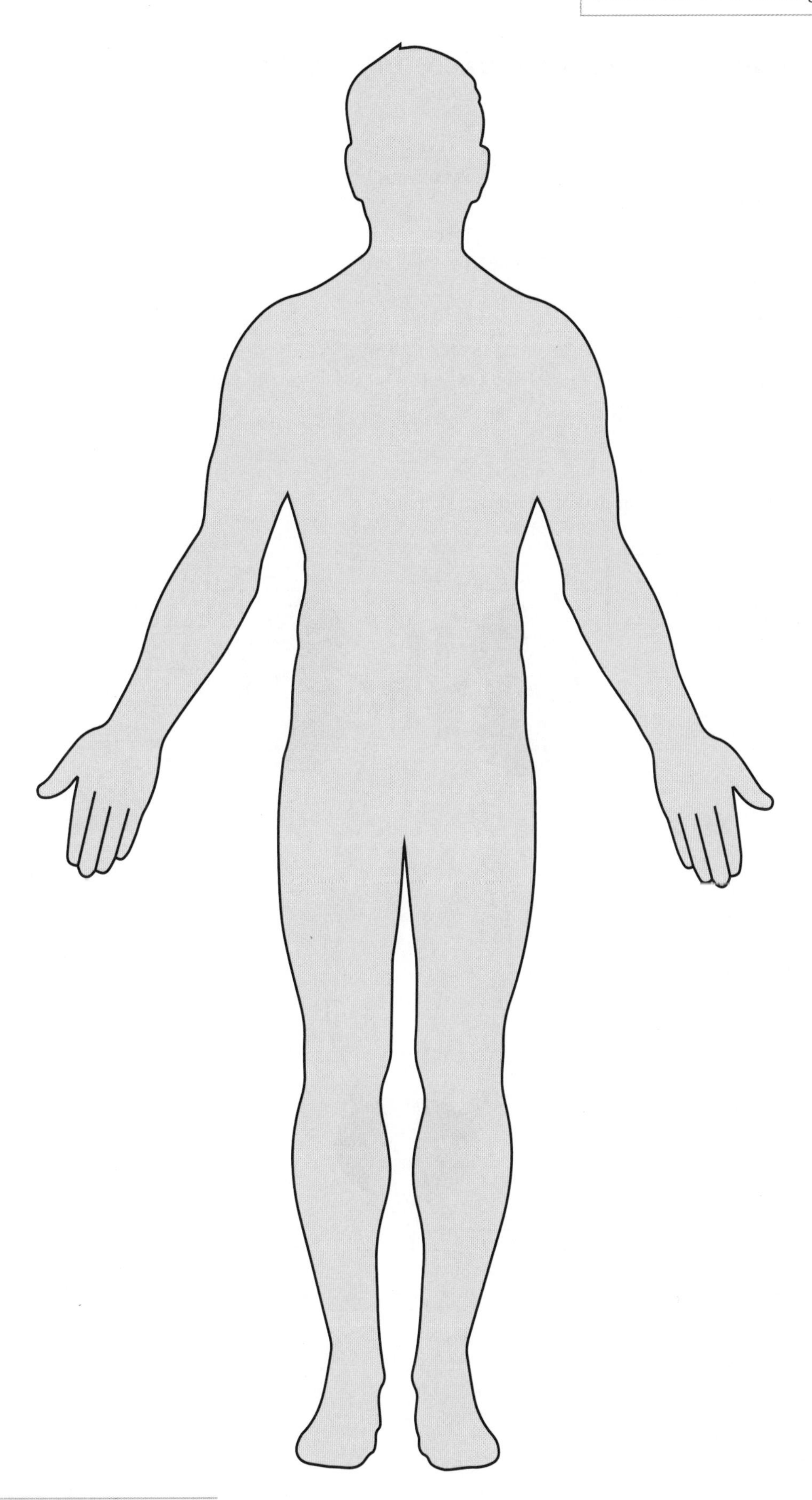

Figure 3.15 Main organs of the human body

3.4 Mammalian systems: excretory system

Key knowledge

Functioning systems

- specialisation and organisation of animal cells into tissues, organs and systems with specific functions: digestive, endocrine and excretory

3.4.1 Structure of the excretory system

PAGE 115

Key science skills

Generate, collate and record data

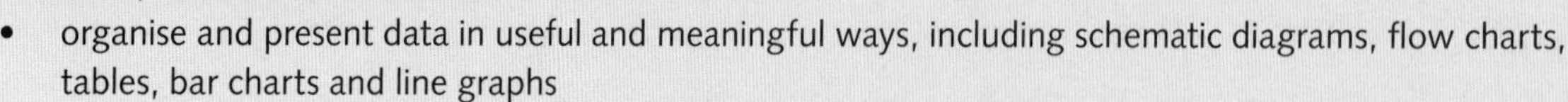

- organise and present data in useful and meaningful ways, including schematic diagrams, flow charts, tables, bar charts and line graphs

1 Figure 3.16 shows a diagram of the human excretory system. Recall the major structures of this system by labelling each structure with the words and phrases from the list provided.

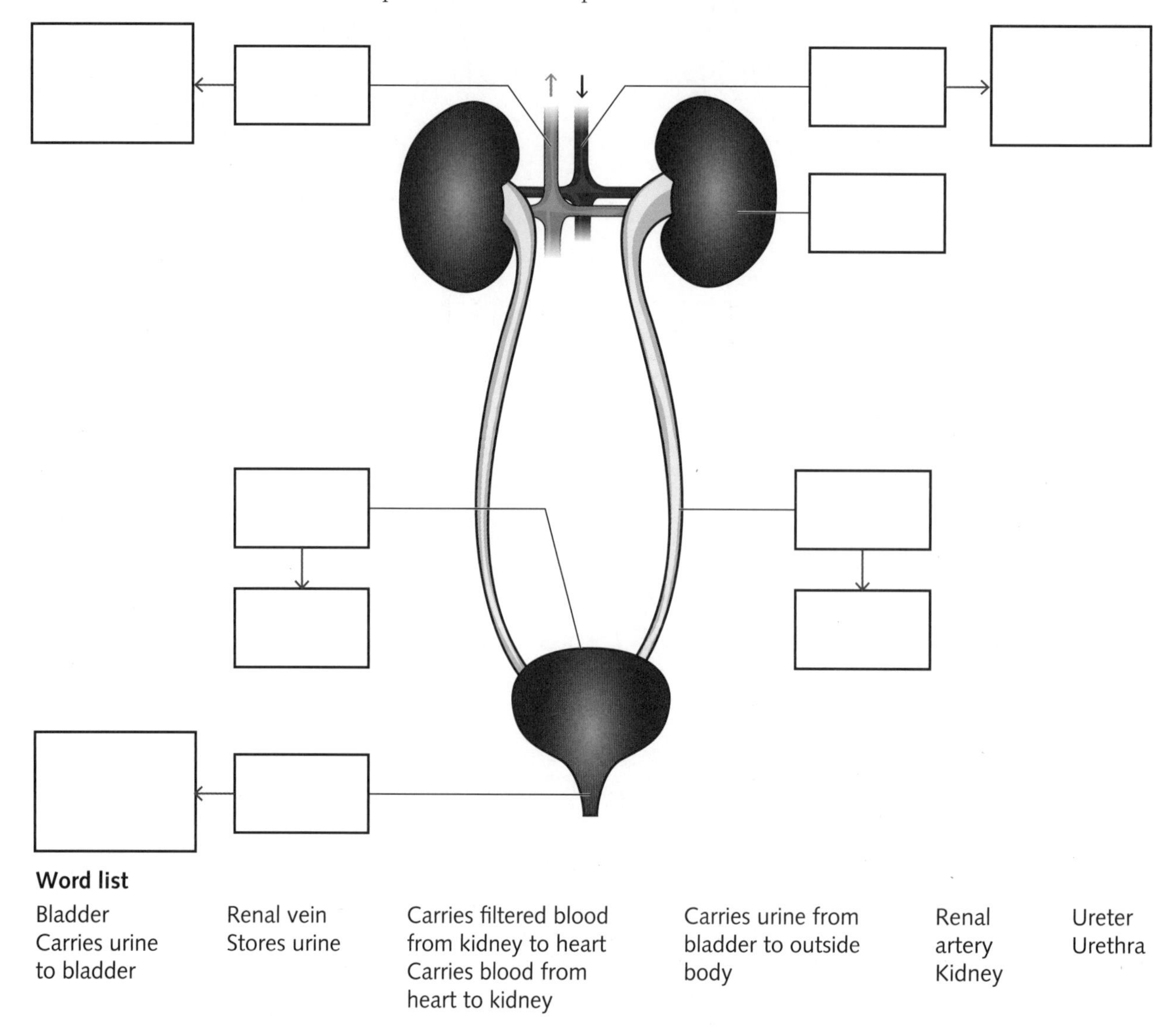

Figure 3.16 General diagram of human excretory system

2 Figure 3.17 shows a diagram drawn by a Senior Biology student summarising the function and structure of the kidney nephron. Use this diagram to draw a flow chart below it that summarises the steps involved in the processing of 50 mL of blood that travels into the kidney through the renal artery and is filtered by the structures in the nephron so that wastes and excess water are eventually removed from the body in urine. Ignore the action of ADH (antidiuretic hormone).

9780170452632

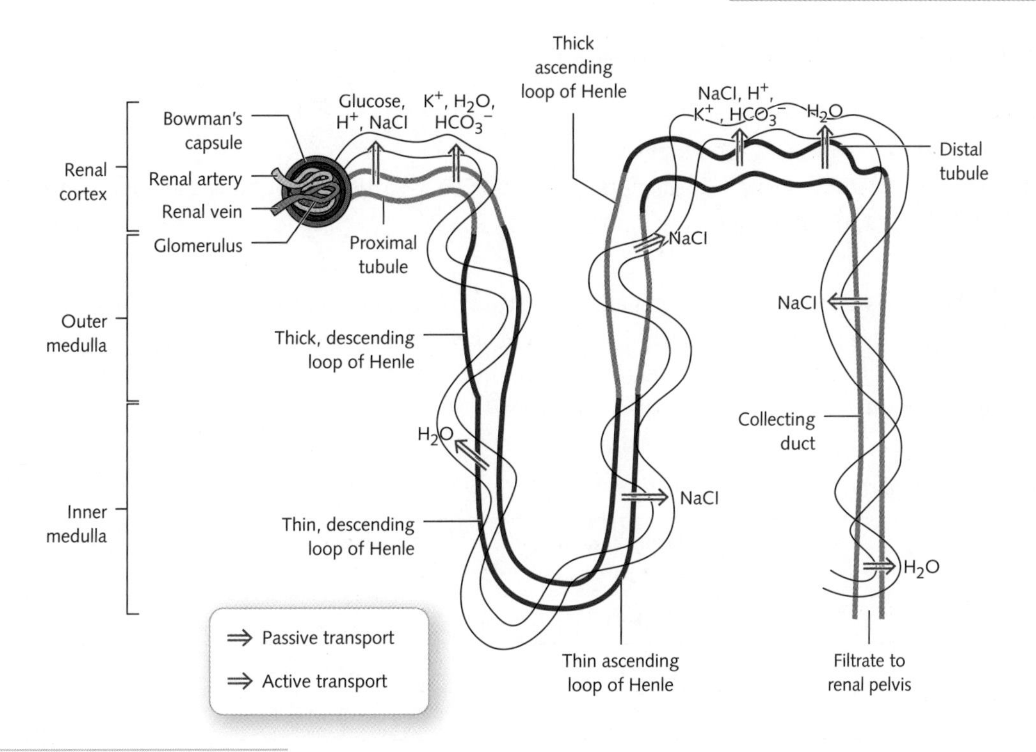

Figure 3.17 Senior Biology student's diagram summarising the structure and function of the kidney nephron

Draw your flow chart here

3.4.2 Composition of fluids in the kidney

PAGE 116

Key science skills

Construct evidence-based arguments and draw conclusions

- use reasoning to construct scientific arguments, and to draw and justify conclusions consistent with the evidence and relevant to the question under investigation

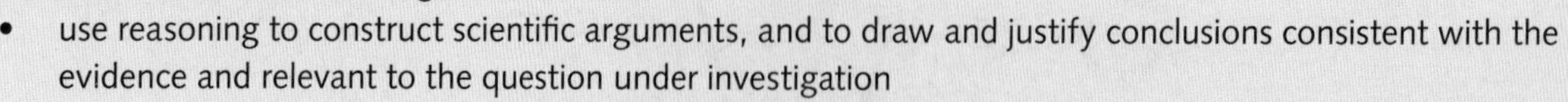

A key skill in VCE Biology is being able to interpret and analyse data in a tabular format. You need to be able to work out what the data is telling you, what the trends in the data are and how to draw conclusions from the data. Table 3.5 shows the composition of fluid as it moves through various structures in the kidney. To avoid being overwhelmed by the data, just look at the specific data you need to answer the questions below.

Table 3.5 Composition of the fluids in the kidney

Component	Blood plasma entering the glomerulus	Filtrate in Bowman's capsule	Urine in collecting duct
Water	90–93	97–99	96
Blood proteins	7–9	some*	0.0
Glucose	0.10	0.10	0.0
Urea	0.03	0.03	2.0
Other N-containing compounds	0.003	0.003	0.24
Ions:			
sodium	0.32	0.32	0.30–0.35
chloride	0.37	0.37	0.60
others (Ca^{2+}, Mg^{2+}, K^{+}, PO_4^{3-}, SO_4^{2-})	0.038	0.038	0.475
pH	7.35–7.45		4.7–6.0 (average 5.0)

*Some of the smallest blood protein molecules only

Analyse the data in Table 3.5 to answer the following questions.

3 a Look at the data for blood proteins in the second row. Account for the lack of blood proteins in the urine.

b The filtrate has a higher concentration of water in the Bowman's capsule (look at the top row of the third column) than in the blood (look at the top row of the second column). Account for this.

c Why does the urea concentration increase from the filtrate to the urine? First, work out where you need to look to get the data that you need to answer this question.

9780170452632

d What does 'Other N-containing compounds' mean?

e Explain the difference in pH (last row) between the blood plasma and urine.

f Glucose (look at the third row) is not normally found in the urine but is found in the filtrate. How can this be?

g Comment on where and how ions (look at sixth row) are reabsorbed in the nephron.

3.5 Chapter review

PAGE 125

3.5.1 Key terms

1 For this activity, use the key terms listed below. Separate the key terms list into terms that refer to vascular plants and terms that refer to animals, but mainly humans. Write them into two columns below. Highlight the terms that appear in both columns.

Key term list	Vascular plants	Animals
absorption adhesion ammonia amylase anus basal metabolic rate bile Bowman's capsule chemical digestion chyme cohesion colon cuticle deamination digestion digestive system distal tubule ductless gland egestion endocrine gland endocrine system epidermis excretion exocrine gland external environment faeces filtrate gall bladder gastric juice gastrointestinal tract glomerulus heterotroph hormones ingestion internal environment kidney lacteal large intestine lignin loop of Henle lymph lymphatic system mechanical digestion microvilli nephron oesophagus organ pancreatic juice parenchyma peristalsis pH phloem polypeptide		

Key term list	Vascular plants	Animals
protease proximal tubule pyloric sphincter rectum renal artery renal pelvis root hair cells root pressure small intestine sphincter stomata terrestrial thyroxine tissue tracheid translocation transpiration transpiration stream transpirational pull urea vascular bundle vascular plant vascular tissue villi xylem xylem vessel element		

2 Separate the list of plant key terms into structural terms (how they are built) and functional terms (what they do) below. Then, do the same with the animal terms.

PLANTS	
Structural terms	Functional terms

ANIMALS	
Structural terms	Functional terms

3 Match the structural key terms to the applicable functional processes in the tables below. Draw a diagram beside each key term to show the general structure.

Table 3.6 Plant key terms

Structural term	Functional term	Diagram

Table 3.7 Animal key terms

Structural term	Functional term	Diagram

Structural term	Functional term	Diagram

3.5.2 Practice test questions

TB PAGE 127

Multiple-choice questions

1 An example of chemical breakdown of food is

A chewing and movement in the mouth by the tongue.
B action of the enzyme trypsin on proteins.
C stomach churning the contents to become chyme.
D bile emulsification of fats in the small intestine.

2 Excess glucose is temporarily stored in the liver as

A fat.
B bile.
C glycogen.
D starch.

3 Which gland is responsible for the fight-or-flight response?

A parathyroid
B thyroid
C adrenal
D testes

Chapter review continued

4 The main pulling force for transpiration is

- **A** when water is taken in at the root surfaces.
- **B** when water molecules adhere to the cells of the xylem.
- **C** when water evaporates at the leaf surface.
- **D** cohesion of water molecules in the transpiration stream.

5 Which one of the following statements about the phloem is correct?

- **A** The contents of the phloem only move from the roots to the leaves.
- **B** Sieve cells contain a nucleus that controls the cell.
- **C** Sucrose and amino acids move in the translocation stream.
- **D** The xylem is made up of living cells, whereas the phloem is made up of dead cells.

6 In the kidney nephron, the function of the Bowman's capsule is to

- **A** filter out water and dissolved materials from the blood.
- **B** reabsorb water and mineral ions from the filtrate.
- **C** filter out urea and proteins from the blood.
- **D** secrete wastes to the outside.

Short-answer questions

7 Xavier is a Year 11 Biology student and as part of his assessment, he investigated the rate of transpiration from geranium leaves. He obtained two sets (X and Y) of ten geranium leaves. He left the ten leaves in set X untreated. He covered the upper surfaces of the ten leaves in set Y with petroleum jelly. He weighed the mass of each set of leaves on an electronic balance and then tied all the leaves in each set to a separate length of thread. The set-up is shown in Figure 3.18a.

He then weighed each set of leaves every 20 minutes over a period of 2 hours and plotted a graph of his results, as seen in Figure 3.18b.

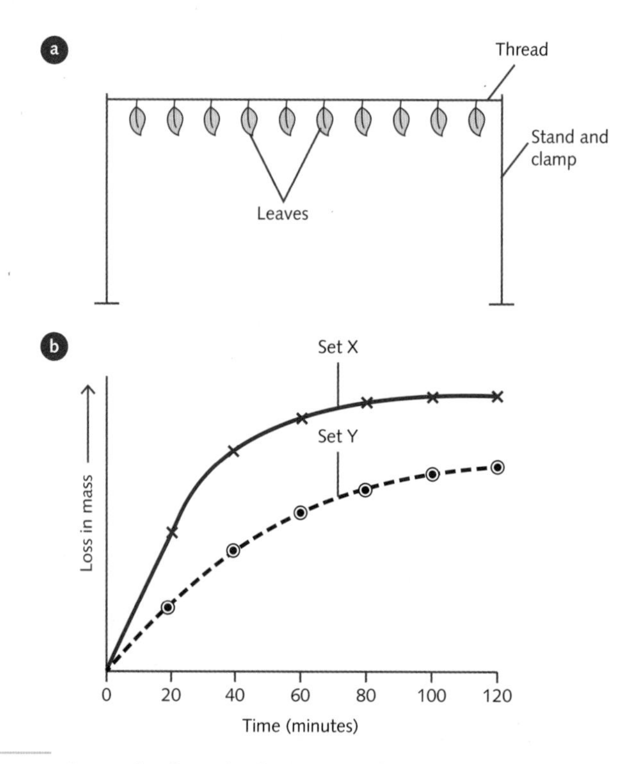

Figure 3.18 a Experimental set-up. **b** Graph of results for Set X and Set Y

a State two environmental conditions that Xavier needed to keep constant during this investigation.

2 marks

b Xavier measured the water loss in milligrams. Explain the advantage of using ten leaves when taking measurements in milligrams.

1 mark

c Explain the change in mass of untreated leaves in Set X shown in the graph.

3 marks

d The results that Xavier obtained for the leaves in Set Y were different from those for Set X. Suggest an explanation for this difference.

2 marks

8 The glomerular filtrate becomes concentrated in the descending loop of Henle and then becomes diluted in the ascending loop of Henle in the nephron. Use appropriate biological terminology and clear and concise expression to explain why this occurs.

4 marks

4 Regulation of systems

Getty Images/Kenny Williamson

Remember

Chapter 4 Regulation of systems will call on content that you have already met in your science studies from previous years or from earlier this year. Take some time to refresh your knowledge of this content before you enter this chapter. Try to answer the following questions from memory. If you cannot do this, then use a reference to assist you.

PAGE 134

1 Describe how water moves from the roots to the leaves of plants.

2 State the roles of the endocrine system.

3 Describe how the excretory system removes excess water from the bloodstream.

4.1 Regulation of water balance in vascular plants

Key knowledge

Regulation of systems

- regulation of water balance in vascular plants

4.1.1 Water balance in vascular plants

PAGE 135

Key science skills

Analyse, evaluate and communicate scientific ideas

- use clear, coherent and concise expression to communicate to specific audiences and for specific purposes in appropriate scientific genres, including scientific reports and posters

A farmer who grew maize crops had a good understanding of how water moved from the water table up into the root zone uptake area and then into the soil solution of the plant. She knew that adequate water needed to be continuously applied through rainfall and irrigation and that the rate of transpiration from the plants was affected by the amount of sunlight and ambient weather conditions in the vicinity of the crop (Figure 4.1). If the crop was without water, it would grow poorly and even fail. This was a risk to the farmer's business.

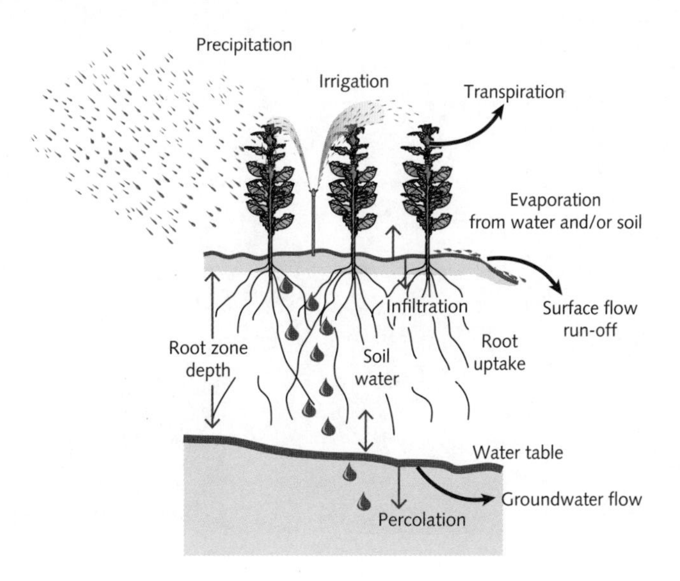

Figure 4.1 Various components of water transport in the vicinity of a crop

What the farmer did not really understand was what went on inside the plants. She decided to attend some classes at a TAFE on this topic. You are the teacher who is instructing students about water balance inside a plant. To explain this, you first need to discuss water movement through the vascular tissue of plants. You are writing your teaching notes and know the following terms need to be included.

adhesion forces	hydrogen bonds	water potential
cohesion forces	tracheid	xylem
evaporation	transpiration	

You also have a diagram one of your colleagues gave you, shown in Figure 4.2.

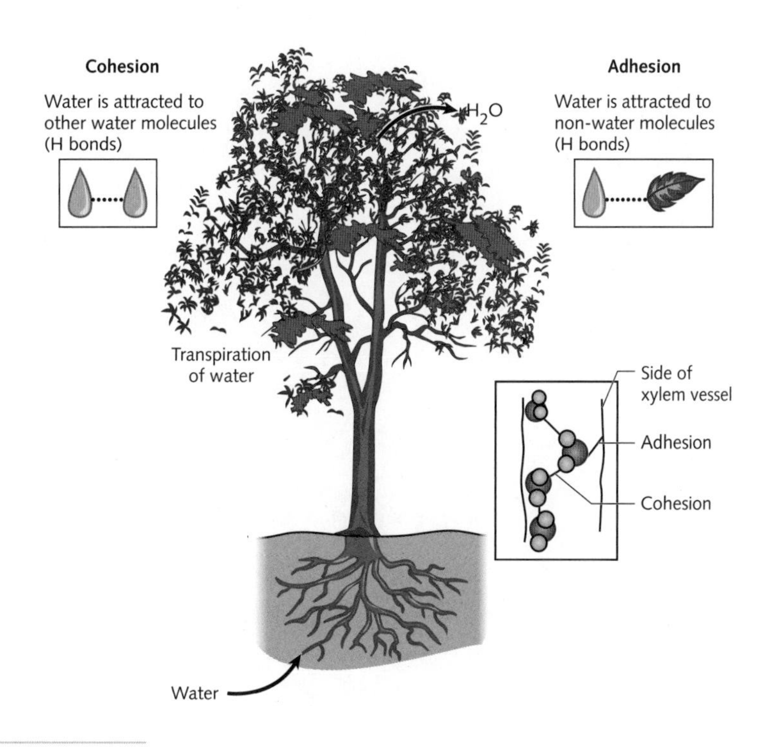

Figure 4.2

Finally, you have a stimulus diagram for the students as shown in Figure 4.3.

Figure 4.3 A lift in a shopping centre

1 In the space below and on the following page, prepare your notes to teach water movement through a plant to the farmer in your class.

- Think carefully about the best order to teach the content for maximum understanding.
- Make sure you draw attention to the effects of dry and humid conditions on rates of water flow.
- Use flow charts and diagrams as relevant.

9780170452632

4.1.2 Stomatal control and water balance

PAGE 136

Key science skills

Develop

Analyse and evaluate data and investigation methods

- identify and analyse experimental data qualitatively, handling where appropriate concepts of: accuracy, precision, repeatability, reproducibility and validity of measurements; errors (random and systematic); and certainty in data, including effects of sample size in obtaining reliable data

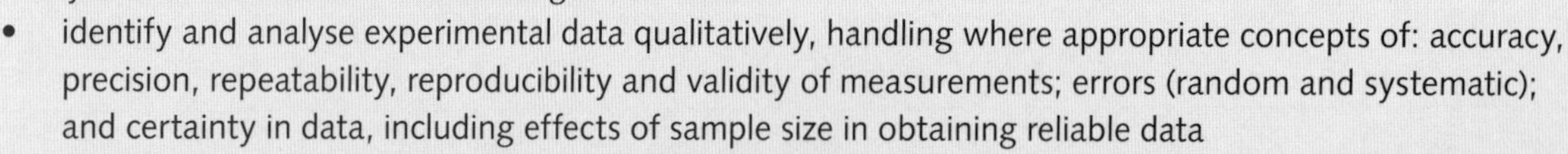

Construct evidence-based arguments and draw conclusions

- use reasoning to construct scientific arguments, and to draw and justify conclusions consistent with the evidence and relevant to the question under investigation

The main driver in water movement from the leaves of plants is the opening and closing of the stomata to enable transpiration to occur. Figures 4.4 and 4.5 summarise the control of stomatal opening and closing. Study the figures carefully to make sure that you understand what they are telling you. Use the information from the figures and your knowledge of stomatal control to answer the questions below.

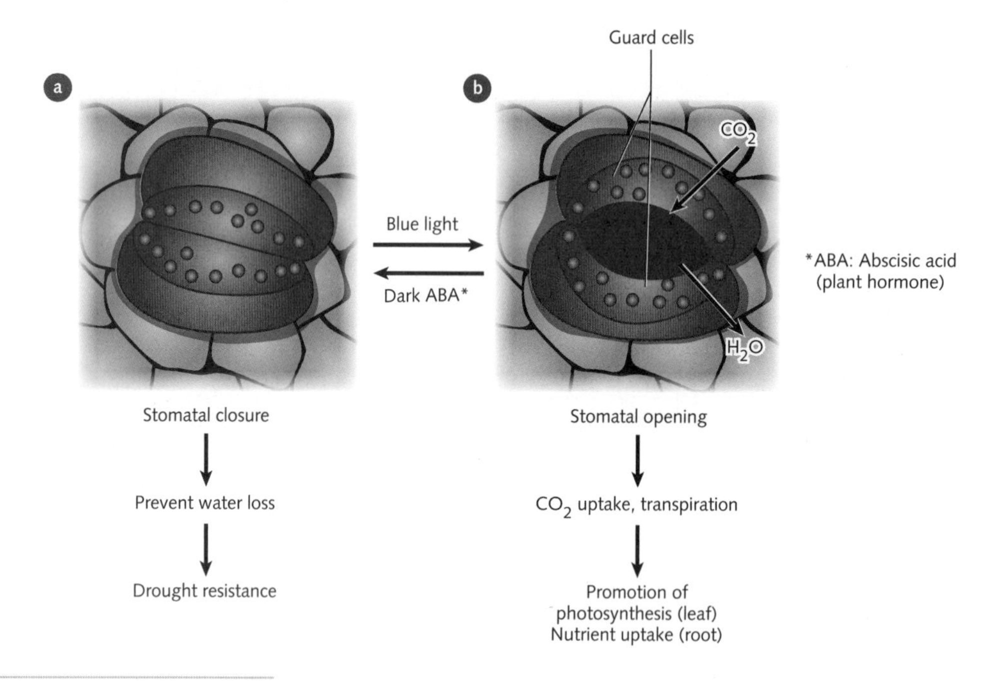

Figure 4.4 Stomata **a** close in the presence of abscissic acid and **b** open in the presence of blue light.

Abscisic acid control of stomatal opening

Inside Outside H_2O K^+ Ca^{2+} ABA

Open stoma **Guard cell plasma membrane** **Closed stoma**

ABA binding leads to influx of Ca^{2+} and the opening of K^+ channels. Water exits guard cells and stoma closes.

Figure 4.5 Abscisic acid controls stomatal opening.

1 a What is a guard cell?

b What is abscisic acid (ABA)?

c What is the main function of ABA?

d Why do stomata open in the presence of blue light, as shown in Figure 4.4?

2 a ABA is specific for a receptor found in the guard cell plasma membrane. When ABA is present, transport proteins in the plasma membrane open, allowing calcium ions to move into the cell and potassium ions to move out of the cell. Refer to Figure 4.5.

Complete the sentence: This entire process causes ...

b Why is this process part of the survival techniques of a vascular plant?

3 The ions of calcium and potassium move 'down a concentration gradient' as they pass through the plasma membrane. What does this mean?

4.2 Human survival in a range of environments

Key knowledge

Regulation of systems

- regulation of body temperature, blood glucose and water balance in animals by homeostatic mechanisms, including stimulus-response models, feedback loops, and associated organ structures

4.2.1 Internal and external environment of the body

PAGE 139

Key science skills

Analyse, evaluate and communicate scientific ideas

- use appropriate biological terminology, representations and conventions, including standard abbreviations, graphing conventions and units of measurement

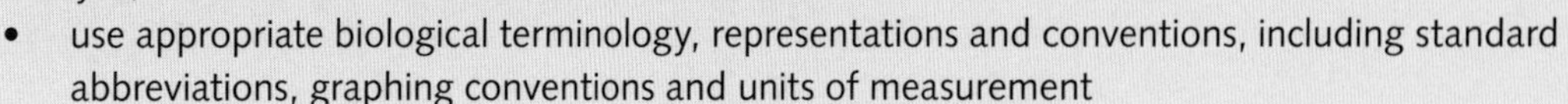

1 Figure 4.6 shows you a simplified diagram of cells, organs and systems in the human body. The skin encloses these but you will notice that the skin has gaps in it. Use your understanding of the external and internal environments of the human body to colour in all the external environment in blue and all the internal environment in red. Note that this drawing is not to scale.

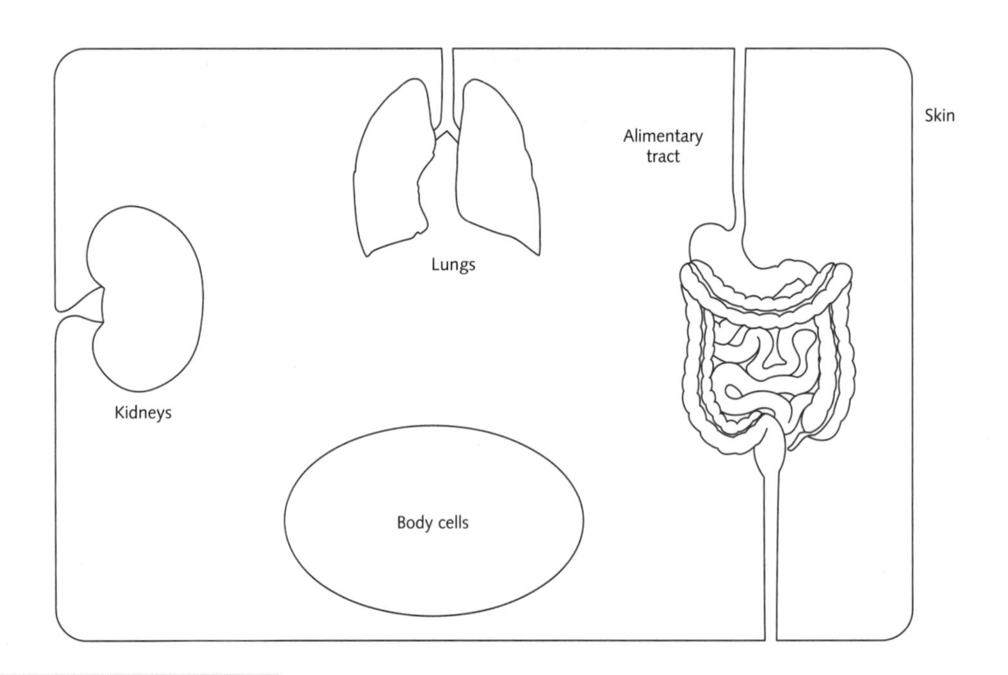

Figure 4.6

2 Annotate Figure 4.6 to show the movement of the following substances between the external and internal environments.

Hint: Consider these factors: (The brackets [...] mean concentration.)

Temperature [Oxygen] [Carbon dioxide] [Water] pH [Na^+]

a oxygen and carbon dioxide in the lungs and body cells

b glucose in the body cells

c products of digestion in the alimentary tract

d wastes and reabsorbed materials in the kidneys

4.2.2 Optimum and tolerance ranges

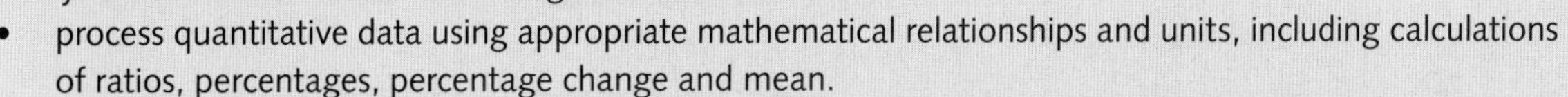

Key science skills

Analyse and evaluate data and investigation methods

- process quantitative data using appropriate mathematical relationships and units, including calculations of ratios, percentages, percentage change and mean.
- identify and analyse experimental data qualitatively, handling where appropriate, concepts of: accuracy, precision, repeatability, reproducibility and validity of measurements; errors (random and systematic); and certainty in data, including effects of sample size in obtaining reliable data

Construct evidence-based arguments and draw conclusions

- use reasoning to construct scientific arguments, and to draw and justify conclusions consistent with the evidence and relevant to the question under investigation

TB PAGE 139

All living organisms live within tolerance ranges of factors (variables) such as temperature, concentration of oxygen and carbon dioxide, and amount of glucose and water in their body fluids. There is a best measure in which the organism functions called the optimum range. This means that if the environmental conditions stray too far out of this range, the organism not only does not prosper, it dies. This is often summarised as shown in Figure 4.7. Take the time to study this figure to work out what it is showing you. Use the information in this figure to answer the questions.

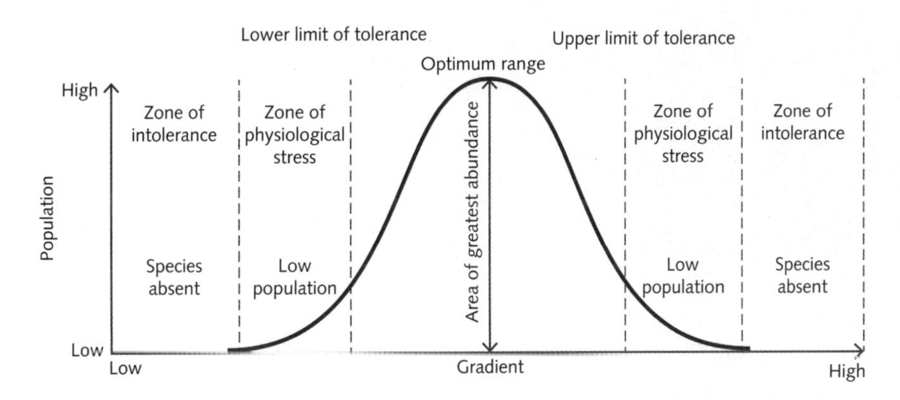

Figure 4.7 Tolerance ranges

1 a Consider the range of tolerance for temperature in the following organisms: humans, polar bear, desert iguana. Would these three animals have the same optimum range? Explain your answer with reference to Figure 4.7.

____

____

____

b Would a desert iguana be able to live in the Arctic? Explain your answer with reference to Figure 4.7.

____

____

c Consider the range of tolerance for oxygen for a human and a shark. Would these two animals have the same optimum range? Explain your answer.

____

____

d State what would happen to the shark if it entered the zone of intolerance for oxygen.

____

____

Figure 4.8 shows the oxygen tolerance ranges for two species of fish that live in freshwater lakes. The shape of the graph is sometimes called the 'Goldilocks curve'; that is, 'not too much, not too little, just right'. Think about how the two graphs differ.

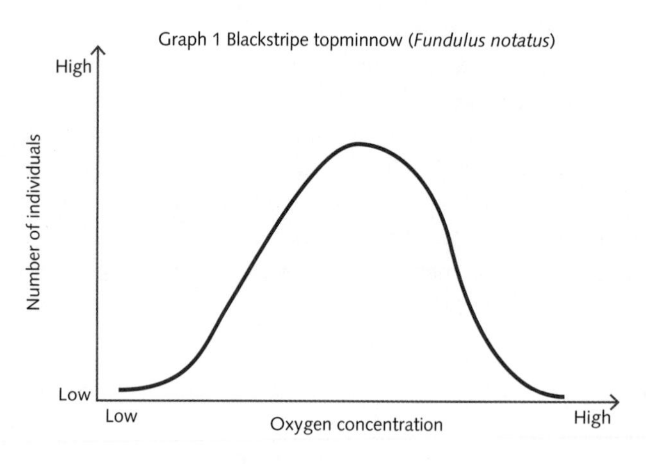

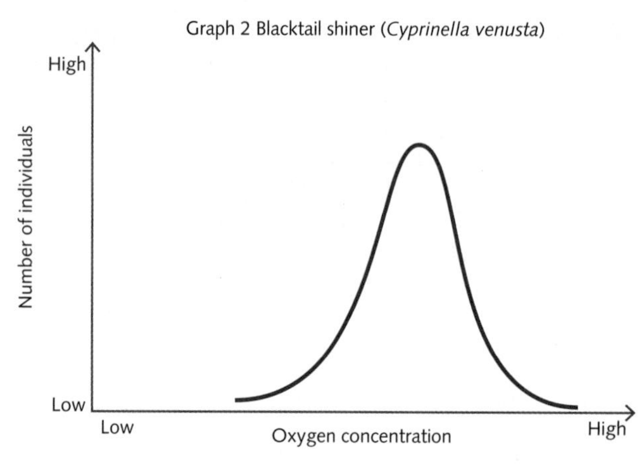

Figure 4.8

2 What do these graphs tell you about the two fish species in terms of the:

a tolerance range of the blackstripe topminnow compared to the blacktail shiner?

b freshwater habitat of each species within the lake?

c requirements for survival of each species?

4.2.3 Stimulus–response model

Key science skills

Generate, collate and record data

- organise and present data in useful and meaningful ways, including schematic diagrams, flow charts, tables, bar charts and line graphs

Analyse, evaluate and communicate scientific ideas

- analyse and explain how models and theories are used to organise and understand observed phenomena and concepts related to biology, identifying limitations of selected models/theories

TB PAGE 140

Figure 4.9 shows the stimulus–response model. The stimulus–response model shows how a stimulus is transferred through receptors and neurons to the effector, where a response is carried out. It enables predictions to be made about what will occur in the body when a stimulus is detected by a receptor.

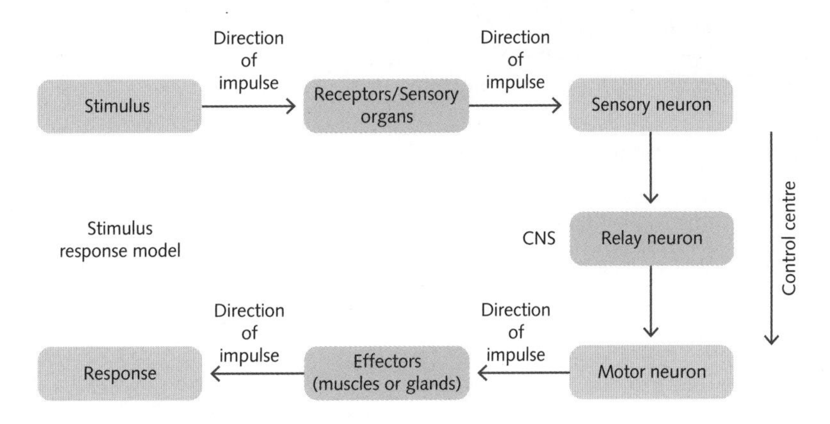

Figure 4.9 Stimulus–response model

1 Figures 4.10–4.12 show examples of stimuli and responses that occur in everyday life. Use the space beside each figure to transform the information in the pictures into a flow chart using the stimulus–response model in Figure 4.9 as a guide. Label each part of the flow chart with the name of the stimulus, receptor, effector and response.

a

Figure 4.10 Heat stimulus

b

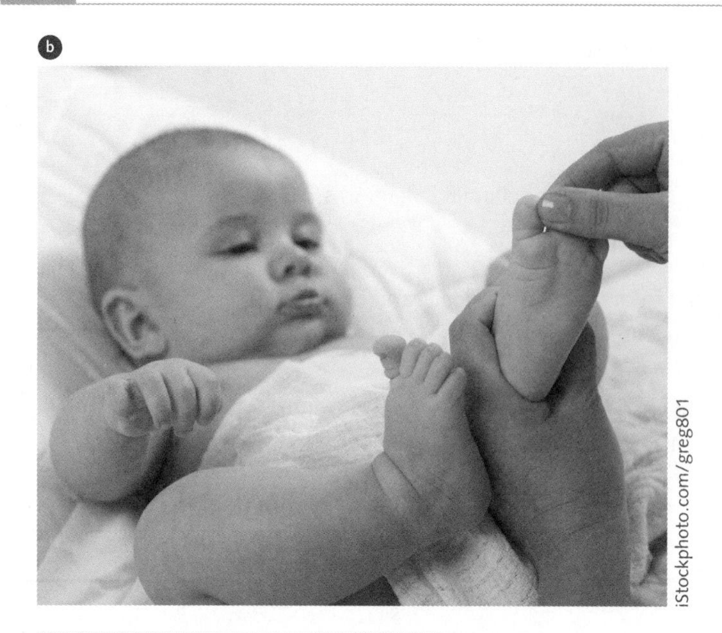
iStockphoto.com/greg801

Figure 4.11 Rubbing the sole of a baby's foot

c

Shutterstock.com/New Africa

Figure 4.12 Response to the smell of food

2 How useful was the stimulus–response model in Figure 4.9 in helping you organise the information in Figures 4.10–4.12?

3 Did you have any difficulty in using this model for any of these examples?

4.2.4 Detection of stimuli from the external and internal environments

The skin is often referred to as the largest organ in your body. The skin is full of sensory receptors making it very receptive to touch. The skin forms the most obvious physical barrier between the external environment and the internal environment.

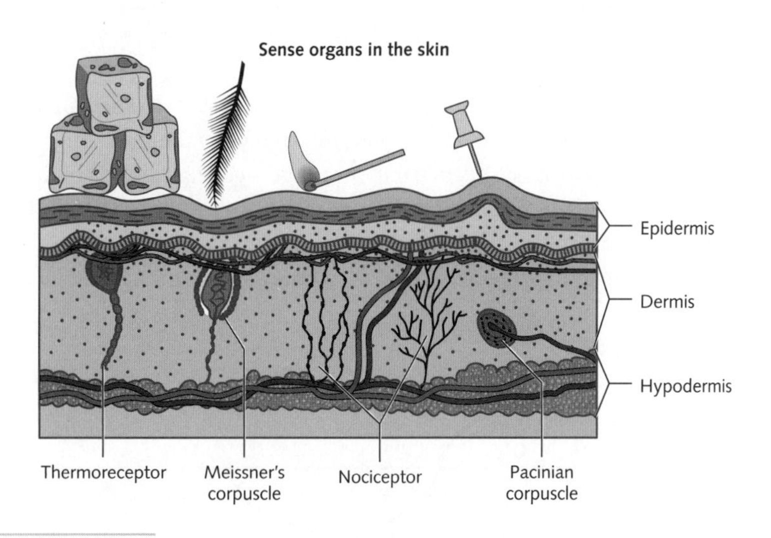

Figure 4.13 Cross-section of the skin showing sensory receptors

1 Considering Figure 4.13, what features of the external environment are being 'sensed' by the following sensory receptors?

a thermoreceptors

b Meissner's corpuscle

c nociceptor

d Pacinian corpuscle

2 How do these receptors contribute towards the survival of the organism?

3 For each type of receptor in question **1**, annotate your responses by categorising them as one of the following: mechanoreceptor, photoreceptor, chemoreceptor, thermoreceptor.

4.2.5 Homeostasis using negative feedback mechanisms

PAGE 143

Key science skills

Generate, collate and record data

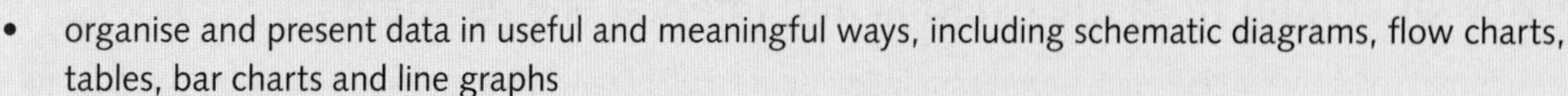

- organise and present data in useful and meaningful ways, including schematic diagrams, flow charts, tables, bar charts and line graphs
- plot graphs involving two variables that show linear and non-linear relationships

Analyse, evaluate and communicate scientific ideas

- analyse and explain how models and theories are used to organise and understand observed phenomena and concepts related to biology, identifying limitations of selected models/theories

Sometimes when an internal stimulus is detected, the response alters the stimulus to such an extent that another internal stimulus is triggered, causing another response. Think about your internal body temperature. You probably know that it is normally 37°C, but in fact it fluctuates around this in a narrow range – getting slightly higher then getting slightly lower. This occurs because of negative feedback and it enables us to maintain a relatively stable internal environment. Maintenance of a stable internal environment is called homeostasis.

Figure 4.14 demonstrates the general principles involved in our understanding of homeostatic negative feedback.

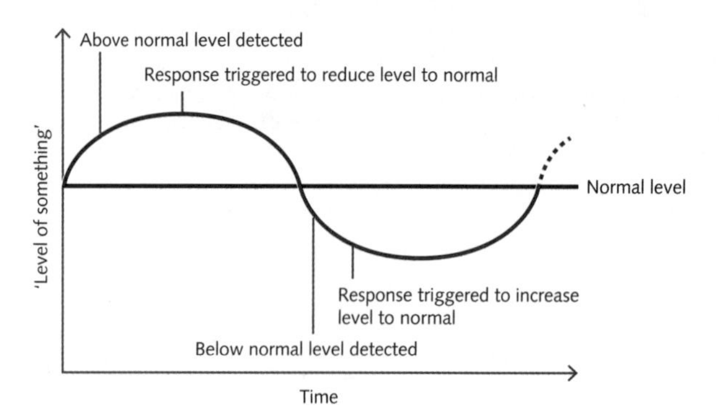

Figure 4.14 Homeostatic negative feedback

1 Use the principles of negative feedback shown in Figure 4.14 to sketch and label a graph for each of the following situations. In each case, time is on the horizontal (x) axis.

a A boy eats an energy bar at recess.

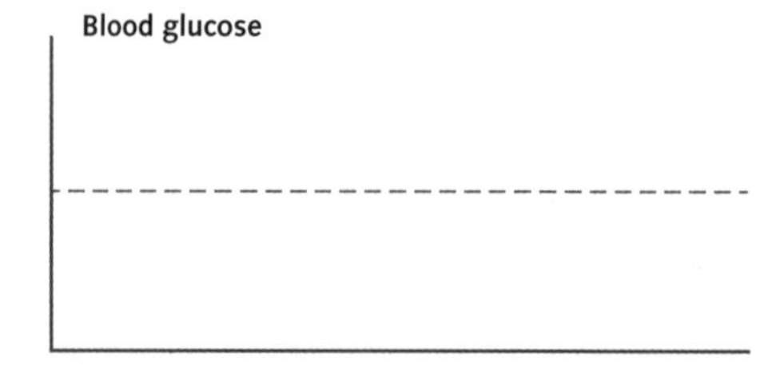

b A girl exhibits a slight fever of a temperature of 38°C. After a sleep, her temperature has returned to normal.

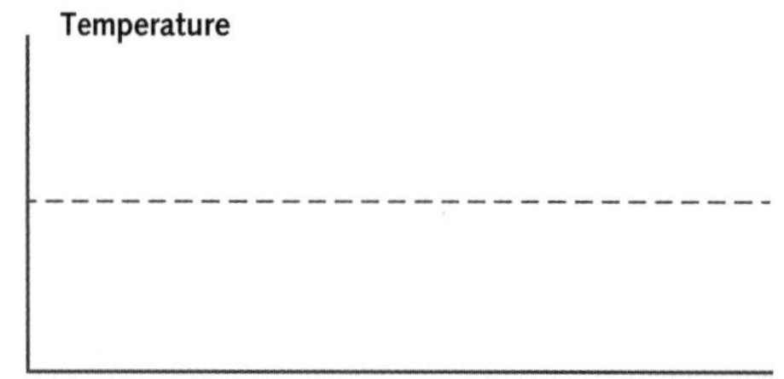

c A student drinks a litre of sparkling water after a run.

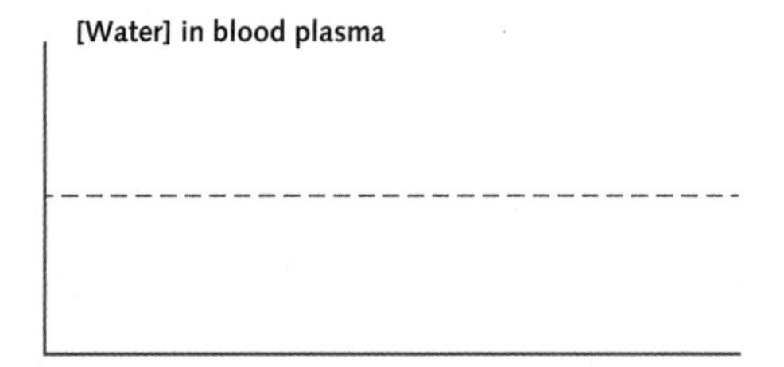

Extension: Negative feedback is not the only form of feedback; there is also *positive* feedback. Figure 4.15 shows an example of positive feedback.

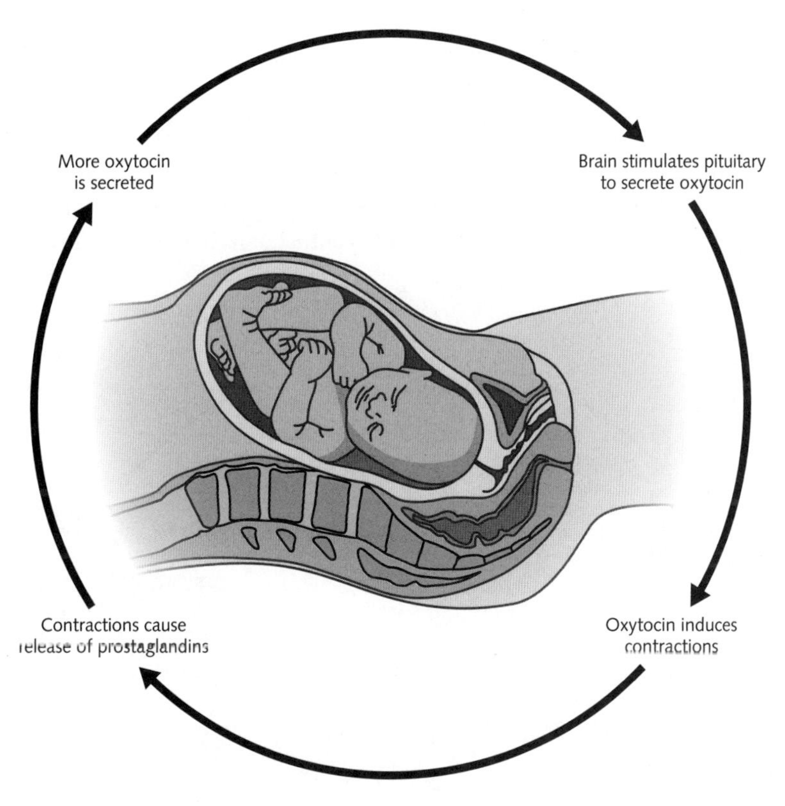

Figure 4.15 An example of positive feedback

2 a Describe what is shown in Figure 4.15.

b Why is this an example of positive feedback?

c What is the messenger in this process?

d What is the effector in this process?

4.3 Temperature regulation in the human body

Key knowledge

Regulation of systems

- regulation of body temperature, blood glucose and water balance in animals by homeostatic mechanisms, including stimulus-response models, feedback loops, and associated organ structures

4.3.1 The hypothalamus as a thermostat

TB PAGE 146

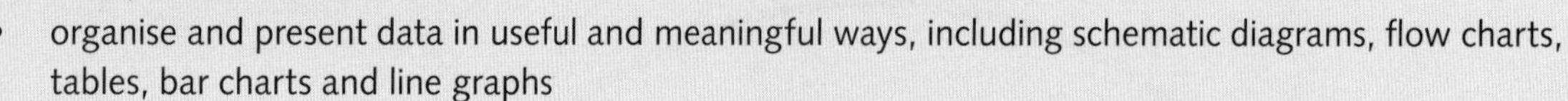

Key science skills

Generate, collate and record data

- organise and present data in useful and meaningful ways, including schematic diagrams, flow charts, tables, bar charts and line graphs

The hypothalamus is a small region located near the base of the brain, near the pituitary gland. It has three main regions: the anterior region, which releases hormones, many of which interact with the nearby pituitary gland; the middle region; and the posterior region. Figure 4.16 summarises the role of the hypothalamus in temperature control.

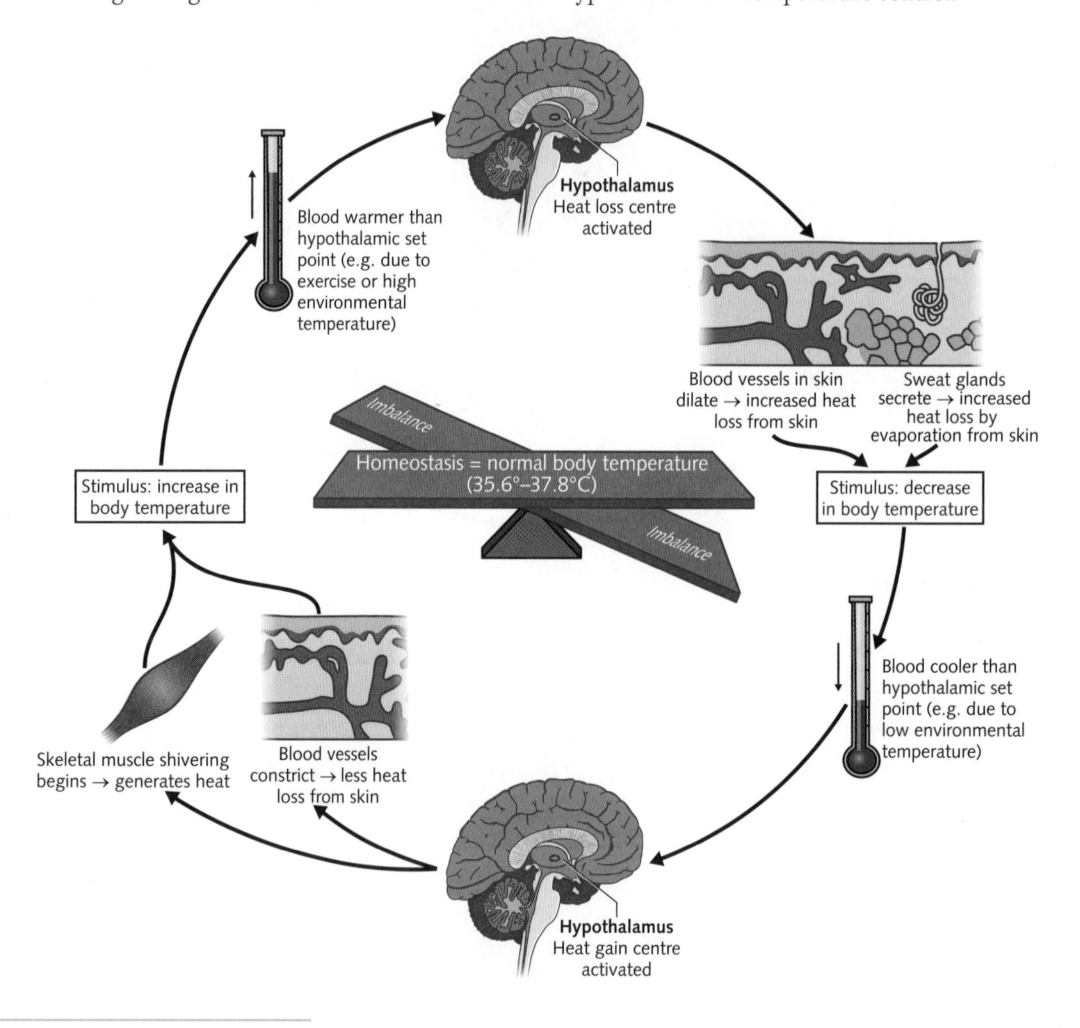

Figure 4.16 The role of the hypothalamus in temperature control

1 Use the information in Figure 4.16 to construct negative feedback loops in the spaces on the following page to demonstrate homeostatic control of temperature in the body for the following events.

9780170452632

a an increase in core temperature above normal

b a decrease in core temperature below normal

4.3.2 Investigating the body's response to a sudden drop in external temperature

TB PAGE 147

Key science skills

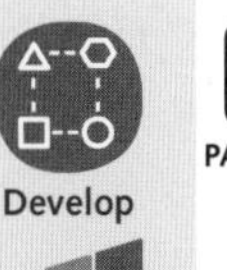

Develop aims and questions, formulate hypotheses and make predictions

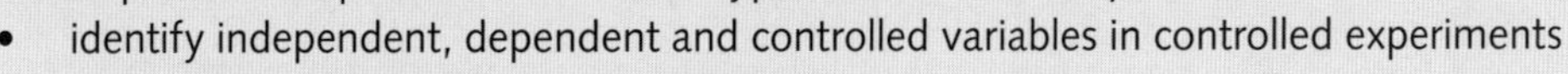

- identify independent, dependent and controlled variables in controlled experiments
- formulate hypotheses to focus investigation

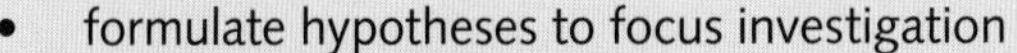

Plan and conduct investigations

- design and conduct investigations; select and use methods appropriate to the investigation, including consideration of sampling technique and size, equipment and procedures, taking into account potential sources of error and uncertainty; determine the type and amount of qualitative and/or quantitative data to be generated or collated

Comply with safety and ethical guidelines

- demonstrate safe laboratory practices when planning and conducting investigations by using risk assessments that are informed by safety data sheets (SDS), and accounting for risks

Generate, collate and record data

- record and summarise both qualitative and quantitative data, including use of a logbook as an authentication of generated or collated data
- organise and present data in useful and meaningful ways, including schematic diagrams, flow charts, tables, bar charts and line graphs
- plot graphs involving two variables that show linear and non-linear relationships

Analyse and evaluate data and investigation methods

- process quantitative data using appropriate mathematical relationships and units, including calculations of ratios, percentages, percentage change and mean

Investigation: investigating the body's response to cold

Mr Xiannidis, your Year 11 Biology teacher, provides you with a worksheet and states that you will be carrying out an investigation in the laboratory and recording the information in your logbook. The completion of a logbook is one of the outcomes of the Year 11 Biology course. You have used your logbook before and feel confident that you know what to do. You will be conducting the investigation with your laboratory partner, remembering your work must be your own.

Introduction

The brain and nervous system play a vital role in the body's response to external stimuli and other events that change internal conditions. The hypothalamus is the control centre for thermoregulation. It receives information about the external environment, interprets that information, and responds to change by sending signals to multiple organ systems.

In this investigation, you will analyse the body's response to a cold stimulus, using a thermometer to measure the surface temperature of the skin, and then relate the data to homeostatic functioning; that is, the maintenance of a regular internal environment.

Research questions

What happens to the temperature of hands when exposed to a significant drop in external temperature? Is the process of thermoregulation observable and reproducible?

Hypothesis

Write a suitable hypothesis for this investigation.

Independent variable:

Dependent variable (include units):

Controlled variables:

Materials

- Ice
- Water
- Waterbath or other similar container
- 110°C thermometer
- Towel

Apparatus set-up

Complete a diagram of the apparatus set-up in the space below, according to scientific drawing conventions.

Risk assessment

Perform a risk assessment for this investigation.

What are the risks in doing this investigation?	How can you manage these risks to stay safe?

Method

Mr Xiannidis has provided you with the following general procedure to follow, but this is not a method. He states that you need to include a control to show that the independent variable is the only variable causing the change in the dependent variable.

Time	Step
0 min	Take baseline temperature measurement of right hand
0–0:20 min	Immerse hand in water for 20 seconds. Dry hand, take first measurement
1:00–2:00 min	Take second measurement
2:00–10:00 min	Take measurements every minute and record in table
10:00–13:00 min	3-minute recovery period
13:00–28:00 min	Switch thermometer to opposite (left) hand and repeat

Using the information provided above, write the method here and on the following page, according to scientific conventions.

Results

The data consists of two types: qualitative and quantitative. Both types of data are to be recorded in a table. You take temperature readings every 60 seconds and record the results, as shown.

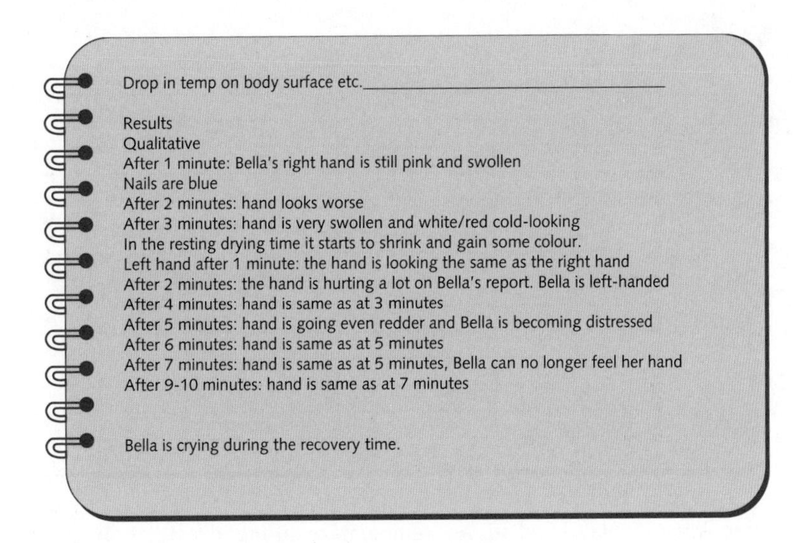
Drop in temp on body surface etc.

Results
Qualitative
After 1 minute: Bella's right hand is still pink and swollen
Nails are blue
After 2 minutes: hand looks worse
After 3 minutes: hand is very swollen and white/red cold-looking
In the resting drying time it starts to shrink and gain some colour.
Left hand after 1 minute: the hand is looking the same as the right hand
After 2 minutes: the hand is hurting a lot on Bella's report. Bella is left-handed
After 4 minutes: hand is same as at 3 minutes
After 5 minutes: hand is going even redder and Bella is becoming distressed
After 6 minutes: hand is same as at 5 minutes
After 7 minutes: hand is same as at 5 minutes, Bella can no longer feel her hand
After 9-10 minutes: hand is same as at 7 minutes

Bella is crying during the recovery time.

Figure 4.17 Isobel's notes of observation of Bella during investigation

1 Tabulate the qualitative data recorded by Isobel , according to scientific conventions.

Measurements

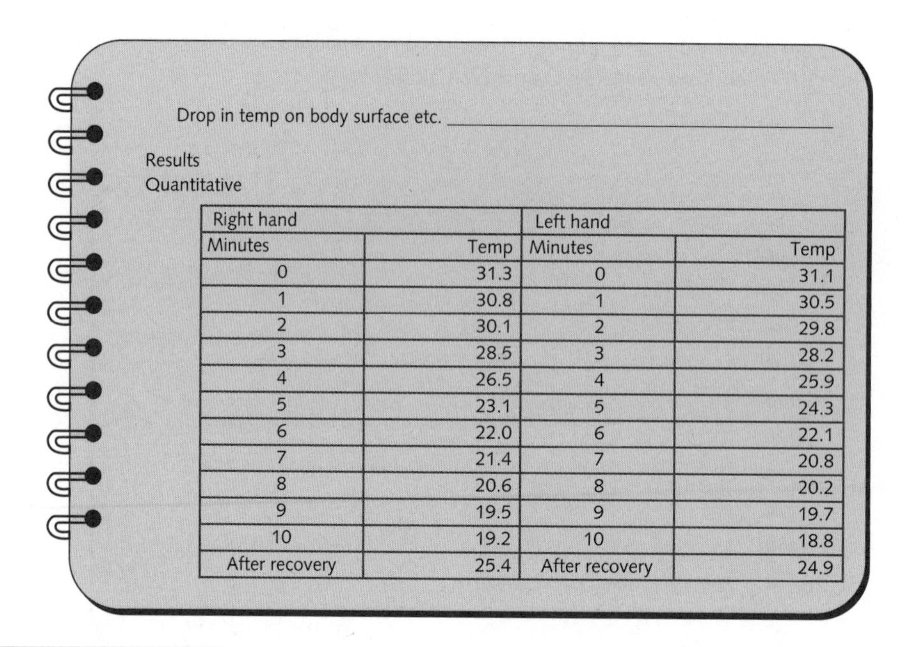

Drop in temp on body surface etc.

Results
Quantitative

Right hand		Left hand	
Minutes	Temp	Minutes	Temp
0	31.3	0	31.1
1	30.8	1	30.5
2	30.1	2	29.8
3	28.5	3	28.2
4	26.5	4	25.9
5	23.1	5	24.3
6	22.0	6	22.1
7	21.4	7	20.8
8	20.6	8	20.2
9	19.5	9	19.7
10	19.2	10	18.8
After recovery	25.4	After recovery	24.9

Figure 4.18 Isobel's measurements of Bella's hand surface temperature during investigation

2 Tabulate the quantitative data according to scientific conventions. For each measurement, calculate the percentage change from the baseline measurement. Show this as separate columns in your data table. The ambient temperature was 22.0°C.

3 Graph the percentage change data for each hand on the graph paper below, according to scientific conventions.

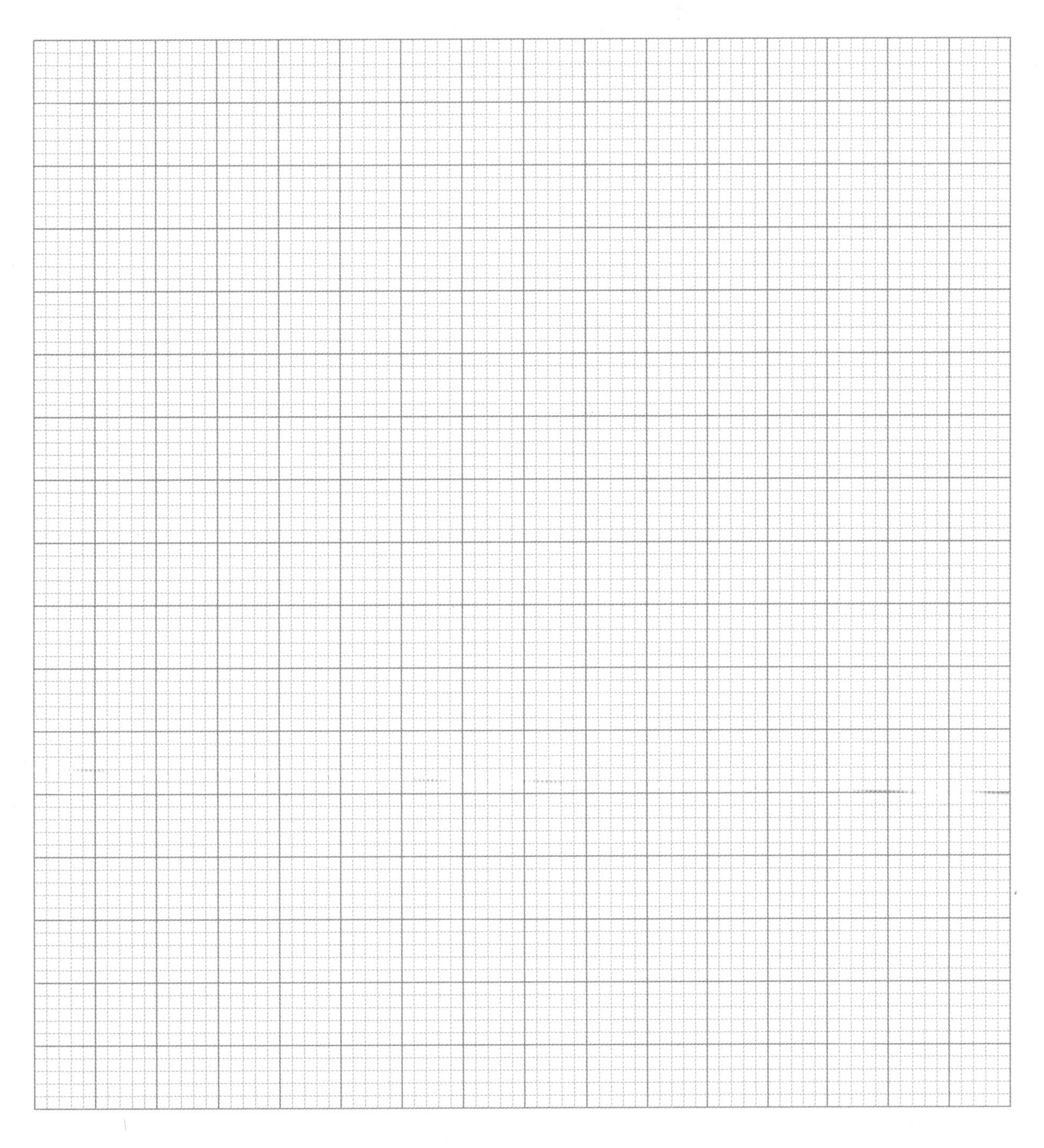

Discussion

Use the questions below to guide your discussion. Normally there would be no guiding questions to help you, but these are the sort of questions that you need to ask yourself when writing a discussion section.

1 Were the initial temperature readings similar between the right hand and the left hand? If not, why not?

2 What did the students set up as a control? Why is a control important in an investigation such as this?

3 Did the temperature of the hands recover to the baseline temperature following the 3-minute recovery period? Explain.

4 What was the maximum percentage change in temperature of each hand over the course of the measurements of temperature? How did the percentage change compare in each hand?

5 How did the independent variable affect the dependent variable? Justify your claim using evidence from the investigation.

6 Based on the evidence collected by Isobel and Bella, explain the results.

7 Is there any evidence in the data or from the observations that experimental errors or other uncontrolled variables affected the results?

8 Identify any new questions that have arisen as a result of this research.

9 If someone is exposed to cold weather for extended periods, where are they most likely to get frostbite? Use the results of this investigation to support your answer.

10 The nervous system plays a critical role in maintaining homeostasis for an organism. The system detects external stimuli, transmits and integrates information about the stimuli, and produces one or more responses. Vasoconstriction occurs when the smooth muscles surrounding arteries contract. Explain how nerves cause muscle contraction. Explain how vasoconstriction protects against the cold.

4.3.3 Temperature regulation in other animals

Key science skills

Analyse and evaluate data and investigation methods

TB PAGE 152

- identify and analyse experimental data qualitatively, handling where appropriate, concepts of: accuracy, precision, repeatability, reproducibility and validity of measurements; errors (random and systematic); and certainty in data, including effects of sample size in obtaining reliable data

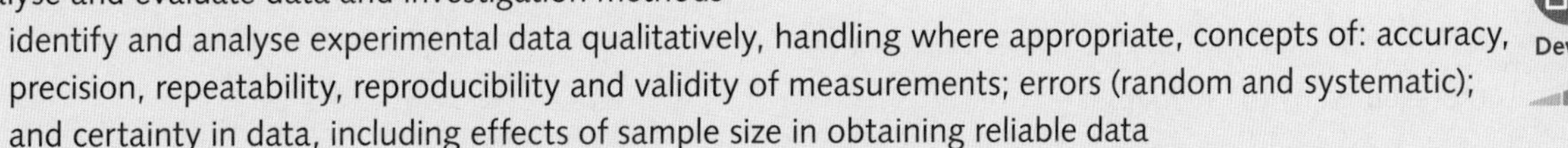

Analyse, evaluate and communicate scientific ideas

- discuss relevant biological information, ideas, concepts, theories and models and the connections between them
- use clear, coherent and concise expression to communicate to specific audiences and for specific purposes in appropriate scientific genres, including scientific reports and posters

Mammals and birds are endotherms and have evolved a wide variety of adaptations to deal with the different challenges to thermoregulation in the world's biomes. A biome is a large area that is characterised by a certain climate. Three biomes are shown in Figure 4.19 with distinctly different climates.

1 For each biome, and the animal that lives there, list at least one structural, physiological and behavioural adaptation that is associated with thermoregulation.

a

b

c

Figure 4.19 **Three biomes and the animals that live there. a Penguins in Antarctica: South Pole with intense weather conditions and full of ice. b Red kangaroos in Australia: grasslands with large open spaces and little shelter. c Lyrebirds in temperate rainforest, Victoria: tall trees with stable to cold weather conditions**

Lizards are ectotherms (their body heat is gained from an external source such as the sun) and spend some time of the day in a burrow. Refer to Figure 4.20 to answer the following questions.

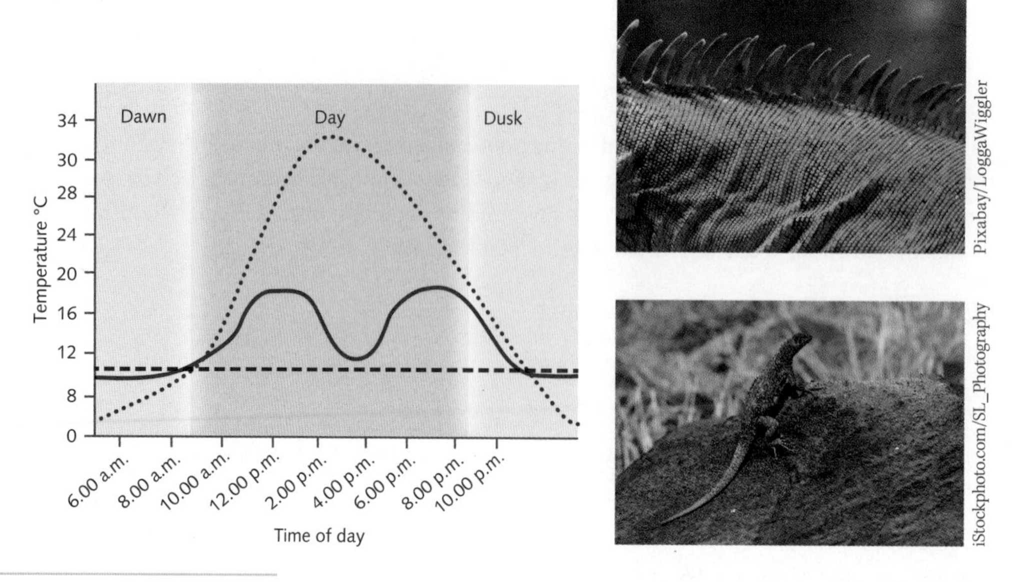

Figure 4.20 Lizard activity during a 14-hour period

2 What is represented by the dotted line?

3 What is represented by the solid line?

4 What is represented by the dashed line?

5 How does this graphical representation explain ectotherms and their adaptations to their habitat?

6 Write a piece of interpretive text of no more than 60 words using data from the graph to tell the story of the lizard over a 14-hour period.

4.4 Regulation of blood glucose levels

Key knowledge

Regulation of systems

- regulation of body temperature, blood glucose and water balance in animals by homeostatic mechanisms, including stimulus-response models, feedback loops, and associated organ structures
- malfunctions in homeostatic mechanisms: type 1 diabetes, hypoglycaemia, hyperthyroidism

4.4.1 Homeostasis for the regulation of blood glucose

Key science skills

Develop aims and questions, formulate hypotheses and make predictions

- identify independent, dependent and controlled variables in controlled experiments

Generate, collate and record data

- organise and present data in useful and meaningful ways, including schematic diagrams, flow charts, tables, bar charts and line graphs
- plot graphs involving two variables that show linear and non-linear relationships

Analyse and evaluate data and investigation methods

- process quantitative data using appropriate mathematical relationships and units, including calculations of ratios, percentages, percentage change and mean

PAGE 156

1 Below are a number of steps that describe the homeostatic control of blood glucose levels within a strict range. These steps have not been written in the order in which they occur.

Steps:

- After a meal, receptors detect an increase in blood glucose.
- Receptors detect a decrease in blood glucose.
- The pancreas secretes insulin.
- The liver removes excess glucose.
- The liver adds glucose to the blood.
- The pancreas secretes glucagon.
- The liver converts stored glycogen into glucose.
- The excess glucose is converted into glycogen, which is stored in the liver.

a Rewrite these steps in the correct order.

b Use the correctly ordered steps from part **a** to create a diagram in the space below showing the negative feedback that is occurring to maintain stable blood glucose levels within the body.

2 A man is being tested for diabetes mellitus. The test of his urine in the surgery showed raised glucose. His fasting blood glucose is measured over a period of time, and the measurements are recorded in Table 4.1. (At time 0, the patient drinks a sugar solution.)
Note: A fasting blood glucose test is performed after a period of 8–10 hours of no food intake.

Table 4.1 Fasting blood glucose over 5 hours

Time (hours)	Blood glucose concentration (mmol L^{-1})
0	5.0
1	6.7
2	6.1
3	5.0
4	4.4
5	4.7

a Graph the data from Table 4.1, using the graph paper on the next page. Provide a title for the graph and label the axes appropriately (including measurement units).

9780170452632

b Determine the type of feedback that the graph is representing. Explain your reasoning.

c Identify the dependent variable.

d Identify the independent variable.

e Calculate the mean (average) for this person's blood glucose.

f Normally, a person's measurement for fasting blood glucose is between 4.0 and 5.4 mmol L^{-1}. Does the patient's blood glucose level always stay within normal limits?

g A fasting blood glucose concentration after 2 hours of 7.0 mmol L^{-1} or higher indicates diabetes. Two tests are required to confirm this. From this dataset, could you diagnose the patient with diabetes?

h Distinguish between hypoglycaemia and hyperglycaemia.

i Does this patient have any apparent issues with blood glucose regulation? Provide evidence from the dataset to support your answer.

j In a normal well-functioning adult, blood glucose concentrations can rise to 7.8 mmol L^{-1}. What does the data tell you that indicates the blood glucose is regulated?

k After 4 hours, the blood glucose concentration has dropped to below the starting level. Describe what happens at 5 hours and explain what this means.

4.4.2 When the glucose homeostasis system malfunctions

TB PAGE 157

Key science skills

Analyse, evaluate and communicate scientific ideas

- discuss relevant biological information, ideas, concepts theories and models and the connections between them

Develop

Figure 4.21 shows the relative concentrations of insulin, glucagon and glucose in the blood in a healthy person.

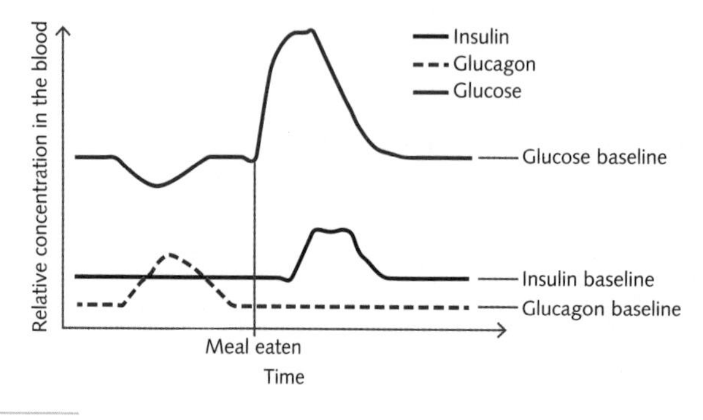

Figure 4.21 Normal hormonal control of glucose in the blood

1 Referring to Figure 4.21, consider what the situation would be if the person suffered from diabetes mellitus. Redraw the graph in the space below. Show the graph for the relative concentration in the blood of each of the three hormones if the person was unmedicated.

2 In the past, medical professionals used to taste a patient's urine to detect sugars in it to indicate if diabetes was present.

a Why would they do this?

b Find out what they do now as a quick test on a urine sample.

4.5 Water balance regulation

Key knowledge

Regulation of systems

- regulation of body temperature, blood glucose and water balance in animals by homeostatic mechanisms, including stimulus-response models, feedback loops, and associated organ structures

4.5.1 Osmoregulation in humans

TB PAGE 159

Key science skills

Generate, collate and record data

- organise and present data in useful and meaningful ways, including schematic diagrams, flow charts, tables, bar charts and line graphs

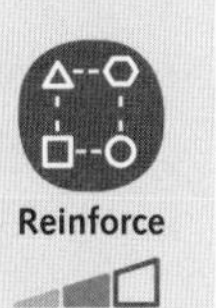

Figure 4.22 shows a representation of the negative feedback that occurs in the body to maintain a constant concentration of water in the blood.

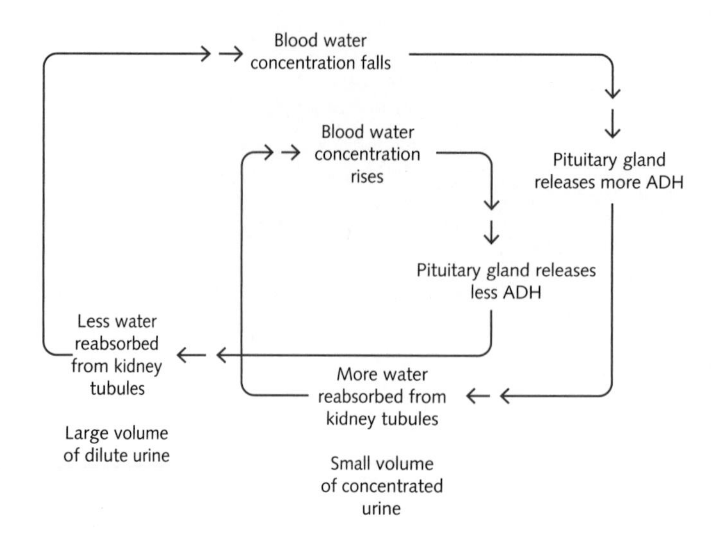

Figure 4.22 Water regulation in the body

Use the information provided in Figure 4.22 to complete Figure 4.23. Add additional annotations to the figure to demonstrate your understanding of water balance in the body and negative feedback.

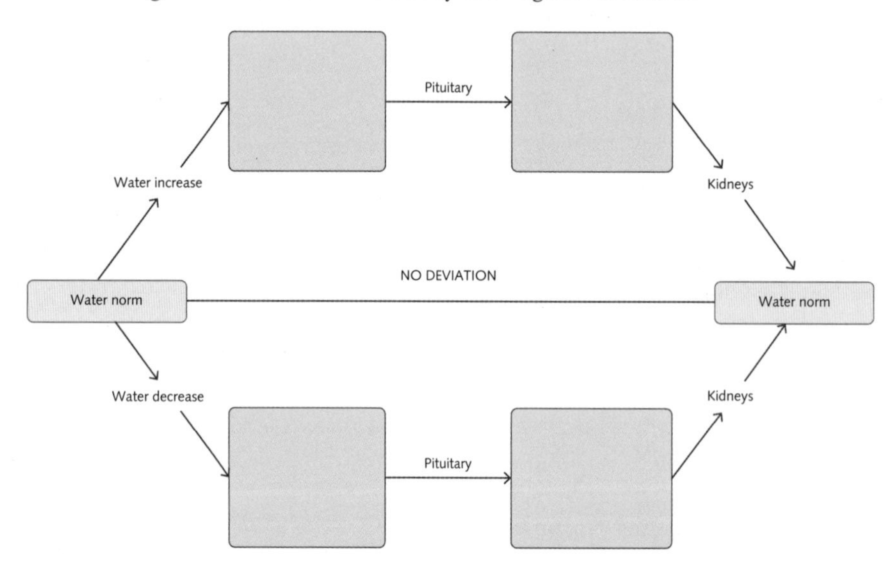

Figure 4.23 Osmoregulation negative feedback

4.5.2 The mechanism of osmoregulation

PAGE 160

Consolidation of knowledge

1 Complete the sentences on the following page using correct terms from the word list below to show your understanding of osmoregulation. Terms may be used more than once.

ADH	hypothalamus	pituitary
antidiuretic hormone	impervious	pituitary gland
brain	more	renal
concentrated	nephron	urine
dilute	osmoreceptors	water
distal tubules	pelvis	less

Osmoregulation is the process used in homeostatic regulation and maintenance of the salts and water balance of an organism's body. In humans, osmoregulation takes place in the kidney units called nephrons and is under the direct control of a hormone called antidiuretic hormone (ADH). Other processes of diffusion and transport across membranes play a major part.

If blood gets too ____________, for example when one drinks a lot of water, the____________ in the hypothalamus of the ____________ detects this. A message is sent to the ____________ gland, causing it not to secrete any ____________. Since no ADH is secreted into the blood, the distal tubes and collecting ducts of the ____________ remain ____________ to water. The result of this is that less or no ____________ is reabsorbed from the filtrate, resulting in a large amount of ____________ ____________ being produced and transported into the ____________ of the kidney.

On the other hand, if the blood is too ____________, osmoreceptors in the ____________ detect this. A message is sent from the hypothalamus to the ____________ ____________ and more ____________ is secreted. ADH reaches the kidney via the ____________ artery. In the nephrons of the kidney, the ____________ ____________ and the collecting duct become ____________ permeable to ____________, so the capillaries absorb much more ____________ from the filtrate. A much more ____________ urine results.

2 A student has handed in the following piece of work to describe osmoregulation in the nephron. The teacher marks the work and notes there are ten mistakes. The student's work is shown in Figure 4.24.

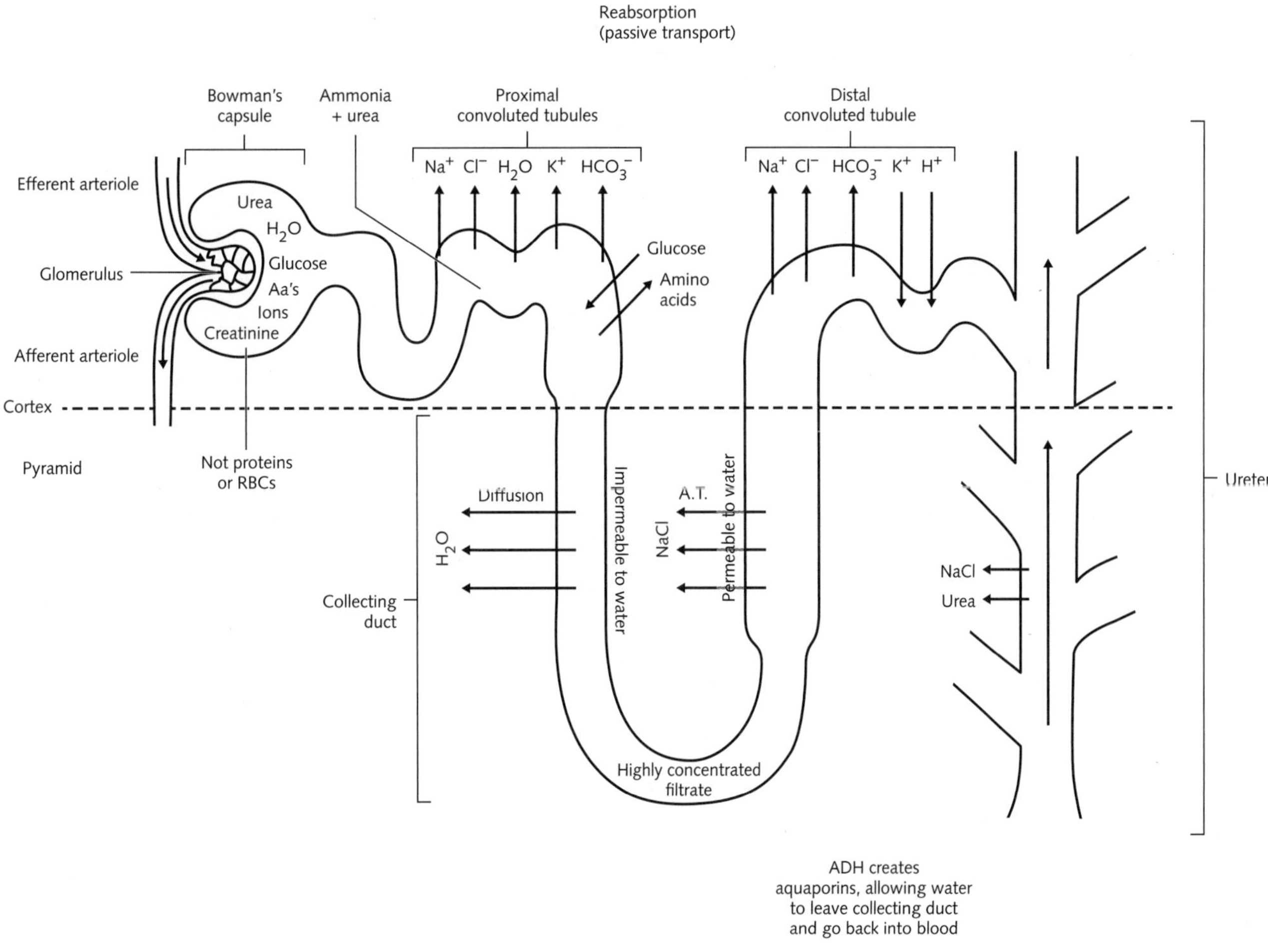

Figure 4.24 Student diagram of osmoregulation

a Circle the mistakes on the diagram in red. Number the mistakes 1 to 10.

b Write down below what the student should have written in each instance to correct the mistake.

1 ____________ 6 ____________

2 ____________ 7 ____________

3 ____________ 8 ____________

4 ____________ 9 ____________

5 ____________ 10 ____________

4.5.3 Water balance in other animals

PAGE 161

Key science skills

Analyse, evaluate and communicate scientific ideas

- analyse and explain how models and theories are used to organise and understand observed phenomena and concepts related to biology, identifying limitations of selected models/theories

At some point in evolutionary history, aquatic vertebrates must have made an evolutionary transition to terrestrial life. One major problem for survival that needed to be overcome is salt and water balance in the terrestrial environment.

1 Animals living on land encounter problems associated with balancing their salt and water concentrations. Describe these problems.

2 **a** Describe the salt and water balance problems that would be encountered by a saltwater aquatic animal transitioning to a terrestrial environment.

b Describe the salt and water balance problems that would be encountered by a freshwater aquatic animal transitioning to a terrestrial environment.

c Compare the problems encountered by saltwater aquatic and freshwater aquatic animals transitioning to a terrestrial environment.

d Do you think it is more likely that freshwater or saltwater aquatic animals evolved into terrestrial animals? Explain your reasoning.

3 Describe two physiological adaptations related to salt and water balance that could enable an aquatic vertebrate to start living in a terrestrial environment.

9780170452632

4.6 Regulation and control of basal metabolic rate and growth by thyroid hormones

Key knowledge

Regulation of systems

- regulation of body temperature, blood glucose and water balance in animals by homeostatic mechanisms, including stimulus-response models, feedback loops, and associated organ structures
- malfunctions in homeostatic mechanisms: type 1 diabetes, hypoglycaemia, hyperthyroidism

4.6.1 Analysing a science journal article

Key science skills

Analyse, evaluate and communicate scientific ideas

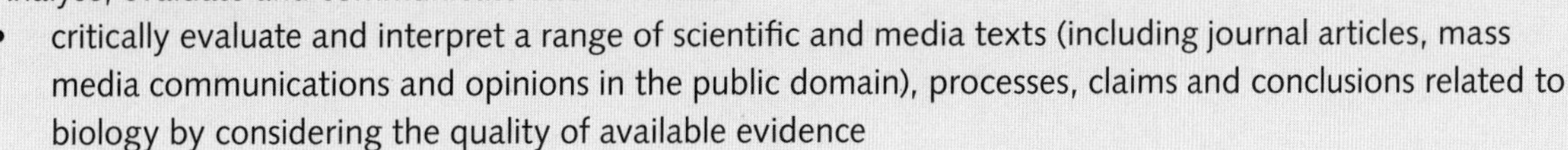

- critically evaluate and interpret a range of scientific and media texts (including journal articles, mass media communications and opinions in the public domain), processes, claims and conclusions related to biology by considering the quality of available evidence

PAGE 163

Carefully read the journal article below.

Human longevity is characterised by high thyroid stimulating hormone secretion without altered energy metabolism

Few studies have included subjects with the propensity to reach old age in good health, with the aim to disentangle mechanisms contributing to staying healthier for longer. The hypothalamic–pituitary–thyroid (HPT) axis maintains circulating levels of thyroid stimulating hormone (TSH) and thyroid hormone (TH) in an inverse relationship. Greater longevity has been associated with higher TSH and lower TH levels, but mechanisms underlying TSH/TH differences and longevity remain unknown. The HPT axis plays a pivotal role in growth, development and energy metabolism. We report that offspring of nonagenarians with at least one nonagenarian sibling have increased TSH secretion but similar bioactivity of TSH and similar TH levels compared to controls. Healthy offspring and spousal controls had similar resting metabolic rate and core body temperature.

by Jansen, S., Akintola, A., Roelfsema, F. et al. *Sci Rep* 5, 11525 (2015). https://doi.org/10.1038/srep11525

1 One of the skills a student of science has to develop is to be able to read and interpret a science journal article. Excerpts from the above journal article have been split up into individual sentences to improve understanding. In the lines below each sentence, rewrite the sentence in simpler words that you understand. You may need to research some of the words to find out what they mean.

Title: Human longevity is characterised by high thyroid stimulating hormone secretion without altered energy metabolism

Few studies have included subjects with the propensity to reach old age in good health, with the aim to disentangle mechanisms contributing to staying healthier for longer.

The hypothalamic–pituitary–thyroid (HPT) axis maintains circulating levels of thyroid stimulating hormone (TSH) and thyroid hormone (TH) in an inverse relationship.

Greater longevity has been associated with higher TSH and lower TH levels, but mechanisms underlying TSH/TH differences and longevity remain unknown.

The HPT axis plays a pivotal role in growth, development and energy metabolism.

We report that offspring of nonagenarians with at least one nonagenarian sibling have increased TSH secretion but similar bioactivity of TSH and similar TH levels compared to controls.

Healthy offspring and spousal controls had similar resting metabolic rate and core body temperatures.

2 Transcribe all your rewritten sentences below to combine them into one coherent paragraph.

3 Reread the paragraph that you wrote above. Are you now able to understand the information that the journal article is trying to convey to the reader?

4.7 Chapter review

PAGE 171

4.7.1 Key terms

1 The following key terms are used in Chapter 4. They have been grouped into sub-thematic areas. Fill in the boxes at the top of each column to show the sub-thematic area.

ammonia	arrector pili muscle	glucagon	effector	basal metabolic rate
antidiuretic hormone	brown fat	glycogen	feedback mechanism	hyperthyroidism
osmoreceptor	conduction	hyperglycaemia	homeostasis	hypothyroidism
osmoregulation	convection	hypoglycaemia	hypothalamus	thyroid-stimulating hormone (TSH)
water potential	endothermic	insulin	exteroceptor	thyrotropin-releasing hormone
urea	evaporation		external environment	thyroxine
vasopressin	hyperthermia		physiological stress	metabolism
	heat balance		receptor	
	hypothermia		response	
	vasoconstriction		set point	
	vasodilation		stimulus	
	radiation		stimulus–response model	
	thermoregulation		tolerance range	
			internal environment	
			interoceptor	
			interstitial fluid	
			negative feedback	
			optimum range	

2 A student looked at these groupings and decided there were too many words to learn in one of them. Choose the grouping that has too many terms in it and rework into smaller themes with headings that would make them easier to learn.

Chapter review continued

4.7.2 Practice test questions

PAGE 172

Multiple-choice questions

The following information applies to Questions 1–4.

©VCAA 2004 EXAM 1 SECTION A Q17 & 18 (adapted) Biologists investigating the regulation of body water in Peking ducks, *Anas platyrhynchos*, put forward the hypothesis that Peking ducks drink more as the saltiness of their drinking water increases. The drinking water was to be supplied in 70-litre wading pools and replenished twice each day. Twelve adult Peking ducks, males and females, were available and two experimental designs were suggested.

Design 1: The same twelve ducks are provided with drinking water of increasing saltiness over a 24-week period.

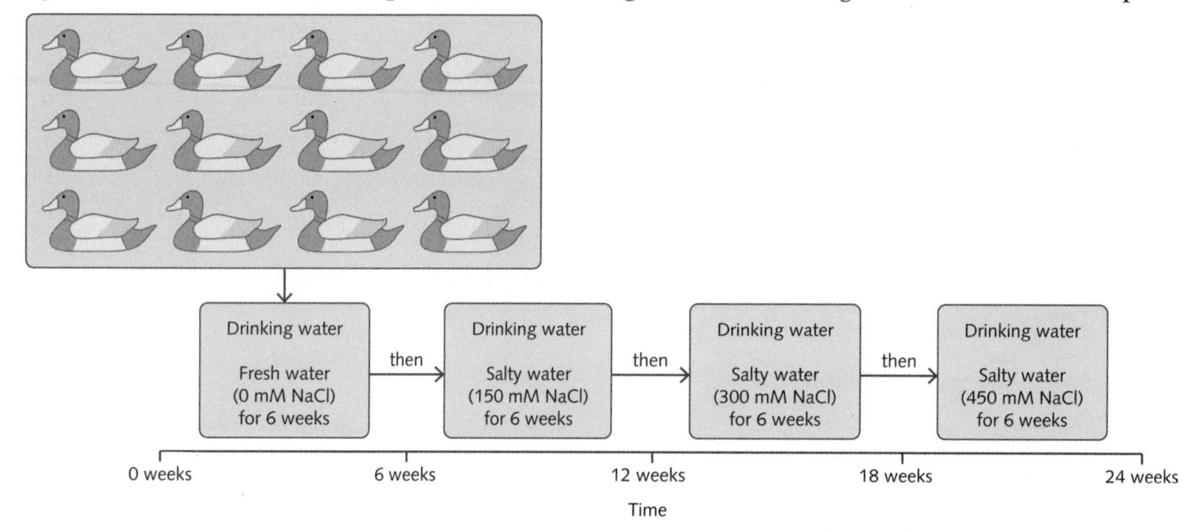

Design 2: The twelve ducks are divided into four groups of three ducks and each group is exposed to drinking water of a different saltiness.

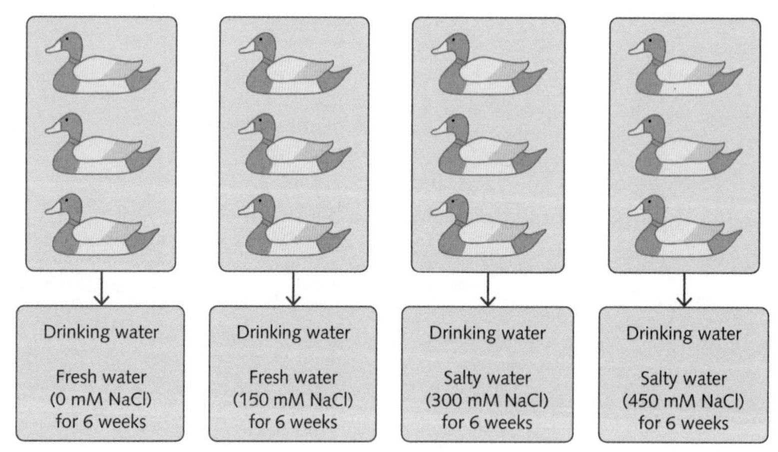

1 The independent variable in this experiment is

- **A** amount of water the ducks drink.
- **B** species of duck.
- **C** gender of the ducks.
- **D** saltiness of the drinking water.

2 The dependent variable in this experiment is

A amount of water the ducks drink.
B species of duck.
C gender of the ducks.
D saltiness of the drinking water.

3 In Design 2 the group of ducks that drinks the freshwater only is called the

A experimental group.
B extraneous group.
C control group.
D variable group.

4 When analysing the expected results of each design, you would be able to say that the

A results gained from Design 1 would be more accurate and valid than those gained from Design 2.
B results gained from Design 2 would be more accurate and valid than those gained from Design 1.
C variability between ducks would not affect results as much in Design 1.
D variability between ducks would not affect results as much in Design 2.

5 **©VCAA** **@2004 EXAM 1 SECTION A Q12 (adapted)** The transpiration rate was measured in the pea plant, *Pisum sativum*. Three identical groups of pea plants were tested. In group **X**, the soil moisture was high, in group **Y** it was medium and in group **Z** it was low.

The graph below shows the results of the experiment.

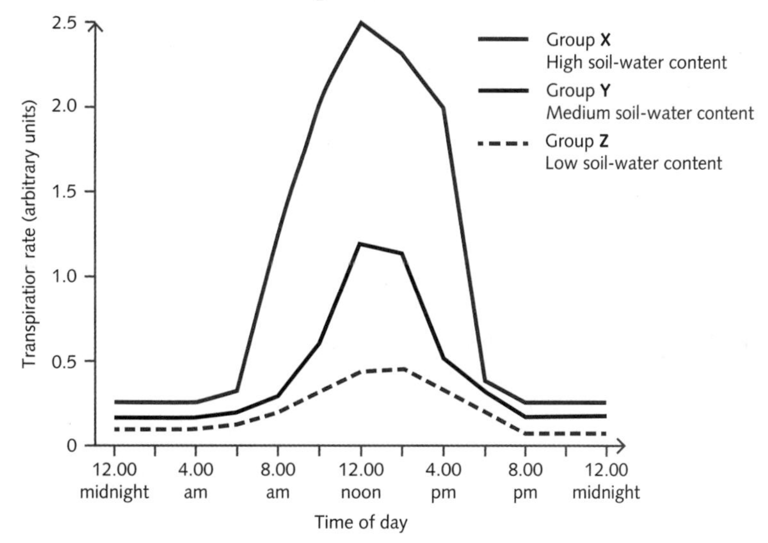

It is reasonable to conclude that

A stomata were closed between 8 a.m. and 5.00 p.m. in group **Z**.
B pea plants that were exposed to higher water content in the soil transpired at a greater rate.
C the transpiration rate of group **Y** at 4 p.m. was twice as much as group **Z**.
D at noon pea plants in group **X** transpired four times as much as group **Z**.

Short-answer questions

6 ©VCAA 2004 VCAA EXAM 1 SECTION B Q4 (adapted) Human blood glucose concentrations are under homeostatic control. When the concentration of glucose in the blood begins, to rise the beta cells in the islets of Langerhans in the pancreas release more insulin hormone and less glucagon. Insulin stimulates the cells to remove glucose from the blood and to convert it into glycogen.

When the concentration of glucose in the blood falls, the alpha cells in the islets of Langerhans in the pancreas secrete less insulin and more glucagon. This stimulates the cells to convert glycogen into glucose and to release it into the bloodstream.

a Use the information given above to draw a flow chart that illustrates the control of blood glucose levels in the human body.

3 marks

b What is the general name given to the control of factors such as blood glucose and salt levels in the body?

1 mark

c What is the general name given to structures that detect the changes in blood glucose concentrations?

1 mark

d The control of blood glucose concentrations involves negative feedback. What is negative feedback?

2 marks

e What is a quick remedy for a person suffering from hypoglycaemia while out on a bike ride?
What does this remedy do?

2 marks

Getty Images/tunart

Scientific investigations 5

Remember

Chapter 5 Scientific investigations will call on content that you have already met in your science studies from previous years or from earlier this year. Take some time to refresh your knowledge of this content before you enter this chapter. Try to answer the following questions from memory. If you cannot do this, then use a reference to assist you.

PAGE 180

1 List the steps of the scientific method.

2 What is a hypothesis?

3 List three different methodologies that scientists can use to answer their research question.

4 What is meant by ethics?

5 Define the following terms in relation to a controlled experiment.
   a control

b independent variable

c dependent variable

6 What types of information do you record in your logbook?

5.1 Choosing your topic

Key knowledge

Investigation design

- biological science concepts specific to the selected scientific investigation and their significance, including the definition of key terms

Part A: Investigation design

5.1.1 Getting started: organisation

PAGE 181

Key science skills

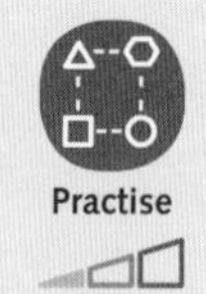

Develop aims and questions, formulate hypotheses and make predictions

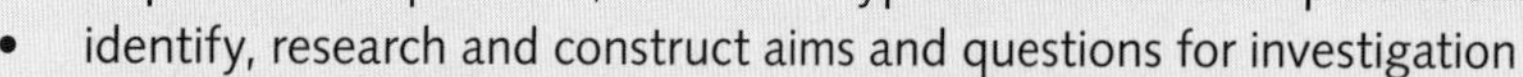

- identify, research and construct aims and questions for investigation

- identify independent, dependent and controlled variables in controlled experiments

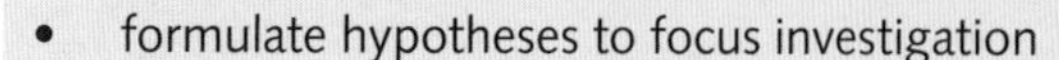

- formulate hypotheses to focus investigation
- predict possible outcomes

For VCE Biology Unit 1, Outcome 3, you will work like a biologist and submit a report of a student-adapted or student-designed scientific investigation using a selected format such as a scientific poster, an article for a scientific publication, a practical report, an oral presentation, a multimedia presentation or a visual representation.

The investigation topic that you choose must be related to the function and/or regulation of cells or systems, and draw a conclusion based on evidence from generated primary data.

When you are planning your scientific investigation, it is important that you get yourself organised early and correctly. By being well organised, you will give yourself the best chance of achieving all the requirements of the outcome in the allocated amount of time and of receiving the best possible marks.

Table 5.1 can be used as a template to help you organise your tasks and due dates. Note that not all items listed in this table are relevant to all research methodologies.

9780170452632

Table 5.1

Completed ✓	Date to be completed by	Task to be completed
		Obtain the assessment criteria that your teacher is going to use to assess this outcome, if it is available.
		Choose an investigation topic related to the function and/or regulation of cells or systems. Get your topic approved by your teacher, if required.
		Set up your logbook so you can start recording your ideas for all the stages of your scientific investigation.
		Complete the secondary research for your chosen topic; make sure you record all sources using the correct referencing format.
		Write the research question for your chosen scientific investigation.
		Decide whether you are going to design an investigation or adapt an existing investigation.
		If you are going to adapt an existing investigation, locate this investigation and record its source.
		Identify the dependent and independent variables and any extraneous variables that need to be controlled.
		Decide if you need to build in a control to your investigation and what this would be.
		Write the aim and hypothesis for your scientific investigation, making sure you follow the 'If … then …' format for writing a hypothesis.
		Predict the results required to either support or refute your hypothesis.
		Decide on which methodology is the best to use to test your research question.
		Decide what type of data you will generate.
		Decide on how much data you will generate and how you will record it.
		Think about how you are going to ensure that your data is accurate, precise, repeatable, reproducible and valid. Build safeguards into your method.
		Think about how you are going to minimise error in your data.
		Think about how you will analyse the data you record.
		Prepare a list of the materials and equipment that you will need to conduct your investigation. Check with your teacher that these will be available.
		Create a risk assessment table to identify any risks with your materials, equipment or method. State how you are going to manage these risks.
		Identify any ethical issues that you need to consider as part of your scientific investigation. If there are any, think about how you are going to deal with them using the ethical guidelines.
		Write the method for your scientific investigation, taking the above items into consideration.
		Conduct your scientific investigation and record all data generated into your logbook.
		Collate your data and present it in the most appropriate format.
		Analyse your data using the appropriate mathematical method.
		Relate your results to your research question and/or hypothesis.
		Write a conclusion, making sure you follow the rules for writing a conclusion.
		Decide on the best presentation format to communicate your scientific investigation, results and conclusion(s).
		Complete the communication piece, making sure that it falls within the word limit, if you have been provided with one.
		Submit your communication piece and logbook for assessment and authentication.
		Congratulate yourself on a job well done!

5.1.2 Getting started: assessment

TB PAGE 184

Before you even start planning your scientific investigation, you need to be aware of how it is going to be assessed. If you know how it is going to be assessed at the start of your planning, then you can incorporate this into your planning. This way you can ensure that you hit all the required criteria and obtain the best possible mark that you can.

Your teacher may use an assessment rubric, such as the one shown in Table 5.2 on page 128, to assess a controlled experiment methodology. The rubric provided in Table 5.2 is a sample rubric and may not accurately reflect how your teacher will assess your scientific investigation. It is provided as an example only so you can complete the activity below. In this activity, you will assess a student's scientific report on a controlled experiment using the rubric in Table 5.2 and decide on a final mark out of 30 (Table 5.3 on page 129).

Table 5.2

Criteria	5 marks	4 marks	3 marks	2 marks	1 mark	0 marks
Use of clear, coherent expression Maximum number of words is 1000; minimum number of words is 300.	Excellent expression of information in a coherent and concise manner. Close to maximum word limit.	Well-expressed information in a coherent and mostly concise manner. Close to maximum word limit.	Generally well-expressed information in a coherent and concise manner. Well under maximum word limit.	Fair expression of information, but lacking coherence and/or conciseness. Closer to minimum word limit.	Poorly expressed information or lack of conciseness and coherency. Too far below minimum word limit.	Not shown
Design or adapt a scientific investigation, including use of correct methodology and use of logbook	Topic chosen carefully to fulfil requirements. Investigation designed or well adapted using the correct methodology to test the research question. All steps recorded clearly in logbook.	Topic chosen mostly fulfils requirements. Investigation designed or adapted using the correct methodology to test the research question. Most steps recorded clearly in logbook.	Topic chosen fulfils most of requirements. Investigation designed or adapted using most of the correct methodology to test the research question. Most steps recorded in logbook.	Topic chosen partially fulfils requirements. Investigation designed or adapted using some of the correct methodology to test the research question. Some steps recorded in logbook.	Topic chosen does not fulfil requirements. Investigation poorly designed or adapted using little of the correct methodology to test the research question. Few to no steps recorded in logbook.	Not shown
Use of scientific method to conduct scientific investigation	Excellent use of scientific method with no errors. Aim, research question, hypothesis and variables clearly defined. Method well laid out and logical.	Good use of scientific method with 1 or 2 errors. Aim, research question, hypothesis and variables defined. Method well laid out but with 1 or 2 lapses in logic.	Good use of scientific method with 3 or 4 errors. Aim, research question, hypothesis and variables partially defined. Method presented but with several lapses in logic.	Poor use of scientific method with more than 5 errors. Aim, research question, hypothesis and variables not all defined. Method presented but with many lapses in coherence and logic.	Little understanding of the scientific method. Aim, research question, hypothesis and variables not defined. Method presented but with no coherence and logic.	Not shown
Generation, collection and display of appropriate qualitative and/or quantitative data, including appropriate use of logbook	Large amount of qualitative and/or quantitative data generated and carefully collected and presented in a well-designed format. All data recorded carefully and clearly in logbook.	Qualitative and/or quantitative data generated and collected and presented in a well-designed format. All data recorded in logbook but not as clear as could be.	Some qualitative and/or quantitative data generated and collected. Format of display could be clearer. Most data recorded in logbook but not as clear as could be.	Some qualitative and/or quantitative data generated and collected. Format of display is disorganised and/or unclear. Some data recorded in logbook.	Little qualitative and/or quantitative data generated and collected. Format of display is disorganised. Little to no data recorded in logbook.	Not shown
Interpretation of data leading to a valid conclusion	Data is very well analysed and interpreted, with no errors. Conclusion drawn from data is valid and refers to hypothesis and data as evidence.	Data is well analysed and interpreted, with a few errors. Conclusion drawn from data is mostly valid and refers to hypothesis and data as evidence.	Data is mostly well analysed and interpreted, with several errors. Conclusion drawn from data is mostly valid and does not refer to both hypothesis and data as evidence.	Data is poorly analysed and interpreted, with many errors. Conclusion drawn from data is almost valid and does not refer to either hypothesis or data as evidence.	Data is not analysed or interpreted. Conclusion drawn from data is not valid and does not refer to either hypothesis or data as evidence.	Not shown
Use of effective science communication	A well-structured, logical and well-presented presentation.	Mostly well-structured, logical and well-presented presentation.	Mostly well-structured and well-presented presentation. Two or more lapses in logic.	Presentation lacks appropriate structure or is not clearly presented. Lacks logical progression.	Poor presentation, lacking logical structure or coherency.	Not shown

9780170452632

Table 5.3

Criteria	Marks out of 5 for each criterion
Use of clear, coherent expression Maximum number of words is 1000; minimum number of words is 300	
Design or adaptation of scientific investigation, including use of correct methodology and use of logbook	
Use of scientific method to conduct scientific investigation	
Generation, collection and display of appropriate qualitative and/or quantitative data, including appropriate use of logbook	
Interpretation of data leading to a valid conclusion	
Use of effective science communication	
Total	/30 marks
Comment:	

What to assess

Students in Mr Myers' Year 11 Biology class were required to design and conduct a controlled experiment into one aspect of plant growth. One student's report and accompanying logbook pages are presented in Figures 5.1 and 5.2 on pages 130 and 131.

What to do

Step 1: Use the rubric in Table 5.2 to assess the student's science report and logbook presented in Figures 5.1 and 5.2.

Step 2: As you read through the report, circle the sentences in the rubric that best represent each criterion (singular of criteria) in the student's report.

Step 3: Once you have read through the student's report and circled all the relevant criteria in the rubric, count the number of marks you will award the student. If you have circled some sentences in different mark columns, decide whether you are going to award half marks for some criteria.

Step 4: Write the marks you are going to award for each criterion in Table 5.3 and calculate the total marks out of 30.

Step 5: Write a comment to the student to let them know where they could have improved to gain further marks.

Investigating plant responses to light from different directions

Introduction

Plants often respond to external stimuli from a particular direction by a bending movement that involves growth. The growth response can be either towards or away from the stimulus. Such growth responses are called tropisms. When plants respond to the external stimuli of light, it is called phototropism.

Aim

To investigate phototropism in plants.

Hypothesis

If plants are exposed to light from different directions, then they will grow towards the light.

Materials

- 3 Petri dishes with damp cotton wool and 20 wheat seedlings (about 5 cm tall) in each
- 2 boxes, one with a slit in the middle of the box and one with a slit at the bottom of the box
- Grow-lux lamp

Risk assessment

What are the risks in doing this investigation?	How can I manage these risks to stay safe?
Mould might grow in the Petri dishes.	Do not inhale near the Petri dishes.
Grow-lux lamps will get hot and could burn skin.	Turn the lamps off and let them cool down before moving them.

Method

1. Find a space that is big enough to fit two boxes with seedlings underneath.
2. Place one Petri dish with wheat under the box with the slit near the middle.
3. Place one Petri dish with wheat under the box with the slit near the bottom.
4. Place the third Petri dish with wheat directly under the Grow-lux lamp.
5. Before the boxes are placed over the seedlings, sketch the seedlings in the results table, indicating colour, height and angle from the horizontal.
6. Return after 3 days and redraw the seedlings in the results, indicating colour, height and angle from the horizontal.

Results

	No box	Middle slit	Bottom slit
Day 1			
Sketch			
Observations	Colour: green Height: 6 cm Angle from the horizontal: 90°	Colour: white-yellow Height: 7 cm Angle from the horizontal: 105°	Colour: white-yellow Height: 7 cm Angle from the horizontal: 125°
After 3 days			
Sketch			
Observations	Colour: green Height: 15 cm Angle from the horizontal: 90°	Colour: white-yellow Height: 19 cm Angle from the horizontal: 138°	Colour: white-yellow Height: 21 cm Angle from the horizontal: 170°

Figure 1 Seedling growth at days 1 and 3

Discussion

The wheat seedlings that grew under the Grow-lux lamp for three days grew straight up (90° to the horizontal). These wheat plants were green, meaning that their chlorophyll had fully developed in the light. The wheat seedlings that grew in the box with the middle slit grew at an angle of 138° to the horizontal. These wheat seedlings were a white-yellow colour, meaning that their chlorophyll had not yet developed due to the lack of light. The wheat seedlings that grew in the box with the bottom slit grew at the greatest angle to the horizontal at 170°. These wheat seedlings were also a white-yellow colour, meaning that their chlorophyll had not yet developed due to the lack of light.

I could improve this method by growing more plants in each condition.

These results support the hypothesis and show that plants will grow towards the light.

Conclusion

The plants with a light source from above grew straight up, but when the light source came in from the side, the plants bent towards the light source. These results support the hypothesis that plants will grow towards the light source.

Figure 5.1 Student science report to be assessed

9780170452632

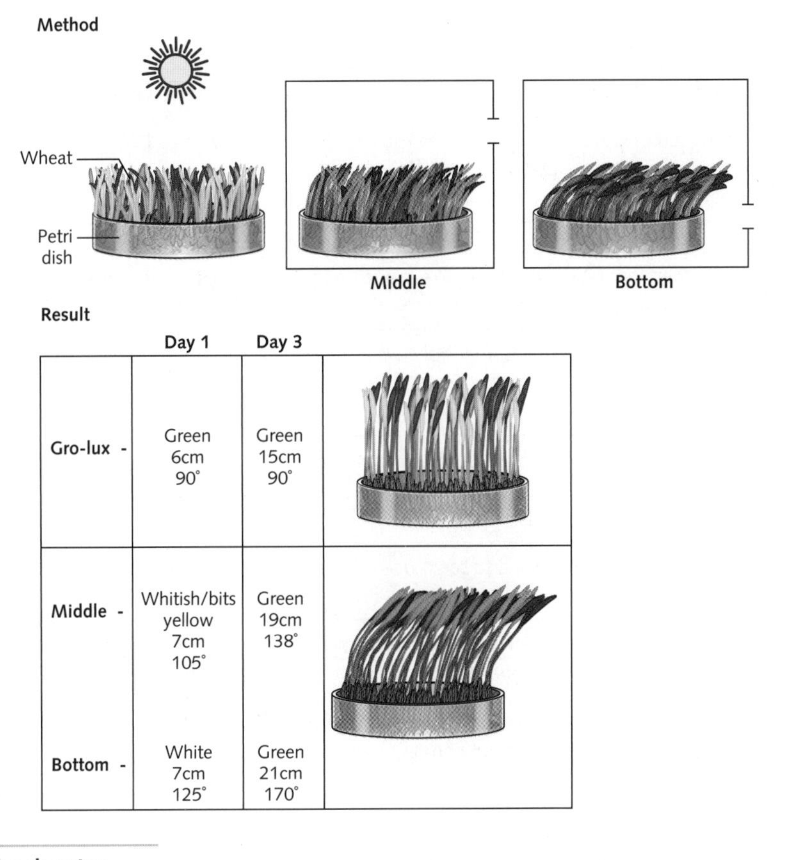

Figure 5.2 Student logbook entry

5.2 Types of scientific methodology

Key knowledge

Investigation design

- biological science concepts specific to the selected scientific investigation and their significance, including the definition of key terms
- scientific methodology relevant to the selected scientific investigation, selected from: classification and identification; controlled experiment; correlational study; fieldwork; modelling; product, process or system development; or simulation

5.2.1 Identifying the different types of scientific methodology

Key science skills

Plan and conduct investigations

- determine appropriate investigation methodology: case study; classification and identification; controlled experiment; correlational study; fieldwork; literature review; modelling; product, process or system development; simulation

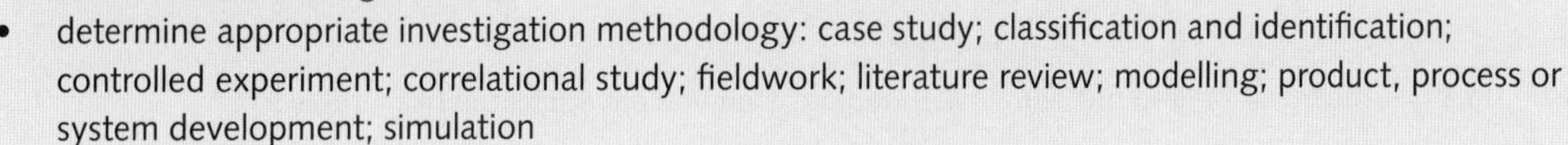

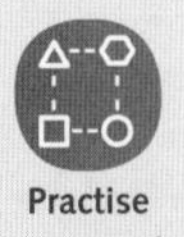

There are many different methodologies to use when undertaking a scientific investigation. Methodology refers to the broader framework of the approach taken in the investigation to test your research question or hypothesis.

Table 5.4 on page 132 shows the seven different types of scientific methodology available to you to choose from to design your scientific investigation. In the first column is the name of the methodology, in the second column is a definition of the methodology and in the third column are research questions. Your task is to decide which methodology would be used to best test each research question. The trouble is that all the cells are mixed up and you need to unmix them.

What to do

Step 1: You will need seven different coloured pencils or highlighters to complete this activity.

Step 2: Use one colour per methodology. Use this to colour in the methodology name, its corresponding definition and the research question that could be best tested using that methodology.

Step 3: Repeat step 2 until you have coloured in all seven methodologies, definitions and research questions.

Table 5.4

Scientific methodology	Definition	Research question
Fieldwork	Making a representation of something usually on a smaller or larger scale so it can be seen and manipulated.	How does DNA replicate itself?
Correlational study	Investigation of a larger group of living things and to classify them by placing them into like groupings and identify them by naming them.	Does the amount of light affect how much fruit my tomato plants produce?
Product, process or system development	Scientific investigation that is undertaken outside the laboratory.	Will the compulsory wearing of facemasks during the 2020 pandemic mean there will also be a drop in the number of cases of influenza?
Controlled experiment	To design an object or product, process or system that will assist a human in meeting their biological demands and requirements.	In what areas of my local parks can blue tongue lizards be found?
Simulation	When two factors are studied to see if one affects the other.	How many different spider species live in the schoolyard?
Modelling	Use a model to simulate real life or part or whole of a system to gain knowledge of its functioning.	How can I best help people who suffer from arthritis and cannot pick up anything?
Classification and identification	Investigation where all variables that could affect the results of an experiment are kept constant except the independent variable.	What does DNA look like?

5.2.2 Designing a scientific investigation

PAGE 194

Key science skills

Plan and conduct investigations

- design and conduct investigations; select and use methods appropriate to the investigation, including consideration of sampling technique and size, equipment and procedures, taking into account potential sources of error and uncertainty; determine the type and amount of qualitative and/or quantitative data to be generated or collated

Comply with safety and ethical guidelines

- demonstrate safe laboratory practices when planning and conducting investigations by using risk assessments that are informed by safety data sheets (SDS), and accounting for risks

Reinforce

Choose one of the research questions from Table 5.4 above. You will then use the methodology that you have paired with this research question and partially develop a scientific investigation that would test that research question. The following questions will provide a guide on what to do.

What to do

Step 1: Identify the research question and matching methodology that you have selected from Table 5.4.

Research question:

Methodology:

9780170452632

Step 2: Write an aim and either a hypothesis or a prediction for your investigation.

Step 3: List the materials and equipment that you will need to carry out your investigation. Construct a risk assessment if required.

Step 4: Write a step-wise method for your investigation.

Step 5: Describe the type of data that your method will generate.

Step 6: Consider how much data your method will generate.

Step 7: Construct a results table to record your data.

5.3 Quantitative and qualitative data

Key knowledge

Investigation design

- techniques of primary qualitative and quantitative data generation relevant to the investigation
- accuracy, precision, reproducibility, repeatability and validity of measurements in relation to the investigation

5.3.1 Primary data

TB PAGE 196

Key science skills

Comply with safety and ethical guidelines

- demonstrate safe laboratory practices when planning and conducting investigations by using risk assessments that are informed by safety data sheets (SDS), and accounting for risks

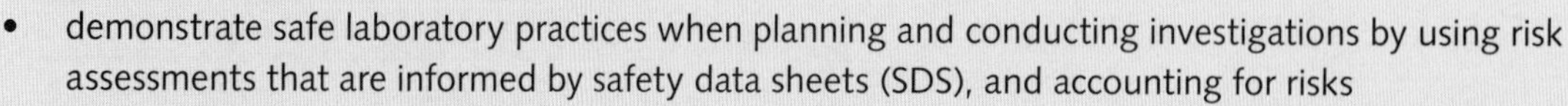

Generate, collate and record data

- systematically generate and record primary data, and collate secondary data, appropriate to the investigation, including use of databases and reputable online data sources
- record and summarise both qualitative and quantitative data, including use of a logbook as an authentication of generated or collated data

As part of your Year 11 Biology class, you have been conducting scientific research to find out where the dirtiest place in the school is. You have exposed five agar plates to five different environments around the school:

1 = staffroom sink

2 = Year 11 locker room

3 = Year 7 classroom

4 = boys' toilets

5 = girls' toilets.

You incubated the five agar plates at 30°C. The plates are shown in Figure 5.3 after two days' growth. Use this information to answer the questions below.

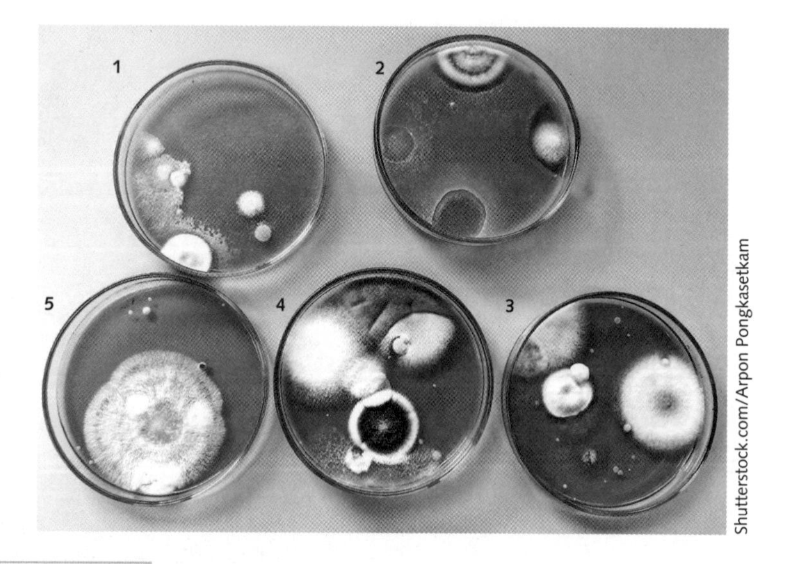

Figure 5.3 Agar plates after 2 days' growth at 30°C

1 Write a research question for this investigation.

______

2 Write a hypothesis for this investigation, making sure that you follow the 'If ... then ...' format for writing a hypothesis.

______

______

3 Identify the methodology used in this investigation.

4 Use the template below to construct a risk assessment for this investigation.

What are the risks in doing this investigation?	How can you manage these risks to stay safe?

5 Determine which qualitative and quantitative data you want to record from your results in Figure 5.3. Design and construct a results table to record this data.

6 Record the data from Figure 5.3 in your results table.

7 The teacher comments that you do not have a control in this experimental set-up. If you were to add a control, what would it be and what purpose would it serve?

5.3.2 Quality of primary data

Key science skills

Analyse and evaluate data and investigation methods

- process quantitative data using appropriate mathematical relationships and units, including calculations of ratios, percentages, percentage change and mean
- identify and analyse experimental data qualitatively, handling where appropriate concepts of: accuracy, precision, repeatability, reproducibility and validity of measurements; errors (random and systematic); and certainty in data, including effects of sample size in obtaining reliable data

Two students were conducting a scientific investigation to find out whether the salt concentration of water affected the amount of light transmitted through the water to the plants. They tested seven different concentrations of salt and measured the transmission of white light through each concentration using a torch and a light meter (Figure 5.4 on page 136). They repeated their measurements for each concentration six times. Their results are shown in Table 5.5 on page 136. Consider these results carefully and answer the questions that follow.

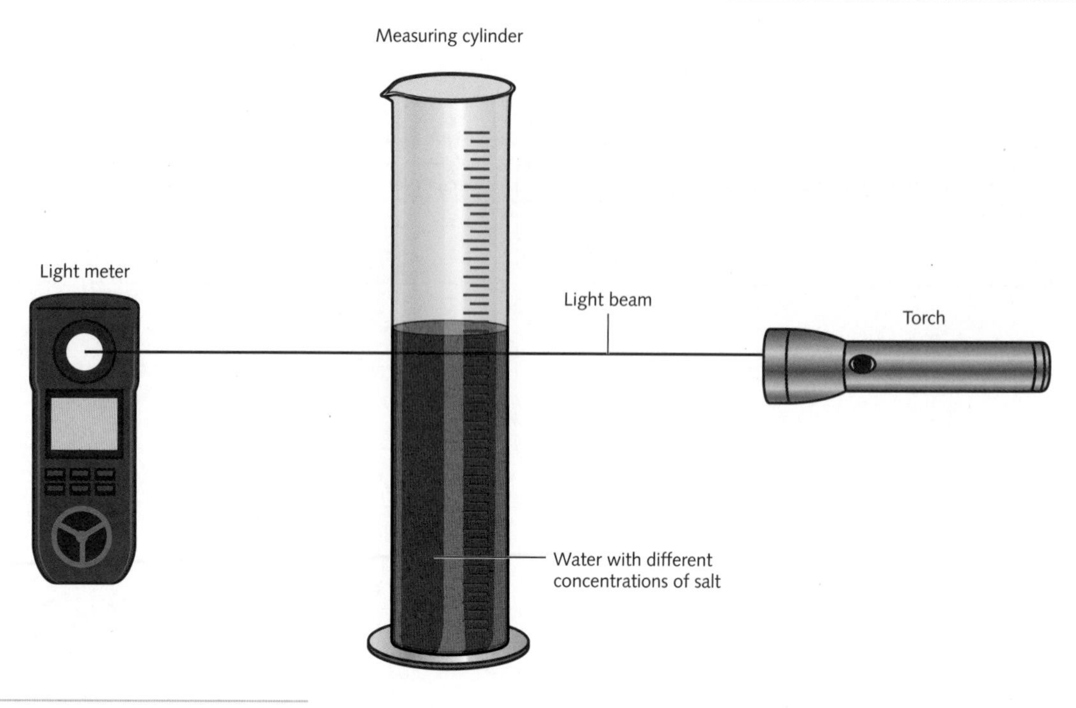

Figure 5.4 Experimental set-up

Table 5.5 Salt concentration and light transmission

Salt concentration (%)	Transmission of light (%)					
	Trial 1	Trial 2	Trial 3	Trial 4	Trial 5	Trial 6
0	78.57	75.27	77.23	78.4	64.88	124.66
3	92.82	69.71	85.23	79.54	88.91	57.96
6	100.05	66.51	88.39	78.29	73.66	61.54
9	110.05	64.91	80.71	109.43	68.29	52.96
12	117.18	59.91	75.66	81.96	65.01	49.95
15	115.46	66.03	72.55	81.06	65.72	55.37
18	120.67	60.48	69.31	74.63	58.43	54.51

1 Calculate the mean results and add them to Table 5.5 by extending the table in the appropriate way.

2 Are these results precise? Remember that precise is how closely individual results are to each other. Explain your answer.

3 Circle in red any results that do not appear to be accurate. Remember that accurate data is when it has been measured and recorded correctly.

4 Circle in blue any outliers. Remember that an outlier is a data point that does not fit the general trend in the data.

5 Are the results valid? Explain your answer.

9780170452632

6 There is one trial that appears to have been measured incorrectly. Identify this trial.

7 Name the type of error (personal, random or systematic) that could have occurred to generate these incorrect results in that trial.

8 Recalculate the mean results excluding this incorrect trial. Add the recalculated mean results to Table 5.5. How does the removal of the incorrectly measured trial affect the mean?

5.4 Health, safety and ethics

Key knowledge
Investigation design
- health, safety and ethical guidelines relevant to the selected scientific investigation

5.4.1 Ethical guidelines

Key science skills
Analyse, evaluate and communicate scientific ideas
- analyse and evaluate bioethical issues using relevant approaches to bioethics and ethical concepts, including the influence of social, economic, legal and political factors relevant to the selected issue

Develop

Ethics are moral principles that govern a person's behaviour. Ethics is considered to be knowing the difference between right and wrong. Consider, for example, that a planning application has been submitted to your local council to build a factory on a site that is currently heavily wooded and home to thousands of native animals. The factory will provide 120 jobs for the local community but at the expense of the trees, shrubs and native animals. What is the right thing to do? This will depend on who you talk to as each person will have a different viewpoint depending on their circumstances – one person might be long-term unemployed, and another person might be a wildlife rescuer. So, you need to consider the ethical dilemma from different viewpoints including:

- Social, e.g. How will the decision affect individuals, families and society in general?
- Economic, e.g. How much will it cost? Who will bear the burden of that cost? Is it affordable at a personal, local or national level? Does anyone benefit financially from the decision?
- Legal, e.g. Is it within the law? Do laws need to be changed to accommodate this issue?
- Political, e.g. Is it an issue that divides the public? Do different political parties have different views on this issue? Will the decision affect one political party more than another? Will it affect how people vote? Will the decision cause demonstrations in the street?

Bioethics are ethics applied to biological dilemmas. For example, in the quest to find a vaccine for the novel coronavirus, SARS-CoV-2, scientists used ferrets and mice as human models (instead of humans). Mice and ferrets are sentient beings, they have a well-developed nervous system and can experience pain, stress and fear. Before the scientists could undertake any research using mice and ferrets, they would have had to present their research to an ethics committee and argue why it was important. They would have argued one of the following five ethical concepts on page 138.

i **Integrity:** the commitment to searching for knowledge and understanding and the honest reporting of all sources of information and communication of results, whether favourable or unfavourable, in ways that permit scrutiny and contribute to public knowledge and understanding.

ii **Justice:** the moral obligation to ensure that there is fair consideration of competing claims; that there is no unfair burden on a particular group from an action; and that there is fair distribution and access to the benefits of an action.

iii **Beneficence:** the commitment to maximising benefits and minimising the risks and harms involved in taking a particular position or course of action.

iv **Non-maleficence:** involves avoiding the causations of harm. However, as positions or courses of actions in scientific research may involve some degree of harm, the concept of non-maleficence implies that the harm resulting from any position or course of action should not be disproportionate to the benefits from any position or course of action.

v **Respect:** involves consideration of the extent to which living things have an intrinsic value or instrumental value; giving due regard to the welfare, liberty and autonomy, beliefs, perceptions, customs and cultural heritage of both the individual and the collective; consideration of the capacity of living things to make their own decisions; and when living things have diminished capacity to make their own decisions ensuring that they are empowered where possible and protected as necessary.

Using this information, read the two scenarios below and answer the questions accompanying each.

Scenario 1

Dr Dom Witt is a biological researcher who is committed to studying the environment. Dr Witt is currently undertaking research on an endangered bird species that lives only in the Otway National Park. Dr Witt is concerned that a plan to build a tourist centre and carpark will further encroach upon this bird's habitat and could cause further loss of individuals. Dr Witt is undertaking a survey of this species to count the individuals and map their distribution throughout the Otway National Park. His preliminary results are showing that there will be very little impact on this species from the visitor centre and carpark. Dr Witt is personally opposed to development in national parks and has decided not to release his results.

1 What is the ethical issue?

2 Which ethical concept(s) applies to this issue? Explain your choice.

3 Identify one social and one economic factor that needs to be considered.

Scenario 2

DNA Data Pty Ltd is a company that provides quick and easy DNA sequencing. If you want to find out if you carry a particular gene for a disease that you may pass onto any potential offspring, you can contact DNA Data and they will send you a kit. The kit comprises a pair of sterile gloves, swab, plastic vial and postage-paid return envelope. All you have to do is put on the plastic gloves, use the swab to remove some cheek cells from inside your mouth, seal the swab in the plastic vial and post it back to the company for analysis. The company will analyse your DNA and have results back to you within two weeks. Your DNA data is readily available to anyone who wants it. It is stored on a server in Russia in case you ever need to access it again.

1 What is the ethical issue?

2 Which ethical concept(s) applies to this issue? Explain your choice.

3 Identify one political and one legal factor that needs to be considered.

5.5 Collecting and analysing data

Key knowledge

Scientific evidence

- the distinction between an aim, a hypothesis, a model, a theory and a law
- observations and investigations that are consistent with, or challenge, current scientific models or theories
- the characteristics of primary data
- ways of organising, analysing, and evaluating generated primary data to identify patterns and relationships including sources of error
- use of a logbook to authenticate generated primary data
- the limitations of investigation methodologies and methods, and of data generation and/or analysis

Part B: Scientific evidence

5.5.1 Analysing your data

Key science skills

Analyse and evaluate data and investigation methods

- process quantitative data using appropriate mathematical relationships and units, including calculations of ratios, percentages, percentage change and mean

PAGE 203

On page 140 are two of sets of data that have been recorded during two different scientific investigations. Inspect each dataset carefully to understand what has happened in the investigation to produce that data. Answer the questions that follow each dataset.

Dataset 1

Table 5.6 Height of seedlings for germinated wheat seeds

Type of water	Trial number	Height of seedling (mm)					Percentage change
		Day 1	Day 2	Day 3	Day 4	Day 5	
Tap water	1	23	43	68	79	96	
	2	21	51	72	81	101	
De-ionised water	1	15	34	48	57	63	
	2	17	38	50	52	61	

Step 1: Calculate the percentage change from day 1 to day 5 in each trial. Enter each result in the appropriate cell in the final column of the Table 5.6.

Step 2: Construct a possible hypothesis that was being tested in this investigation.

Step 3: Does the data supplied in Table 5.6 support or refute this hypothesis? Explain why, using data to support your explanation.

Dataset 2

Table 5.7 Effect of different concentrations of sucrose solution on raw potato

Concentration of sucrose solution (mol dm^3)	Initial length of potato strip (mm)	Final length of potato strip (mm)	Percentage change in length of potato strip	Mean percentage change in length of potato strip
0	49.5	51.5		
	49.0	52.5		
0.2	50.0	51.0		
	50.5	51.5		
0.4	50.0	51.0		
	49.5	51.5		
0.6	50.0	49.0		
	51.0	49.5		
0.8	49.0	48.5		
	50.5	49.0		
1.0	49.5	48.0		
	50.5	47.5		

Step 1: Calculate the percentage change in the length of each potato strip for each concentration of glucose. Enter each result in the cell in the appropriate column in Table 5.7.

Step 2: Calculate the mean percentage change in the length of the potato strips for each concentration of glucose. Enter each result in the cell in the appropriate column.

Step 3: Construct a possible hypothesis that was being tested in this investigation.

Step 4: Does the data supplied in Table 5.7 support or refute this hypothesis? Explain why, using data to support your explanation.

5.6 Communicating your results

Key knowledge

Science communication

- the conventions of scientific report writing including scientific terminology and representations, standard abbreviations and units of measurement
- ways of presenting key findings and implications of the selected scientific investigation.

Part C: Scientific communication

5.6.1 Presenting your work as an article

Key science skills

Analyse, evaluate and communicate scientific ideas

- use clear, coherent and concise expression to communicate to specific audiences and for specific purposes in appropriate scientific genres, including scientific reports and posters

Develop

TB PAGE 211

Unit 1 Outcome 3 gives you the opportunity to present your scientific research findings in a number of different ways: as a scientific poster, an article for a scientific publication, a practical report, an oral presentation, a multimedia presentation or a visual presentation. You are going to practise writing an article of no more than 400 words for scientific publication. You are going to source your information from one of the scientific investigations presented in sections 5.3.1 (page 134) and 5.3.2 (page 135).

Use the template below to structure your scientific article. Remember, it is not a scientific report but an informative article for a journal.

Title: Author:	
Abstract: brief summary of the article	

Introduction: what was the purpose of the study?	
Material and method: what was used and how was it used?	
Results: what was discovered?	
Discussion: what do the results mean? How do these results add to scientific knowledge?	
Conclusion: what can you learn from the investigation?	

9780170452632

5.7 Chapter review

PAGE 218

5.7.1 Key terms

Differentiate (explain how they are different) between each pair of terms shown below

accurate; precise

quantitative data; qualitative data

dependent variable; independent variable

aim; research question

ethics; bioethics

random error; personal error

methodology; method

modelling; simulation

observation; hypothesis

integrity; non-maleficence

social; political

repeatable; reproducible

Chapter review continued

5.7.2 Practice test questions

Multiple-choice questions

1 In a scientific investigation, the one variable that is manipulated is called the

A controlled variable.
B extraneous variable.
C independent variable.
D dependent variable.

2 A scientist conducted a scientific investigation to determine how the concentration of salt in a body of water affects the type of plants that can live in that body of water. In this scientific investigation, the independent variable is the

A temperature of the water.
B size of the body of water.
C amount of salt in the water.
D number of plants in the water.

The following information applies to Questions 3 and 4.

A scientist wants to study the effects of fertiliser on tomato plants, so they set up a scientific investigation where plant A receives no fertiliser, plant B receives 5 mg of fertiliser each day, and plant C receives 10 mg of fertiliser each day.

3 Which plant is the control plant?

A Plant B
B All of them
C Plant C
D Plant A

4 In order to make the data collected valid, the scientist would need to

A ensure that there is only one independent variable affecting the dependent variable.
B ensure that there is only one dependent variable affecting the independent variable.
C add more trials at each condition.
D remove any outliers from the data before calculating the means.

5 When writing a scientific report, the order of the headings in the report is

A introduction, aim, materials, results, method, discussion, conclusion.
B introduction, aim, materials, method, results, conclusion, discussion.
C aim, introduction materials, method, results, discussion, conclusion.
D introduction, aim, materials, method, results, discussion, conclusion.

Short-answer questions

6 Mrs Barnes decides to set up a scientific investigation to test if potatoes cook faster in a microwave or on the stovetop. She peels and dices 500 grams of potatoes and separates this into two lots of 250 grams each. She places one lot of potatoes on a microwave-safe dish and places it into the microwave. She cooks it on high for 30 seconds and repeats until the potatoes are cooked. She records the total time it took to cook the potatoes. She places the other 250 grams into a saucepan with water and places it on high heat on the stovetop. She records how long it takes to cook the potatoes. She records her results.

Microwave: 2 minutes 30 seconds
Stovetop: 5 minutes 13 seconds

a What is the independent variable in this investigation?

1 mark

b What is the dependent variable in this investigation?

1 mark

c List one extraneous variable.

1 mark

d Write a hypothesis for this investigation.

2 marks

e Identify and discuss one limitation in Mrs Barnes' method.

2 marks

7 Sarah and Mosey undertook field work to find out how many different species of eucalypt occurred in their local area. They divided their local area into eight equal segments based on compass bearings and used a eucalypt key to classify each eucalypt tree within each segment. They recorded their results in their logbooks and these are

South east 3
North east 5
South west 7
North west 9
East 4
West 1
South 8
North 11

On the following page, construct a graph of their results using the graph paper. Use correct scientific conventions for drawing a scientific graph.

4 marks

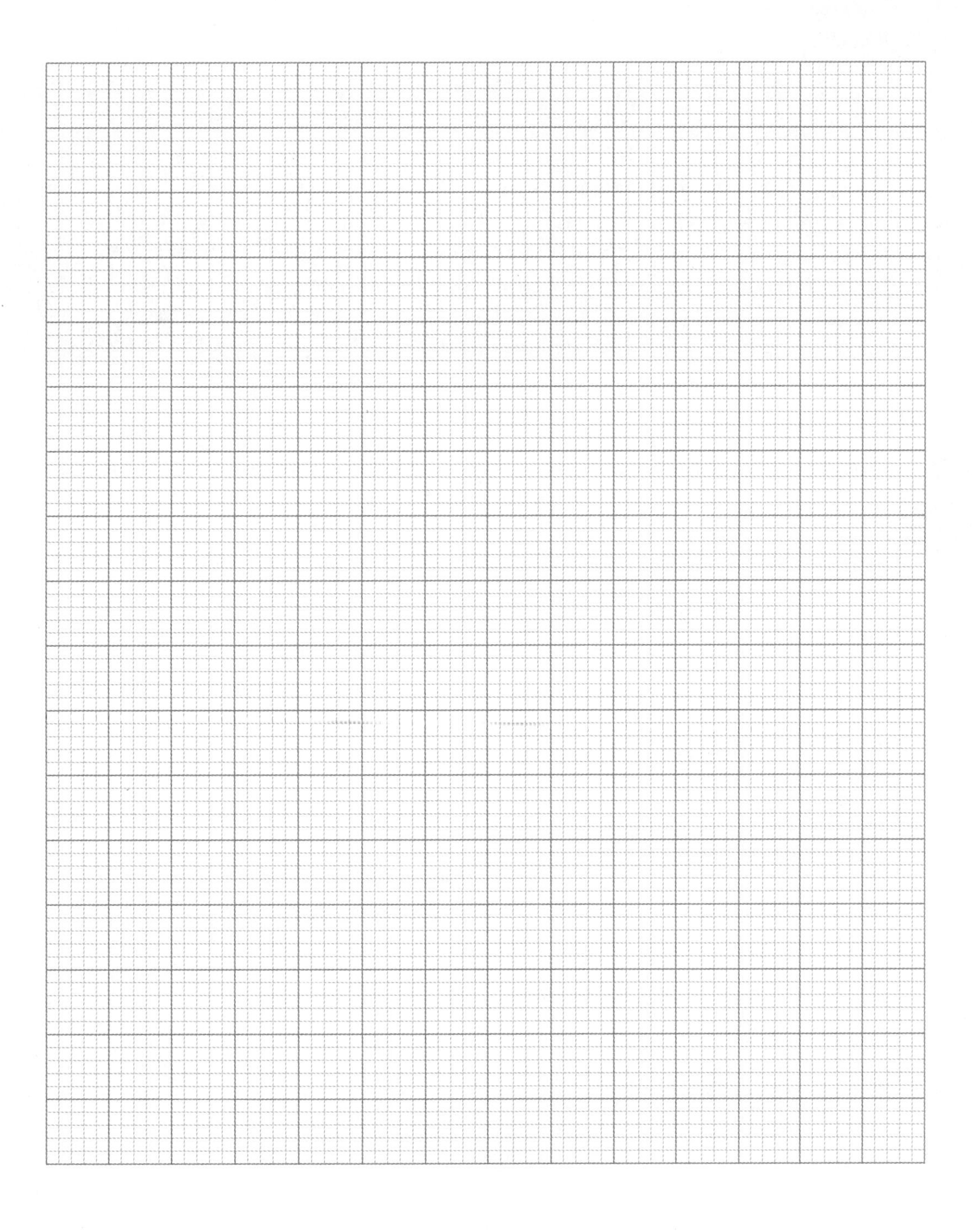

6 Chromosomes to genomes

Getty Images/SCIEPRO

Remember

Chapter 6 Chromosomes to genomes will call on content that you have already met in your science studies from previous years or from earlier this year. Take some time to refresh your knowledge of this content before you enter this chapter. Try to answer the following questions from memory. If you cannot do this, then use a reference to assist you.

PAGE 224

1 Why is DNA called the molecule of heredity?

2 Where in prokaryotic cells is DNA located?

3 Where in eukaryotic cells is DNA located?

4 What is the union of two haploid cells to form a diploid zygote called?

5 What is the division of the nucleus of a eukaryotic cell to form two identical daughter cells called?

6.1 Distinction between genes, alleles and a genome

Key knowledge

From chromosomes to genomes

- the distinction between genes, alleles and a genome

6.1.1 Structure of DNA: making a model

PAGE 226

Key science skills

Analyse, evaluate and communicate scientific ideas

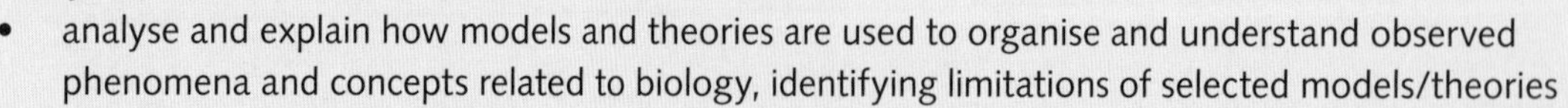

- analyse and explain how models and theories are used to organise and understand observed phenomena and concepts related to biology, identifying limitations of selected models/theories

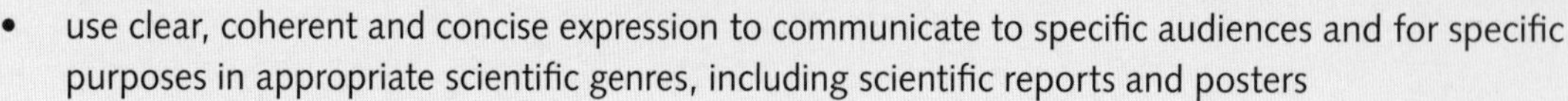

- use clear, coherent and concise expression to communicate to specific audiences and for specific purposes in appropriate scientific genres, including scientific reports and posters

DNA is the molecule of inheritance. It is passed between generations and carries the code for production of proteins. You are going to make a model of DNA. Modelling in science enables you to create a representation of an object that is too small, too large or can only be represented using mathematics. In this instance, you are going to make a physical model of DNA using the information that is already known about DNA.

What to do

To complete this activity, you will need 1 sheet of white paper, a holepunch, scissors, coloured pencils or markers, a biro and 2 cm wide sticky tape.

Part 1: Making DNA

Colour areas of the white paper in the colours listed below. Use the holepunch to make dots of each colour so that you have:

12 × brown = deoxyribose sugar

12 × pink = phosphate

6 × orange = adenine

6 × yellow = thymine

6 × green = guanine

6 × blue = cytosine

Step 1: Cut two pieces of 2 cm wide sticky tape, each 20–30 cm long.

Step 2: Turn the tapes so they are sticky side up and place them onto the desk in front of you.

Step 3: Label one strand 5′ and 3′ using a biro. Label the other strand 3′ and 5′.

Step 4: Construct the 'backbone' of each DNA strand with alternating brown and pink dots along the outside edge on one strip and the opposite on the second strip (Figure 6.1).

Step 5: Create a base sequence by placing orange, yellow, green or blue dots in any order along the opposite edge on one of the strands (Figure 6.1).

Step 6: Make the complementary DNA strand by placing the matching dots on the second strand as per the complementary base pairing rule.

Step 7: Annotate as applicable using a biro and put in the hydrogen bonds by using small lines of paper as in the diagram below. Carefully stick another piece of sticky tape on top to seal the dots on each strand. Your model should now look like Figure 6.1 below.

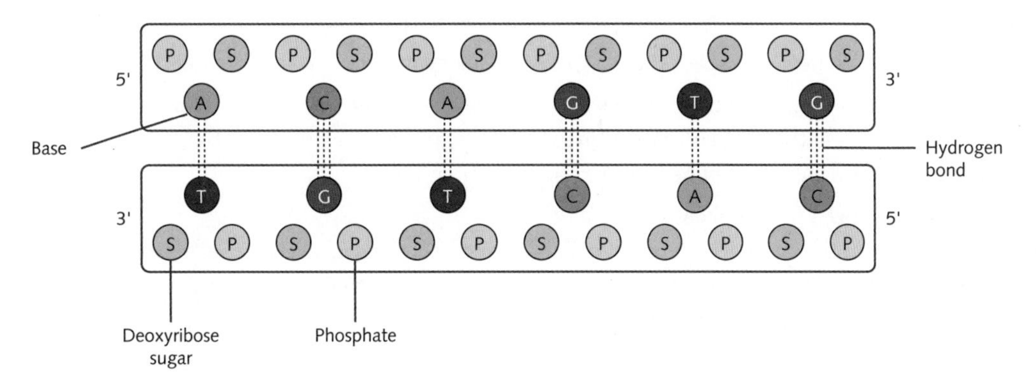

Figure 6.1 The completed DNA strand

Part 2: DNA replication

Use the double-stranded DNA that you made in Part 1.

Step 8: Unzip the two DNA strands created in Part 1, as shown in Figure 6.2. Do this by cutting the hydrogen bonds between the complementary bases.

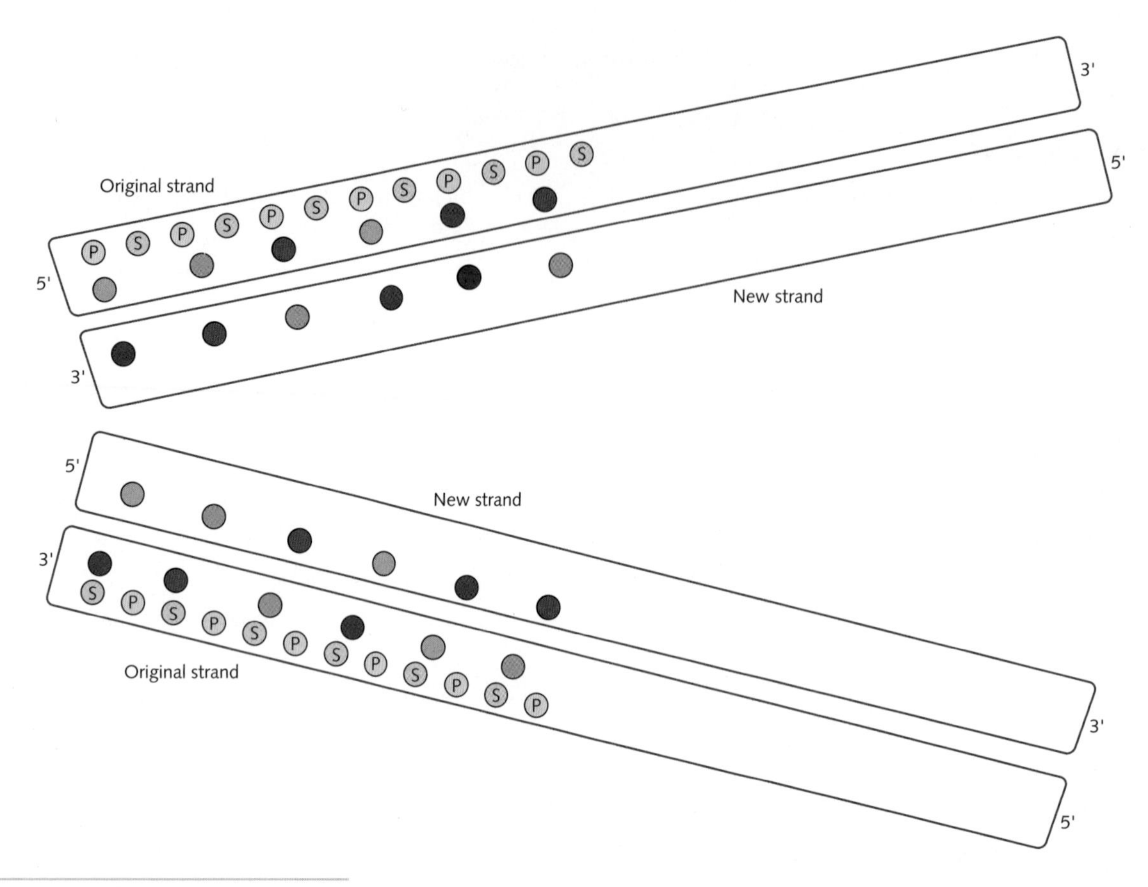

Figure 6.2 Unzipping the DNA double strand

Step 9: Place two new pieces of tape sticky side up next to the separated original pieces.

Step 10: Label the ends of the new strands of sticky tape 5′ → 3′ as appropriate to the original strands.

Step 11: Replicate each strand by placing the complementary bases – orange, yellow, green or blue dots on the new stands from the 5′ end to the 3′ end. (You don't need to set up the 'backbone' on the new strands, although you need to understand that a backbone would exist on the new strands.)

The DNA sequence has now been fully replicated.

Step 12: Cover the new strands with sticky tape to seal in the dots. You now have two new double strands of DNA that are semi-conserved (when one strand of DNA serves as a template to create a new strand of DNA). Use more sticky tape to stick the completed strands onto page 151.

Replicated DNA strand

Use sticky tape to stick your DNA strands here. Draw in hydrogen bonds between your DNA strands and annotate with labels from the list below.

Original strand
New strand
Sugar–phosphate backbone
Nitrogenous base
Deoxyribose sugar
Phosphate
Complementary bases
Hydrogen bonds

6.1.2 The ABC of DNA

TB PAGE 227

1 a What is DNA?

b 'Each strand is made up of two zones or regions.' Explain what this means.

c What holds the two strands together in the double helix?

2 a When do cells make accurate copies of DNA?

b Cells duplicate their DNA. Where does this happen?

c What does the term 'semi-conservative' mean in this context?

d What is the complementary base-pairing rule?

3 a What information is coded into DNA?

b What are nitrogenous bases?

c What is the 'backbone' of DNA constructed from?

4 Find out what 3′ and 5′ mean.

5 What is a nucleotide?

6 a What is the basic unit of DNA?

b What does DNA stand for? (Be careful with your spelling.)

7 A nucleic acid was analysed and found to contain 36% adenine and 14% guanine. What percentage of thymine and cytosine would this nucleic acid contain?

8 a How does the hydrogen bonding between adenine and thymine bases differ from the hydrogen bonding between cytosine and guanine bases?

b The percentage of cytosine in a double-stranded DNA molecule is 18. What is the percentage of thymine in that DNA molecule?

6.1.3 Genes, alleles and genomes

Key science skills

Analyse, evaluate and communicate scientific ideas

- critically evaluate and interpret a range of scientific and media texts (including journal articles, mass media communications and opinions in the public domain), processes, claims and conclusions related to biology by considering the quality of available evidence

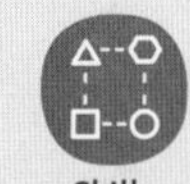

TB PAGE 228

Read the article below, concentrating on the information about genes and alleles. Use this information to answer the questions that follow.

To labradoodle, or not to labradoodle? That is the question.

Shutterstock.com/Leoniek van der Vliet

Figure 6.3 A litter of labradoodle puppies

There are more than 350 dog breeds in the world – many of them hybrids or crosses between breeds. Only 200 of these 350 breeds are recognised by the Australian National Kennel Council as true breeds. Intense breeding and hybridisation can lead to breeds that inherit diseases from one or either parent. Take brachycephalic which occurs in flat-faced dogs like pugs and bulldogs. The shape of the head, throat and muzzle can flatten, which shrinks breathing passages and causes respiratory problems.

The Australian labradoodle is a designer dog (Figure 6.3 on the previous page) much sought after due to its steady nature and non-shedding and hypoallergenic coat qualities. The labradoodle was originally bred by crossing labrador retrievers and standard or miniature poodles. Labradors were selected as they provided the alleles that create a docile and easily trainable dog, whereas the poodle contributes the alleles for coat qualities. These qualities were deemed to be perfect for assistance dogs for the Royal Guide Dogs Association of Australia.

Since 1989, there have been many generations of breeding labradoodles, which have been selected for these desirable qualities while also minimising the risk of continuing inherited disorders found in each of the parental breeds. Since then, breeders have run breeding programs to produce puppies that reflect their newly established standard. But they have been unable to persuade dog registration authorities to accept the Australian labradoodle as a new breed.

A study by Professor Elaine Ostrander's group at the National Institutes of Health in Maryland, US, analysed more than 150 000 locations on the genomes from modern Australian labradoodles as well as individuals from each of the breeds that were part of its ancestry. The goal was to investigate how a new dog breed develops and how the choices made by the breeders fix the defining characteristics of that breed within the genome over subsequent generations.

Professor Ostrander's team found that genes inherited from labrador retrievers are still detectable in modern labradoodles, but, overall, the genome is dominated by genes inherited from the different poodle breeds.

The report said this appears to have resulted from breeders focusing on just one main factor – hair quality – in making decisions on which dogs to use for matings, saying, 'Today's Australian labradoodle is largely poodle with an excess of poodle alleles related to coat type.'

Adapted from: https://www.vettimes.co.uk/news/labradoodle-largely-poodle-due-to-breeder-choices-study/

https://www.inverse.com/science/dog-genome-study-finds-surprising-truth-about-poodle-mixes

1 What genes were desired from the labrador and the poodle in creating the labradoodle?

2 What alleles did each breed contribute to the labradoodle?

3 What was the original reason for creating the labradoodle?

4 What is meant by '... analysed more than 150 000 locations on the genomes from modern Australian labradoodles ...'?

5 What is the evidence provided that refutes the labradoodle as a new breed? Comment on the quality of this evidence.

6 On the basis of this evidence, do you think the labradoodle should be recognised as a new breed? Provide your reasons.

6.1.4 Genomics

Key science skills

Analyse and evaluate data and investigation methods

- identify and analyse experimental data qualitatively, handling where appropriate, concepts of: accuracy, precision, repeatability, reproducibility and validity of measurements; errors (random and systematic); and certainty in data, including effects of sample size in obtaining reliable data

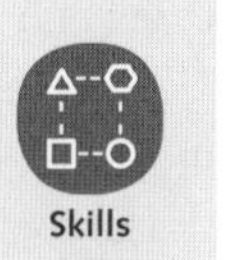

TB PAGE 229

Genomics is the study of genomes of organisms. Figure 6.4 shows two lengths of genetic sequencing of genes in two people, A and B.

To complete this activity, you will need yellow, green, blue and orange highlighters or pencils.

Person A

1	GGATGCGAAG	GCTGCGGCGT	CCTGGGGCGA	GGCGCTGACG	TGAGCTCGGC	GCACCTGGGC
61	TGGGCAGGTA	AGGGCTGGTG	CGGGACGGGG	AGAGGAACCT	GCAGTCCCTA	CTTGGGTAGA
121	GCCAGGCGCC	CCTTGGCTAA	GACGTCGAGG	AGCGTGGTAG	CGACGGGTGA	TCTTCGCTGC
181	GGACTTGGTT	CGGAGGGACG	TCCGCTTCTG	GTGGACAGAT	TGAGCAAAGG	CCTGGGCTGT
241	AGAGACAGGG	AAGTACCAGG	AAGGGGTGGA	TGACCCTGAC	CCAGCTAAAT	GGAAGGCCCA
301	TCTTATACTC	ATGAAATCAA	CAGAGGCTTG	CATGTATCTA	TCTGTCTATC	TATCTATCTA
361	TCTATCTATC	TATCTATCTA	TCTATCTATC	TATCTATCTA	TGAGACAGGG	TCTTGCTCTG
421	TCACCCAGAT	TGGACTGCAG	TGGGGGAATC	A		

Person B

1	GGATGCGATG	GCTGCGGCGT	CCTGGGGCGA	GGCGCTGACG	TGAGCTCGGC	GCACCTGGGC
61	TGGGCAGGTA	AGGGCTGGTG	CGGAACGGGG	AGAGGAACCT	GCAGTCCCTA	CTTGGGTAGA
121	GCCAGGCGCC	CCTTGGCTAA	GACGTCGAGG	AGCGTGGTAG	CGACGGGTGA	TCTTCGCTGC
181	GGACTTGGTT	CGGAGGGACG	TCCGCTTCTG	GTGGACACAT	TGAGCAAAGG	CCTGGGCTGT
241	AGAGACAGGG	TTGTACCAGG	AAGGGGTGGA	TGACCCTGAC	CCAGCTAAAT	GGAAGGCCCA
301	TCTTATACTC	ATGAAATCAA	CAGAGGCTTG	CATGTATCTA	TCTGTCTGTC	TATCTATCTA
361	TCTATCTATC	TATCTATCTA	TGAGACAGGG	TCTTGCTCTG	TCACCCAGAT	TGGACTGCAG
421	TGGGGGAATC	A				

Figure 6.4 DNA sequences of Person A and Person B

What to do

Person A's sequence is regarded as a reference sequence. You are to compare the sequences from Person A and Person B. You are looking for differences in Person B's sequence from the reference sequence (Person A). Make all your coloured marks onto Person B's sequence.

Step 1: Highlight any single nucleotide changes in yellow.

Step 2: Highlight any inserted nucleotides in green.

Step 3: Highlight any places where nucleotides have been deleted in blue.

Step 4: Short tandem repeats are sequences of 2–5 bases that are repeated in sequence a number of times (e.g. TACTACTACTAC). Highlight in orange any short random repeats.

6.1.5 Human genomics research

PAGE 229

Key science skills

Analyse, evaluate and communicate scientific ideas

- analyse and evaluate bioethical issues using relevant approaches to bioethics and ethical concepts, including the influence of social, economic, legal and political factors relevant to the selected issue

Many people said it couldn't be done. Not only was it done, but it was done years before schedule. What is being talked about? The Human Genome Project or HGP was a 13-year-long publicly funded initiative that started in 1990 with the aim of sequencing the entire human genome. The project was initially headed by James D Watson, one half of the Watson and Crick pair of scientists who originally discovered the structure of DNA in 1953.

The project started with two early goals. The first goal was to build genetic and physical maps of the human and mouse genomes, and the second goal was to sequence the smaller yeast and worm genomes as a test run for sequencing the larger, more complex human genome. When the yeast and worm genomes were successfully completed, the sequencing of the human genome proceeded.

The two key principles for the Human Genome Project were:

1. to establish an all-inclusive effort aimed at understanding our shared molecular heritage, and to benefit from diverse approaches
2. that all human genome sequence information be freely and publicly available within 24 hours of its assembly. The entire human genome sequence is now available online for all to access (Figure 6.5).

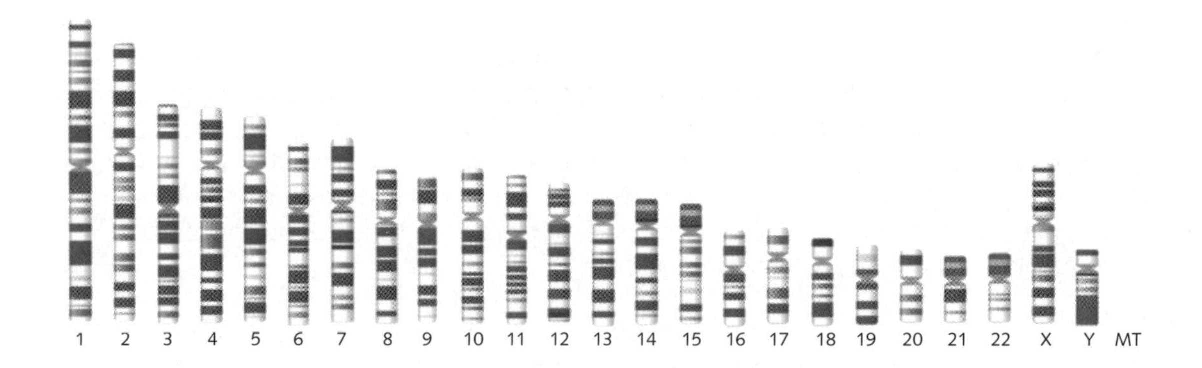

Figure 6.5 Many online sites allow you to search each human chromosome for its sequence.

These founding principles ensured unrestricted access for all scientists in academia and in industry, and it provided the means for rapid and novel discoveries by researchers of all types. The Human Genome Project has so far discovered more than 18 000 disease genes, including 33 different types of cancer, leading to improvements in diagnosis, treatment and prevention. Genetic tests are now available for more than 2000 different diseases, and more personalised and targeted drug treatment has been developed.

You can get your own genome mapped for a small cost, but you need to be aware of any legal and bioethical issues. A few are mentioned below.

- Who owns your DNA sequencing? You or the company that mapped it?
- Where will your DNA information be stored and who has access to it?
- If the company finds out that you have a disease that will kill you in five years, should they tell you?

1 Add two more bioethical issues that could result from DNA mapping.

2 Choose one of the bioethical issues listed above and decide which approach(es) to bioethics could be applied to that issue. (See page 49.) Provide reasons for your choice.

3 Consider any other factors that may influence your chosen issue by completing the table below. Complete the factors applicable to your issue. (Refer to section 5.4.1 for a discussion of social, economic, legal and political factors.)

Factor	How this factor may influence your issue
Social	
Economic	
Legal	
Political	

6.2 Chromosomes

Key knowledge

From chromosomes to genomes

- the nature of a pair of homologous chromosomes carrying the same gene loci and the distinction between autosomes and sex chromosomes
- variability of chromosomes in terms of size and number in different organisms

6.2.1 Chromosomes

PAGE 231

Key science skills

Analyse, evaluate and communicate scientific ideas

- use appropriate biological terminology, representations and conventions, including standard abbreviations, graphing conventions and units of measurement

Chromosomes are made of tightly wound DNA. Chromosomes can appear in many forms throughout the cell cycle. They are referred to differently depending on the stage of the cell cycle, so it is important that you know these names and how to use them.

Figure 6.6 represents a chromosome in various forms. Draw circles around each of the structures listed below and label this diagram to show:

- double-stranded DNA
- chromatid
- chromosome
- homologous chromosomes
- chromatin.
- histone proteins
- centromere

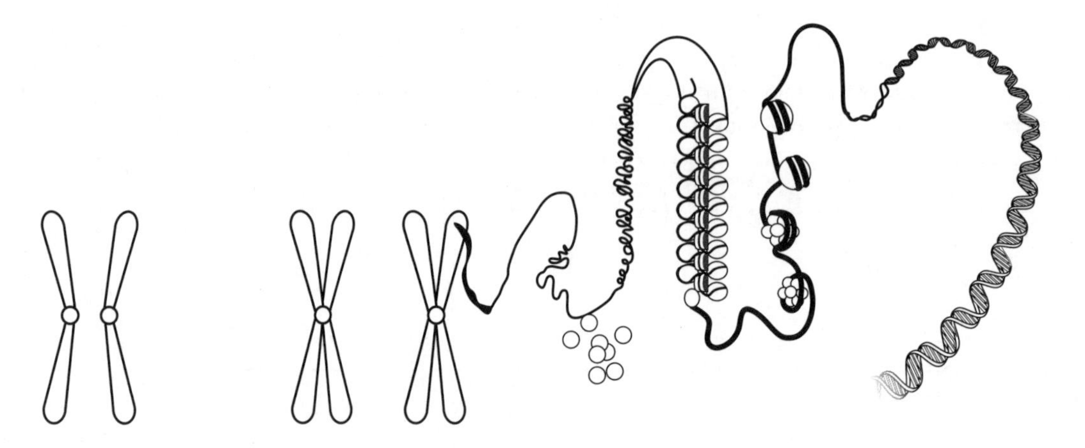

Figure 6.6 Chromosome in various forms

6.2.2 Autosomes and sex chromosomes

PAGE 232

Figure 6.7 shows a karyotype of a human male and a human female. Karyotypes are formed by matching chromosomes on size, position of centromere and banding pattern to form homologous pairs. Each homologous pair of autosomes is given a number from 1 to 22, from the largest to the smallest. The sex chromosomes follow. Use information from Figure 6.7 and your knowledge to answer the questions that follow.

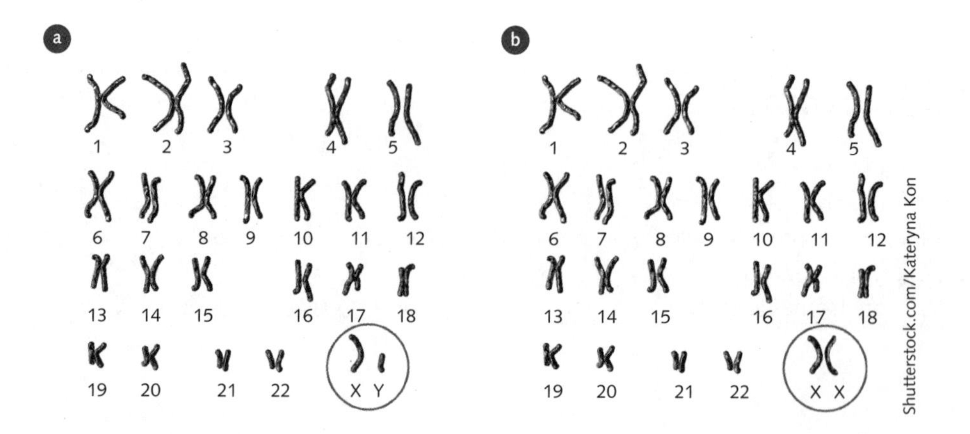

Figure 6.7 Human karyotypes

9780170452632

1 How many chromosomes does a human normally have in each of their body cells?

2 Using Figure 6.7:

a list the numbers of the autosomes

b identify the sex chromosomes. What is the sex of each individual?

3 What type of sex chromosomes do males have?

4 What type of sex chromosomes do females have?

5 Who contributed the X chromosome to a male child?

6 Who contributed the Y chromosome to a male child?

7 Who contributed the X chromosomes to a female child?

6.2.3 Variations in nuclear chromosomes of eukaryotes

Key science skills

Analyse, evaluate and communicate scientific ideas

- use appropriate biological terminology, representations and conventions, including standard abbreviations, graphing conventions and units of measurement

Develop

TB PAGE 233

1 Define the following terms.

a ploidy

b diploid

c haploid

d monoploid

e polyploid

f aneuploid

2 Complete Table 6.1 to show the haploid and diploid state in different species.

Table 6.1

Species	Number of chromosomes in haploid cells (*n*)	Number of chromosomes in diploid cells (2*n*)
Human	23	
Fruit fly		8
Chimpanzee	24	
Bat		44
Koala		16
Kangaroo	8	
Tasmanian devil	7	
Rice		24
Platypus	26	
Chicken	39	
Spinach		12
Cucumber	7	

6.2.4 The endosymbiotic theory

TB PAGE 236

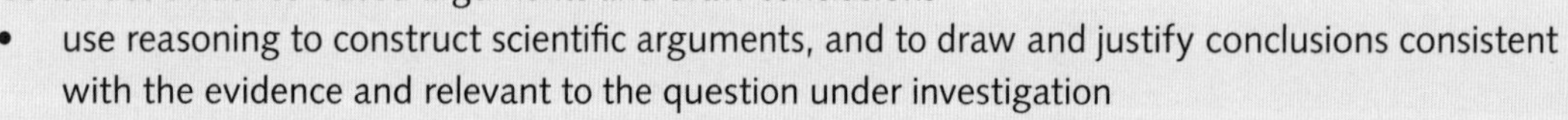

Key science skills

Construct evidence-based arguments and draw conclusions
- use reasoning to construct scientific arguments, and to draw and justify conclusions consistent with the evidence and relevant to the question under investigation

The endosymbiotic theory states that mitochondria and chloroplasts were once bacteria that were living inside larger host cells.

The following similarities are evidence that supports this theory.

1 Prokaryotic bacteria contain their own DNA in the form of circular plasmids.
2 Mitochondria and chloroplasts contain circular DNA.
3 Mitochondria and chloroplasts are about the same size as prokaryotic bacteria.
4 Mitochondria, chloroplasts and prokaryotic bacteria are surrounded by a double lipid bilayer.
5 Mitochondria, chloroplasts and prokaryotic bacteria all divide by binary fission.

Use your knowledge of biology and the evidence for the endosymbiotic theory presented above to explain the endosymbiotic theory. Do this by annotating the diagram in Figure 6.8. Using the space provided, start at the top of the diagram and explain what is happening at each step.

9780170452632

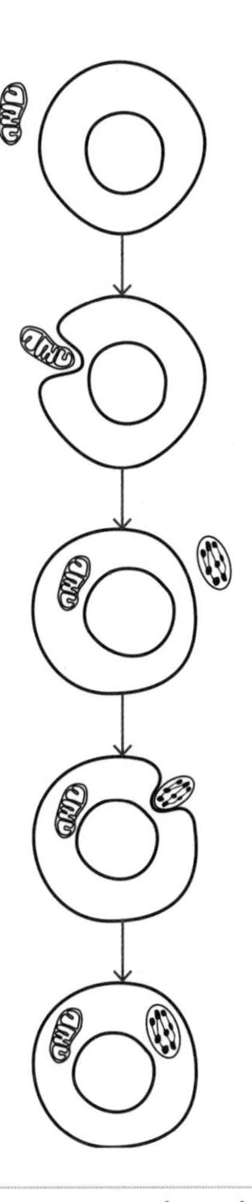

Figure 6.8 The endosymbiotic theory

6.3 Karyotypes for identifying chromosomal abnormalities

Key knowledge

From chromosomes to genomes

- karyotypes as a visual representation that can be used to identify chromosomal abnormalities

6.3.1 Chromosomal abnormality

Key science skills

Construct evidence-based arguments and draw conclusions

- use reasoning to construct scientific arguments, and to draw and justify conclusions consistent with the evidence and relevant to the question under investigation

A person with an unknown syndrome presented at a genetics clinic. The person underwent a cheek swab to obtain cheek cells, which were then viewed under a microscope. The chromosomes in one of the cells were used to prepare a karyotype. Figure 6.9 on page 162 shows the karyotype of this person, known as Person Z. The karyotype was prepared at a genetic screening laboratory by matching homologous chromosomes by their size, location of centromere and banding patterns. Autosomes are organised in size order from the largest (chromosome 1) to the smallest (chromosome 22). The sex chromosomes are placed at the end, either XX or XY. A chromosome count in a normal person is 46. Some people have extra or missing autosomes or sex chromosomes, as shown in Table 6.2 on page 162.

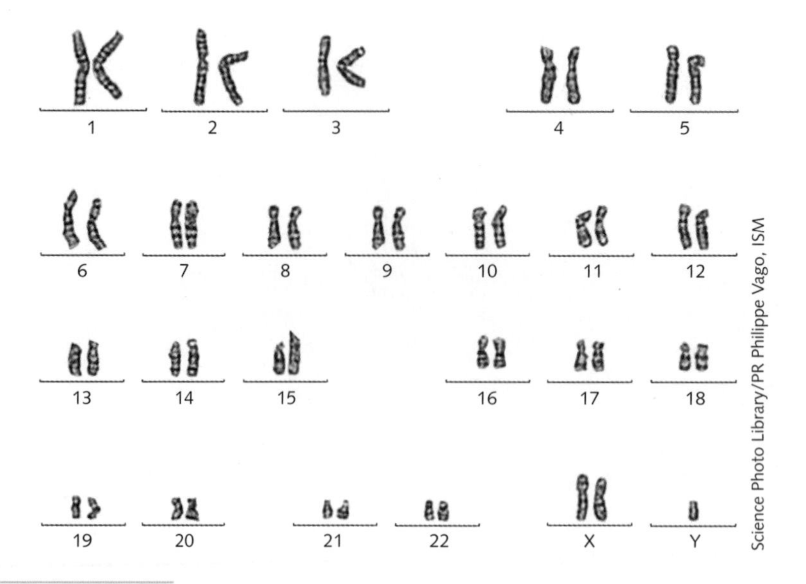

Figure 6.9 Human karyotype of Person Z

Table 6.2

Syndrome	Chromosomal abnormality
Some common chromosomal disorders Normal number of chromosomes (46)	The patient's condition is not due to an abnormal number of chromosomes
Down syndrome	Trisomy 21 (three copies of chromosome 21)
Klinefelter syndrome	More than two sex chromosomes (e.g. XXY; XXXY)
Turner syndrome	Only one sex chromosome (X)
Patau syndrome	Trisomy 13 (three copies of chromosome 13)
Trisomy 8 syndrome	Trisomy 8 (three copies of chromosome 8)

1 How many autosomes does Person Z have?

2 How many sex chromosomes does Person Z have?

3 What syndrome would you diagnose Person Z as having? Provide evidence for your diagnosis.

4 Research the symptoms of this syndrome.

6.4 Production of gametes and sexual reproduction

Key knowledge

From chromosomes to genomes

- the production of haploid gametes from diploid cells by meiosis, including the significance of crossing over of chromatids and independent assortment for genetic diversity.

6.4.1 Introducing variation through meiosis

Key science skills

TB PAGE 243

Generate, collate and record data

- organise and present data in useful and meaningful ways, including schematic diagrams, flow charts, tables, bar charts and line graphs

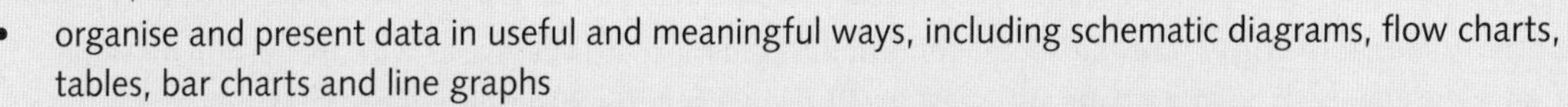

- plot graphs involving two variables that show linear and non-linear relationships

Analyse and evaluate data and investigation methods

- process quantitative data using appropriate mathematical relationships and units, including calculations of ratios, percentages, percentage change and mean

Analyse, evaluate and communicate scientific ideas

- analyse and explain how models and theories are used to organise and understand observed phenomena and concepts related to biology, identifying limitations of selected models/theories

The production of gametes (haploid cells) occurs through the process of meiosis. It is a two-step process that results in a diploid ($2n$) cell dividing into four haploid (n) gametes.

The process of meiosis is divided into two stages: meiosis I and meiosis II.

Meiosis I halves the number of chromosomes in the cell from $2n$ to n. Crossing over of parts of the paired homologous chromosomes occurs during prophase I and is the basis of variation within a species.

Meiosis II resembles mitosis except the starting cell is haploid rather than diploid. Four haploid gametes are produced at the end of this division.

During anaphase I, the homologous chromosomes are separated to either end of the cell. During anaphase II, the chromatids are separated into gametes. This separation of chromosomes or chromatids is called disjunction. If the chromosomes or chromatids do not pull apart properly during these divisions, a non-disjunction event occurs (Figure 6.10). If a zygote is formed from a gamete that has experienced a non-disjunction event, the resulting zygote and then offspring will have an extra or a missing chromosome in every cell of their body.

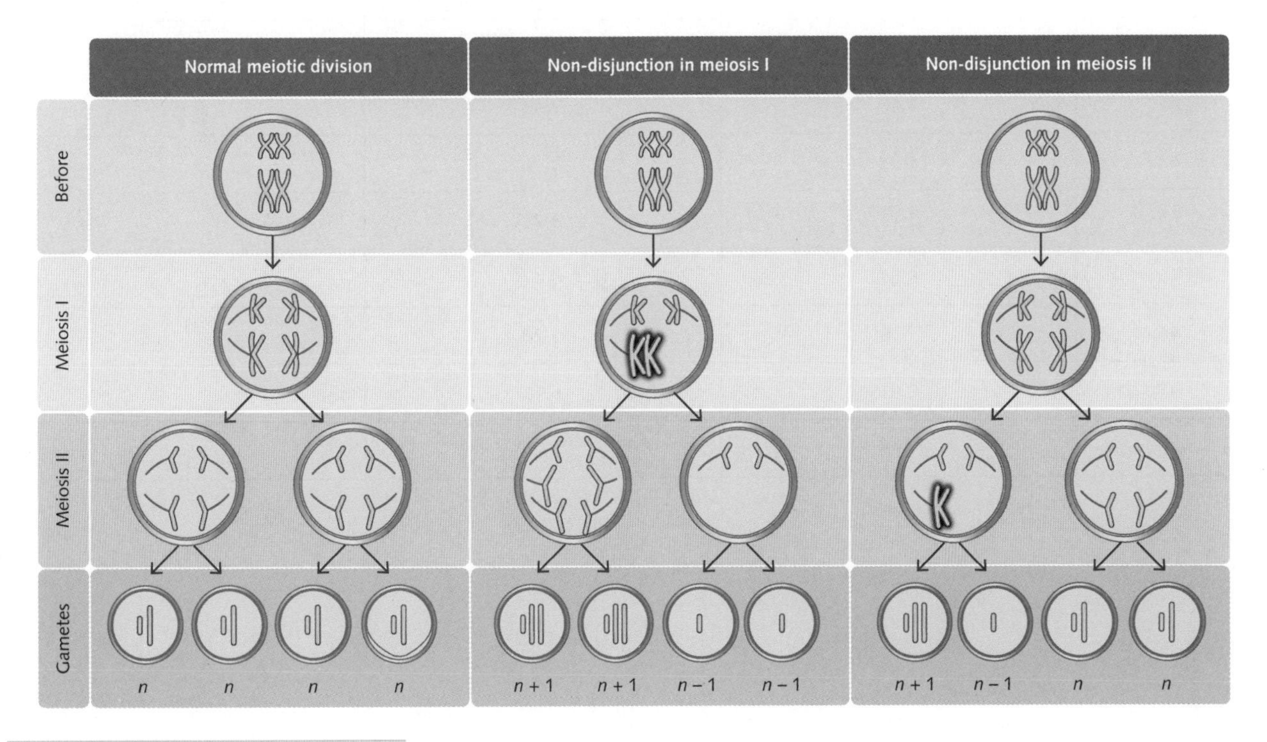

Figure 6.10 Disjunction and non-disjunction results in gametes formed through meiosis.

1 If disjunction produces four haploid gametes, state the possible result of non-disjunction of one chromosome when it occurs in the following phases.

a Meiosis I

b Meiosis II

It has been hypothesised that the risk of chromosomal abnormalities in the foetus increases with increasing maternal age. Table 6.3 shows the chance of having a baby with trisomy 21 (Down syndrome) or with one of the three most common trisomy syndromes (trisomy 21, 18 or 13) per 100 pregnancies. This data was compiled from more than 17 000 amniocentesis tests. An amniocentesis test is when amniotic fluid is removed from the uterus using a long needle and the cells contained in it are tested. Amniotic fluid surrounds the developing foetus which constantly sheds cells into the fluid.

Table 6.3 Risk of foetal chromosomal abnormalities with increasing maternal age

Maternal age (years)	Percentage of pregnancies having a baby with trisomy 21, 18 or 13	Percentage of pregnancies having a baby with trisomy 21	Percentage of pregnancies having a baby with either trisomy 18 or 13
34	0.45	0.26	
35	0.46	0.29	
36	0.57	0.39	
37	0.78	0.55	
38	1.10	0.79	
39	1.52	1.10	
40	2.04	1.49	
41	2.67	1.94	
42	3.40	2.46	
43	4.23	3.06	
44	5.17	3.73	
45	6.21	4.47	
46	7.36	5.28	
47	8.61	6.16	
48	9.96	7.11	

2 Use the data in Table 6.3 to plot the chance of common trisomy syndromes (trisomy 21, 18 or 13) against maternal age on the graph grid below.

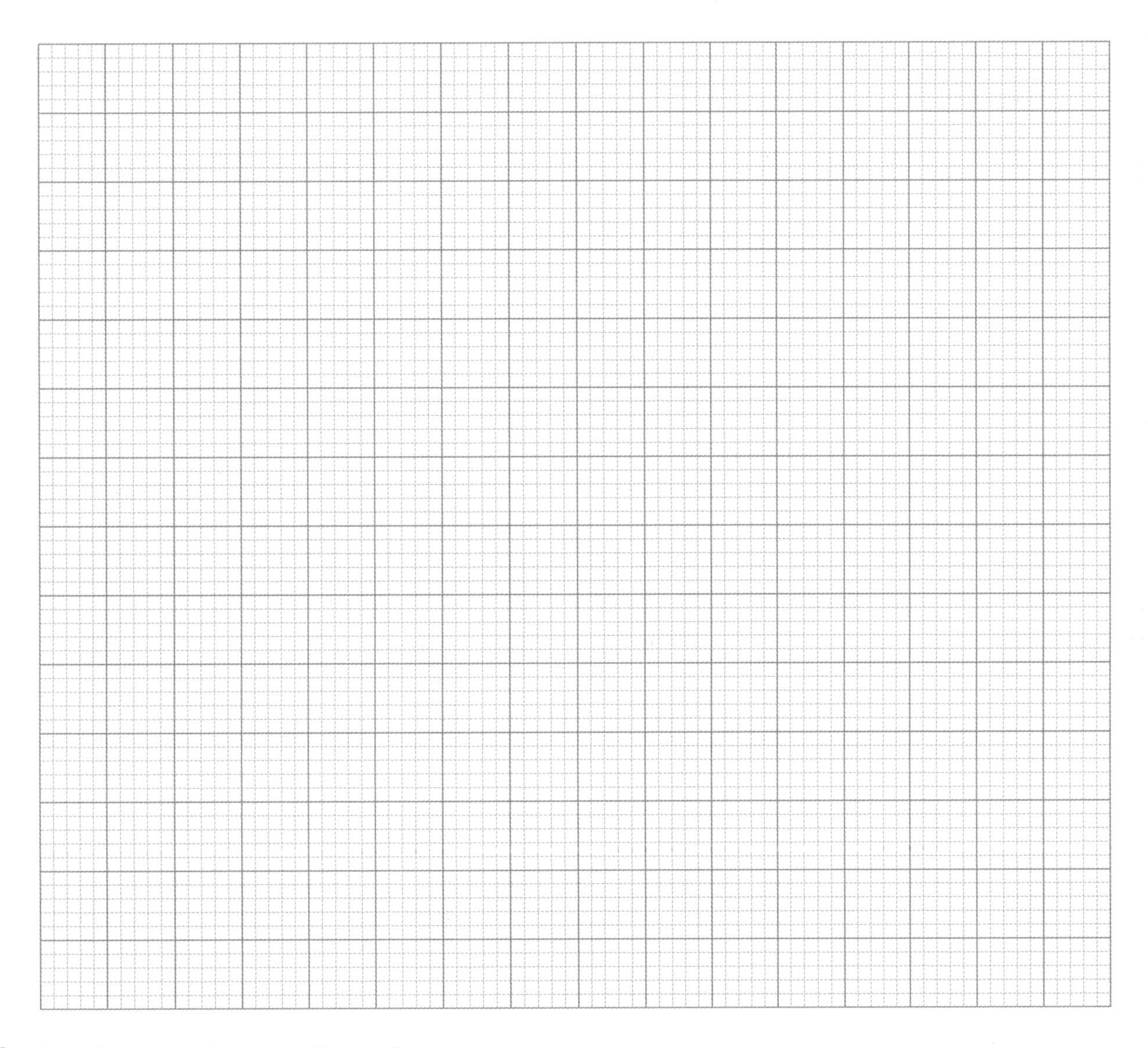

3 Describe the trend shown in this graph.

4 Complete the fourth column in Table 6.3. Add trisomy 21 to your graph as a separate line.

5 Using the information provided in Table 6.3 and the graph, draw two conclusions about trisomy 21, 18 or 13. Make sure each conclusion is supported by evidence from the data.

6.4.2 Meiosis, fertilisation and mitosis

Key science skills

Analyse, evaluate and communicate scientific ideas

- discuss relevant biological information, ideas, concepts theories and models and the connections between them

PAGE 244

Meiosis is the division of a diploid cell to produce four haploid gametes for the purposes of reproduction.

Fertilisation is the uniting of two haploid gametes (sperm and ova) to restore the diploid state in the production of a zygote.

Mitosis is the division of a diploid cell to produce an identical diploid cell for the purposes of growth, repair and replacement.

This activity explores how these three processes fit together.

What to do

Step 1: Meiosis: Using the diploid cell in Figure 6.11, complete the two divisions of meiosis by drawing the chromosomes onto Figure 6.12.

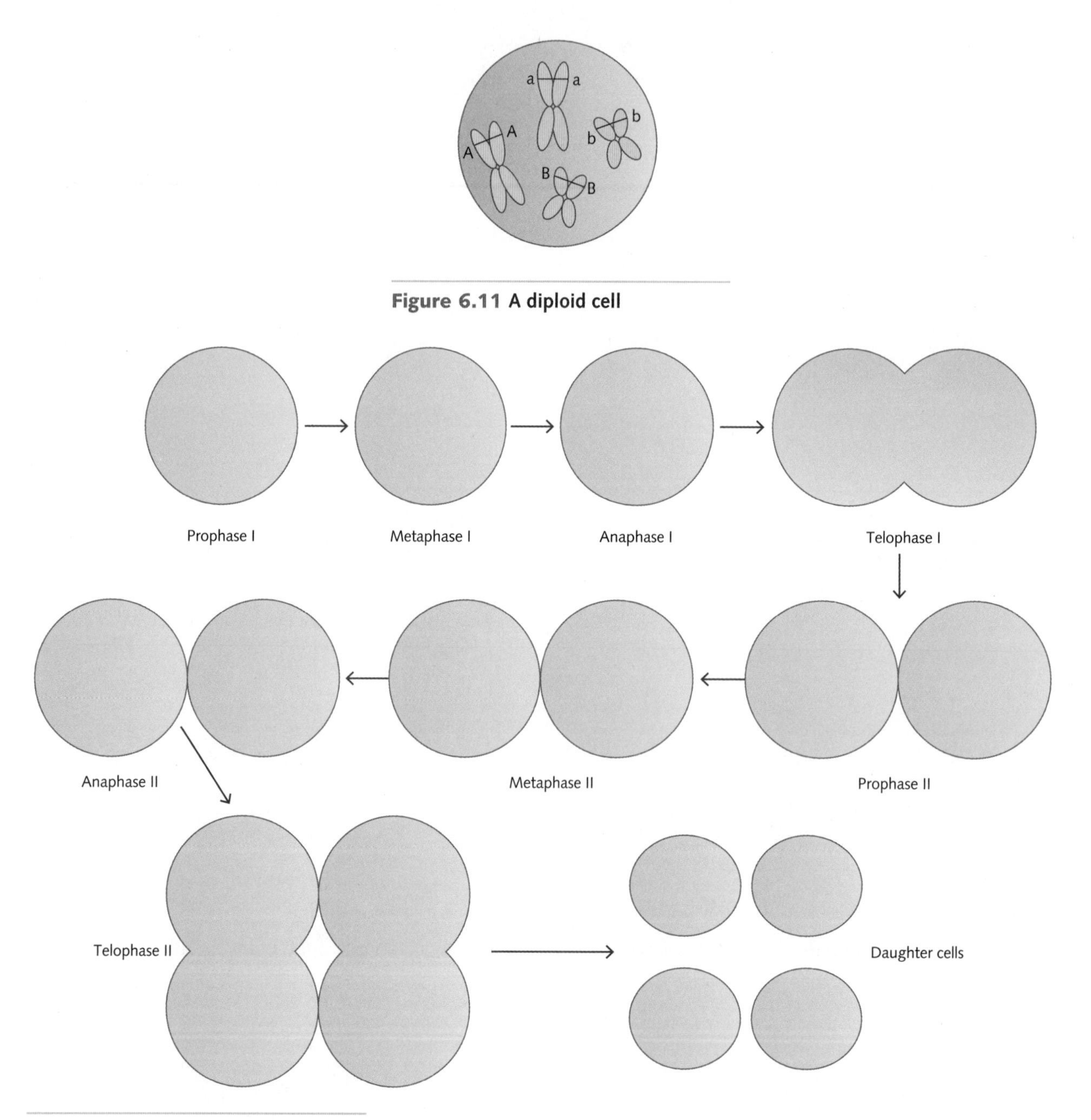

Figure 6.11 A diploid cell

Figure 6.12 Two divisions of meiosis

Step 2: Fertilisation: Choose one of the haploid gametes resulting from the two meiotic divisions and draw it into the blank circle next to the other provided gamete in Figure 6.13. Join the two gametes in the process of fertilisation to produce a diploid zygote.

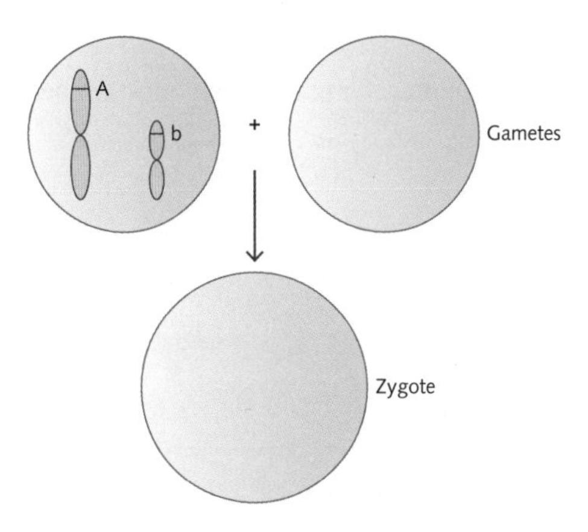

Figure 6.13 Two gametes in the process of fertilisation

Step 3: Mitosis: Using the diploid zygote you created in Step 2, start growing the zygote into a foetus by the process of mitosis by completing Figure 6.14.

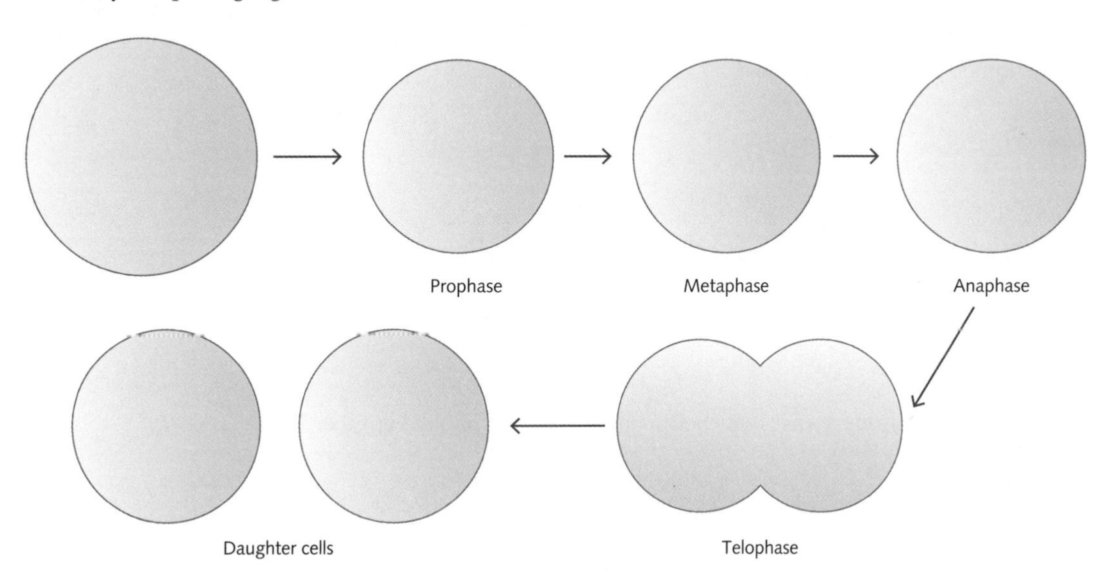

Figure 6.14 Growing a zygote into a foetus by the process of mitosis

Describe how the processes of meiosis, fertilisation and mitosis work to maintain a stable chromosome number of 46 in human body cells.

6.5 Chapter review

PAGE 251

6.5.1 Key terms

Use your knowledge of the key terms in this chapter to answer each question below.

1 What are the two types of reproduction?

2 What are somatic cells?

3 What is a chromatid?

4 What are homologous chromosomes?

5 What is the difference between cells produced by meiosis and those produced by mitosis?

6 An organism has 18 chromosomes in its body cells. What is its haploid number?

7 What is non-disjunction?

8 What is the significance of crossing over?

9 What is DNA made of?

10 What is an autosome?

9780170452632

Chapter review continued

PAGE 253

6.5.2 Practice test questions

A Year 11 Biology student sat an exam and submitted the following answers 1–4 indicated with a circle. In each case, state whether you agree or disagree with the student's answers and state why.

1 The building blocks of RNA and DNA are

A amino acids.
(B) monosaccharides.
C nucleotides.
D fatty acids and glycerol.

2 marks

2 How would the shape of a DNA molecule change if thymine paired with guanine and cytosine paired with adenine? The DNA molecule would

A be longer.
B be wider.
C have regions where no base-pairing would occur.
(D) have different widths along its length.

2 marks

3 When one DNA molecule is copied to make two DNA molecules, the new DNA molecules contain

(A) 25% of the parent DNA.
B 50% of the parent DNA.
C 75% of the parent DNA.
D 100% of the parent DNA.

2 marks

4 A gene is

(A) a piece of DNA.
B three nucleotides that code for a glucose molecule.
C a sequence of nucleotides that codes for a functional product.
D a sequence of monosaccharides that codes for a functional product.

2 marks

5 Circle the errors in the segments of double-stranded DNA below.

Segment 1	Segment 2	Segment 3	Segment 4
A-C-G-G-C	G-G-T-G-A	G-A-T-T-A	C-A-A-T-T
T-T-C-C-G	C-C-A-C-T	C-C-A-A-T	G-T-T-A-C

3 marks

6 Figure 6.15 shows a karyotype from a person with a chromosomal abnormality called Patau syndrome.

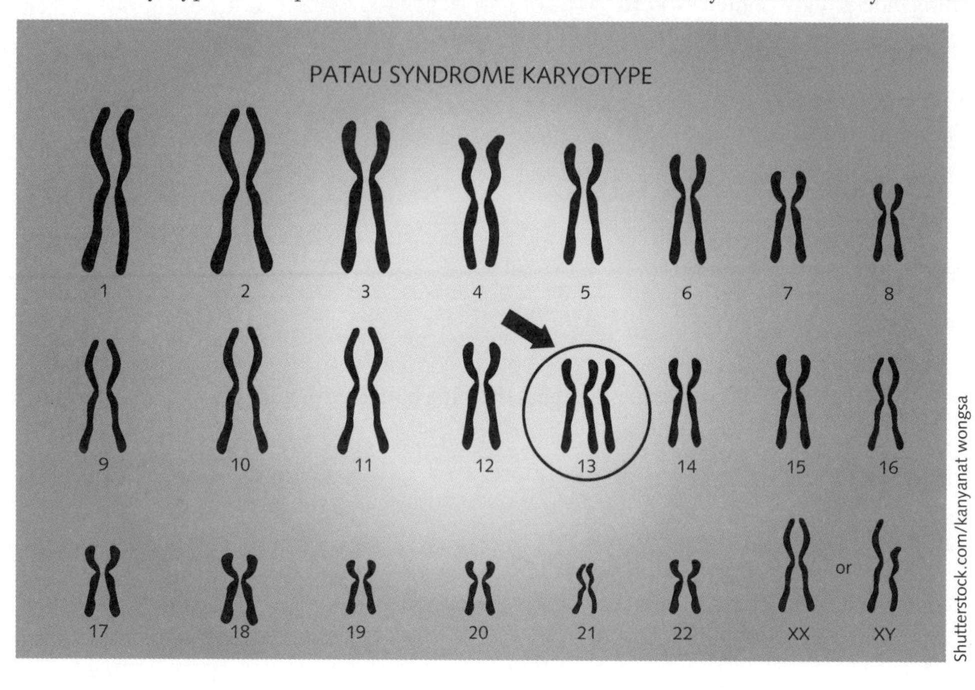

Figure 6.15 The karyotype of a person with Patau syndrome

a Explain how this abnormality came about.

2 marks

b How many autosomes does this person have?

1 mark

c If this person's cells contained XX chromosomes, what sex would they be?

1 mark

Shutterstock.com/nobeastsofierce

Inheritance

7

Remember

TB
PAGE 260

Chapter 7 Inheritance will call on content that you have already met in your science studies from previous years or earlier this year. Take some time to refresh your knowledge of this content before you enter this chapter. Try to answer the following questions from memory. If you cannot do this, then use a reference to assist you.

1 What is a gene and what does it code for?

2 What term is used for different forms of the same gene located at the same locus on a chromosome?

3 What are non-sex chromosomes called?

4 What are the two types of sex chromosomes in humans?

5 What is meant by the term 'homologous chromosomes'?

6 Distinguish between haploid and diploid cells. Where would you find each type of cell in the human body?

7.1 Patterns of inheritance

Key knowledge

Genotypes and phenotypes

- the use of symbols in the writing of genotypes for the alleles present at a particular gene locus
- the expression of dominant and recessive phenotypes, including codominance and incomplete dominance

7.1.1 Mendel's peas

PAGE 261

Key science skills

Analyse and evaluate data and investigation methods

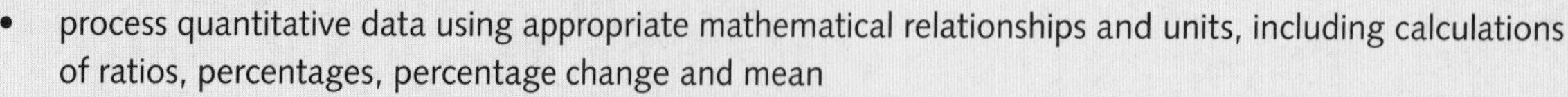

- process quantitative data using appropriate mathematical relationships and units, including calculations of ratios, percentages, percentage change and mean

Analyse, evaluate and communicate scientific ideas

- use appropriate biological terminology, representations and conventions, including standard abbreviations, graphing conventions and units of measurement

Gregor Mendel (1822–1884) carried out breeding experiments on pea plants. The conclusions he drew from these experiments form the foundation of the modern study of heredity. The principles that underly this study are called Mendelian genetics.

Gregor studied seven contrasting characteristics in pea plants (Table 7.1). These characteristics were pure breeding; that is, they were always passed onto offspring. Thus, a tall pea plant crossed with a tall pea plant always produced a tall pea plant. Mendel controlled his plant crosses by carefully pollinating the flowers by hand.

Table 7.1

Characteristic	Seed (endosperm)		Seed shape		Flower colour		Pod colour		Pod shape	
Different forms of the characteristic	Yellow	Green	Round	Wrinkled	Purple	White	Yellow	Green	Inflated	Constricted

Characteristic	Flower position		Stem length	
Different forms of the characteristic	Axial (along stem)	Terminal (at tip of stem)	Tall	Short

The convention in biology is to use a letter of the alphabet to represent the characteristic that is being followed in a cross. The upper case of this letter represents the dominant form and the lower case of this letter represents the recessive form. The dominant form masks the recessive form. For example, for the characteristic of stem length, T = tall and t = short.

1 Using this convention, write the letters that represent the alleles for the dominant and recessive characteristics of Mendel's pea plants in Table 7.2. (Hint: In Table 7.1, the dominant form is always shown first.) One has been done for you as an example.

Table 7.2

Characteristic	Dominant form	Recessive form
Stem length	T = tall	t = short
Flower position		
Seed (endosperm)		
Seed shape		
Flower colour		
Pod colour		
Pod shape		

Mendel performed thousands of crosses to draw his conclusions about heredity. He crossed pure-breeding plants (parental generation or P) for each dominant and recessive characteristic. In the first generation (first filial generation or F_1), he always obtained plants that displayed the dominant characteristic. When he crossed individuals from the first filial generation (F_1) for each characteristic, he obtained the results in the second filial generation (F_2), as shown in Table 7.3. The first column shows the characteristic under consideration and the second column shows the number of plants in the F_2 that display the dominant or recessive characteristic.

2 Complete the last column of Table 7.3 by calculating the ratio for each characteristic to its smallest whole number. To calculate the ratio, divide the smaller number into the larger number, then round each number to its nearest whole number, e.g. stem length = $\frac{787}{277} = \frac{2.84}{1} = 3:1$

Table 7.3 F_2 results for seven characteristics in pea plants

Characteristic	F_2 dominant : recessive	Ratio
Stem length	787 : 277	3 : 1
Flower position	651 : 207	
Seed (endosperm)	6022 : 2001	
Seed shape	5474 : 1850	
Flower colour	705 : 224	
Pod colour	428 : 152	
Pod shape	882 : 299	

3 What do you notice about the ratios of each characteristic?

7.1.2 The relationship between genes, alleles and traits

PAGE 263

Key science skills

Analyse, evaluate and communicate scientific ideas

- use appropriate biological terminology, representations and conventions, including standard abbreviations, graphing conventions and units of measurement

Mendel thought that there were *factors* that carried characteristics from one generation to the next. We now know the factors that Mendel referred to as genes. Chromosomes, genes and alleles were not described as the material of heredity until later in the nineteenth and early twentieth centuries, well after Mendel had completed his work.

In every one of our body cells we have 46 chromosomes, 23 inherited from our mother and 23 inherited from our father. These chromosomes are made up of DNA and a segment of DNA can code for the production of a protein. A segment of DNA that codes for the production of a protein is called a gene. We also know that genes can come in different forms, called alleles. For example, the gene for pea plant pod shape has two alleles, which result in inflated or constricted pods. These alleles occur at a particular place or locus on the chromosome (Figure 7.1).

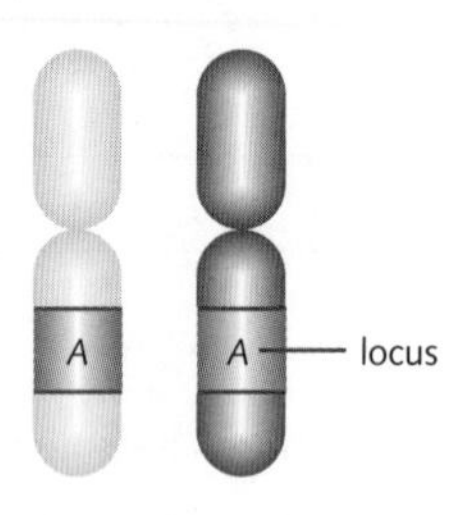

Figure 7.1 Homologous chromosomes showing the locus for allele A

Figure 7.1 shows that each chromosome contains one gene at a particular position or locus. As we have a pair of each chromosome, it follows that we have a pair of each gene. For the chromosomes shown in Figure 7.1 for gene *A*, the individual is said to be *AA* (homozygous dominant); if the alleles were *aa* the individual is said to be homozygous recessive, and if the alleles were *Aa* the individual is said to be heterozygous for that gene.

Use this information to complete the tasks below.

1 Using the letters you have chosen to represent the different genes in Table 7.2, write the letters that represent the following pairs of alleles in pea plants.

a homozygous dominant for seed shape

____

b homozygous recessive for pod shape

____

c heterozygous for flower colour

____

d homozygous recessive for flower position

____

e homozygous dominant for pod colour

____

9780170452632

2 **a** Use the space below to show a parental cross of a pure-breeding round seed and a pure-breeding wrinkled seed plant. Show the predicted genotype(s) and phenotype(s) of the F_1 generation and their probability of occurring.

b Cross two of the F_1 generation from question **2a** to produce the F_2 generation. Show the predicted genotype(s) and phenotype(s) of the F_2 generation and their probabilities of occurring.

3 Use the Punnett square provided below to show a cross between two pea plants that are both heterozygous for flower colour. Show the predicted genotype(s) and phenotype(s) of the F_1 generation and their probability of occurring.

Gametes		

4 In foxes, red (*R*) coat colour is dominant to silver-grey (*r*) coat colour. What phenotypic ratio would you expect in the offspring of a cross between a male fox that is heterozygous for coat colour and a silver-grey female fox? Show your working in the space below.

5 In humans, cystic fibrosis (CF) is a common, inherited, single-gene disorder. Carriers are heterozygous for the affected gene (*Cc*), they carry the recessive allele but this is masked by the dominant allele. They can, however, pass the recessive allele on to their offspring. The disease is only expressed in people with the genotype (*cc*). Complete the following single gene (monohybrid) crosses for CF. Show the predicted genotype(s) and phenotype(s) of the F_1 generation and their probability of occurring.

a Parents: *Cc* × *Cc*

b Parents: *Cc* × *CC*

7.1.3 Dominance is not always clear cut

PAGE 267

Key science skills

Analyse, evaluate and communicate scientific ideas

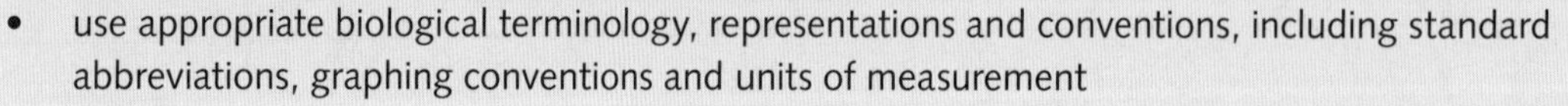

- use appropriate biological terminology, representations and conventions, including standard abbreviations, graphing conventions and units of measurement

Incomplete dominance

Sometimes, depending on the gene in question, the expression of the dominant allele does not always mask the expression of the recessive allele; instead they can blend together to display an intermediate form. Consider, for example, the gene that controls the colour of snapdragon flowers. When a pure-breeding red snapdragon plant is crossed with a pure-breeding white snapdragon plant, pink snapdragon flowers are expressed in the offspring (red + white = pink). This is known as incomplete (or partial) dominance and is expressed as a mixing of the two pure-breeding parents. As neither allele is dominant, the letters representing each allele are both shown as a capital letter representing the gene and then a superscript letter that represents the allele, e.g. red snapdragon flowers (C^R) and white snapdragon flowers (C^W). As each cell has two chromosomes, each carrying the gene, then the genotype for a red snapdragon flower will be shown as C^RC^R.

Consider the following cross, which shows a pure-breeding red snapdragon plant crossed with a pure-breeding white snapdragon plant.

Parents:	C^RC^R × C^WC^W
Gametes:	all C^R all C^W
Genotype of F_1:	all C^RC^W
Phenotype of F_1 plants:	all will display pink flowers.

Incomplete dominance also occurs in other species.

Chickens have a gene that controls their feather type, which can be represented by the letter *T*. The alleles for this gene are incompletely dominant. Use the letter *F* to represent the allele for the trait that is expressed as frizzled feathers, and the letter *S* to represent the allele for the trait that is expressed as straight feathers. When the alleles are inherited together, they express as slightly frizzled chicken feathers. Use the above notation to answer the following questions.

1 Use the space below to show the following crosses. In each cross state the phenotypic and genotypic ratios in the F_1.

a frizzled feathers × straight feathers (Hint: Parents will be $T^FT^F \times T^ST^S$)

b slightly frizzled feathers × slightly frizzled feathers

c frizzled feathers × slightly frizzled feathers

Codominance

Codominance is another type of dominance relationship where both alleles are fully expressed in the heterozygote. Consider feather colours in turkeys. When a turkey with black feathers breeds with a turkey with white feathers, the offspring all grow both black and white feathers. The alleles for feather colour are codominant and both are expressed in the offspring, not blended to be grey like in incomplete dominance. Again, the notation uses the superscript, so black feathered turkeys will have a genotype of F^BF^B, white feathered turkeys will have a genotype F^WF^W and black-and-white-feathered turkeys will have a genotype F^BF^W.

2 Predict the offspring of a cross between a white-feathered turkey and a black-and-white feathered turkey. What ratio of feather colours would you expect in the F_1?

Camellia plants have white or red flowers. Alleles for flower colour are codominant in that when a white-flowered camellia is crossed with a red-flowered camellia, the offspring have flowers that are red and white blotched. The genotype for flower colour in a camellia can be expressed as red (F^RF^R) and white (F^WF^W).

3 State the genotype of a red-and-white-blotched camellia.

4 Use the space below to show the following crosses. In each cross, state the phenotypic and genotypic ratios in the F_1.

a red camellia × white camellia

b white camellia × red-and-white-blotched camellia

c red-and-white-blotched camellia × red-and-white-blotched camellia

7.2 Genetic material, environmental factors and epigenetic factors

Key knowledge

Genotypes and phenotypes

- proportionate influences of genetic material, and environmental and epigenetic factors, on phenotypes

7.2.1 Epigenetics

Key science skills

Construct evidence-based arguments and draw conclusions

- evaluate data to determine the degree to which the evidence supports the aim of the investigation, and make recommendations, as appropriate, for modifying or extending the investigation

Carefully read the following article, paying attention to the aim of the investigation and whether the evidence supports the aim. Answer the questions that follow.

Obesity and epigenetics

Obesity is a disease like any other disease. It has causes, symptoms that affect the functioning of the body, and cures. It is a disease that is characterised by too much body fat. It has multiple causes, including excessive intake of kilojoules at some stage in the person's life. Obesity is associated with many other debilitating diseases such as cardiovascular disease, type-2 diabetes, hypertension and cancer.

Researchers into obesity from the University of Sao Paulo in Brazil studied the link between DNA methylation patterns and obesity. They studied both obese women and women who had a normal body mass index (BMI). DNA methylation is when a methyl group is added onto the cytosine group on DNA molecules. This addition changes how the gene is expressed and can often turn off the gene expression.

Women that fell within the normal range for BMI were found to have an average of 11% more methylation marks on their DNA than obese women. The team wanted to find out if a low kilojoule diet would restore the DNA methylation levels in obese women so they would equate to those in women with normal BMI.

Women with a BMI of around 58.5 (the normal BMI range for women is between 18 and 25) were placed on a kilojoule-controlled diet for 6 weeks. At the end of the 6 weeks, they had not only lost weight but had changed their DNA methylation at more than 16 000 sites associated with 9200 genes. Many of these genes had pathways to cancer. Around 650 genes did not change and remained different between the obese and normal-BMI groups. The benefit of the intervention therefore was only partial.

Adapted from https://www.whatisepigenetics.com/low-cal-diet-could-change-epigenetic-patterns-in-obesity-related-disease/

1 State the aim of this experiment.

2 What methodology did the researchers employ in the design of this experiment?

3 What were the results of this experiment?

4 Do the results of the experiment support the aim of the experiment? Explain.

5 How could this experiment be extended to gain more conclusive results?

6 How is obesity an example of epigenetics?

7.3 Patterns of inheritance

Key knowledge

Patterns of inheritance

- pedigree charts and patterns of inheritance, including autosomal and sex-linked inheritance
- predicted genetic outcomes for a monohybrid cross and a monohybrid test cross

7.3.1 Monohybrid crosses

PAGE 272

Key science skills

Analyse, evaluate and communicate scientific ideas

- use appropriate biological terminology, representations and conventions, including standard abbreviations, graphing conventions and units of measurement

A monohybrid cross is a cross between two individuals where the inheritance of one gene is followed. Consider the homologous chromosomes from a guinea pig shown in Figure 7.2. They carry a gene for fur colour at a particular locus. Let's represent the gene with the letter B for the dominant allele, which is expressed as black fur, and b for the recessive allele, which is expressed as white fur.

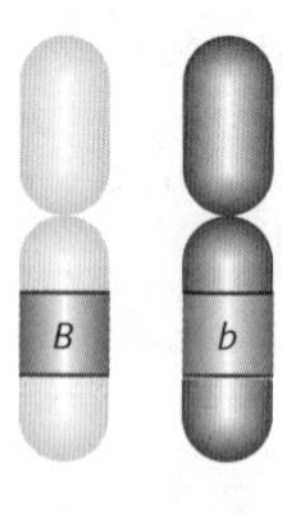

Figure 7.2 Guinea pig homologous chromosomes displaying the locus for fur colour

1 What is the genotype of the guinea pig shown in Figure 7.2 in relation to fur colour?

2 a Show a cross between two guinea pigs of the same genotype as that in question **1** with respect to fur colour.

9780170452632

b What is the probability that this cross will produce a guinea pig with white fur?

c Why is this an example of a monohybrid cross?

3 Six black guinea pigs were produced from the cross in question **2a** but the owner did not know if they were homozygous dominant or heterozygous in relation to the gene for fur colour.

a Show the crosses that the owner could do to determine the genotypes of these six black guinea pigs for certain.

b What is this type of cross called?

c What result from these crosses would show that the black guinea pigs were either homozygous dominant or heterozygous for fur colour?

4 *Drosophila melanogaster*, the fruit fly, is an animal much studied by geneticists. It only has four pairs of chromosomes and a very short generation time, so it is ideal to study in the laboratory. It also has some very distinct features that are under the control of single genes, such as wing shape and eye colour.

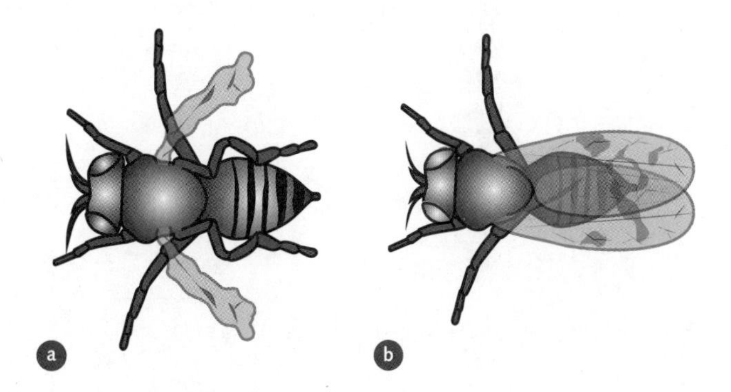

Figure 7.3 ***Drosophila melanogaster*** **showing a vestigial wings and non-red eyes, b long wings and red eyes**

a Long wings are dominant over vestigial (short stumpy) wings. Assume the fruit flies shown in Figure 7.3 on the previous page are both homozygous for wing shape. What is the probability of the two fruit flies shown in Figure 7.3 producing a vestigial wing offspring in the F_1 generation? In the F_2 generation?

b Red eyes are dominant over non-red eyes. Cross two fruit flies that are heterozygous for eye colour and state the expected ratio in the F_1 offspring.

7.3.2 Sex-linked inheritance

PAGE 277

Key science skills

Analyse, evaluate and communicate scientific ideas

- use appropriate biological terminology, representations and conventions, including standard abbreviations, graphing conventions and units of measurement

Human body cells have 46 chromosomes. These are made up of 22 pairs of autosomes and one pair of sex chromosomes. In humans, the sex chromosomes are X and Y, as seen in Figure 7.4. Note the size difference between the two chromosomes, the X chromosome being much larger than the Y chromosome. Remember that females have two X chromosomes, whereas males have one X and one Y chromosome. You can see that in males, many genes on the X chromosome do not have a matching gene on the Y chromosome. When you are considering crosses that involve genes situated on the X and Y chromosomes, there is a special notation that is used. For example, the gene that controls red-green colour blindness is situated on the X chromosome, so this is shown as X^R to represent the allele for normal colour vision and X^r to represent the allele for red-green colour blindness. There is no matching locus on the Y chromosome, so males only have one allele for colour vision, this is shown as X^RY or X^rY. Use this information to answer the questions below.

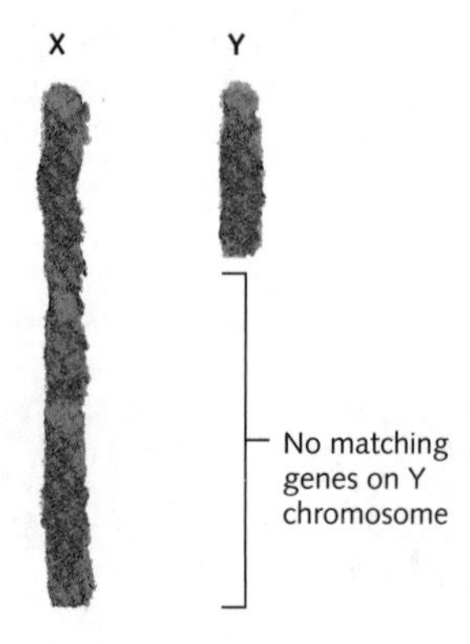

Figure 7.4 **X and Y human sex chromosomes. Note the size difference.**

9780170452632

1 Write the notation to represent the following individuals with respect to their colour vision.

a female, homozygous for normal colour vision

b female, carrier for red-green colour blindness

c female with red-green colour blindness

d male with normal colour vision

e male with red-green colour blindness

2 Use the notation you have written in question **1** to show the probability of producing a child with red-green colour blindness from the following crosses.

a Female, carrier for red-green colour blindness × male with red-green colour blindness

b Female, homozygous for normal colour vision × male with red-green colour blindness

3 The gene for Duchenne muscular dystrophy is located on the X chromosome. The disease is caused by the recessive allele, represented by the letter *d*. The dominant allele, represented by the letter *D*, masks the recessive allele. The Punnett square below shows a typical pairing of parents for muscular dystrophy.

Gametes	X^D	Y
X^D	$X^D X^D$	$X^D Y$
X^d	$X^D X^d$	$X^d Y$

a What is the chance that a son from these two parents will suffer from Duchenne muscular dystrophy?

b Use a Punnett square to show a cross between a homozygous dominant mother and a father with Duchenne muscular dystrophy.

i What is the probability that their next daughter will be a carrier?

ii What is the probabillity that their next son will suffer from the disease?

4 Two parents who do not exhibit a genetic condition called phenylketonuria (PKU) have a daughter with PKU. PKU is an inborn error of metabolism that results in decreased metabolism of the amino acid phenylalanine. Untreated, PKU can lead to intellectual disability, seizures, behavioural problems and mental disorders.

A student determined there were four different rationales for this occurring in the child of two 'non' PKU parents. For each rationale below, justify why or why not it could be the reason for the PKU offspring.

a The allele for PKU is located on the Y chromosome.

b PKU is an autosomal dominant trait.

c PKU is an autosomal recessive trait.

d PKU is an X-linked recessive trait carried by the father.

7.4 Pedigree charts for autosomal and sex-linked inheritance

Key knowledge

Patterns of inheritance

- pedigree charts and patterns of inheritance, including autosomal and sex-linked inheritance

7.4.1 Pedigree charts

PAGE 280

Key science skills

Analyse, evaluate and communicate scientific ideas

- use appropriate biological terminology, representations and conventions, including standard abbreviations, graphing conventions and units of measurement

A pedigree chart is a visual representation of the inheritance of a particular trait. Analysing pedigree charts provides information about the inheritance of that trait over several generations. Consider the pedigree chart shown in Figure 7.5. Use this pedigree chart to show your understanding of the symbols used to create these charts by answering the questions below.

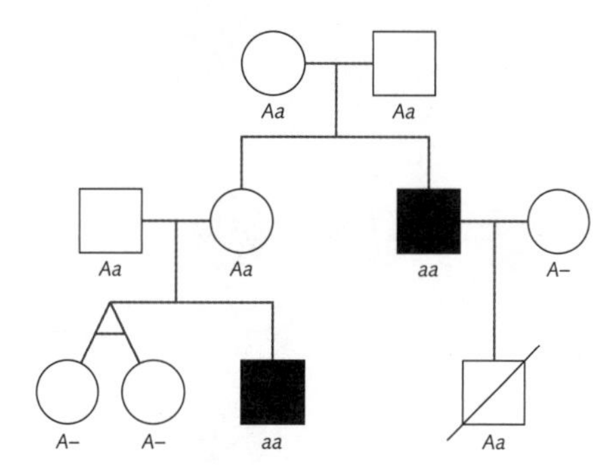

Figure 7.5 Pedigree chart

1 On Figure 7.5:

a write the generation numbers

b write numbers for each individual shown in the chart

c state the generation and number of a male

d state the generation and number of a female

e label a marriage line

f label an offspring line

g state the generation and numbers of identical twins

h state the generation and number of a deceased person

i state the generation and number of a person showing the trait under investigation.

2 Determine the type of inheritance shown in Figure 7.6. Provide the reasoning behind your answer by referring to particular individuals in the pedigree chart. Write the genotype of each individual in the spaces provided in the chart.

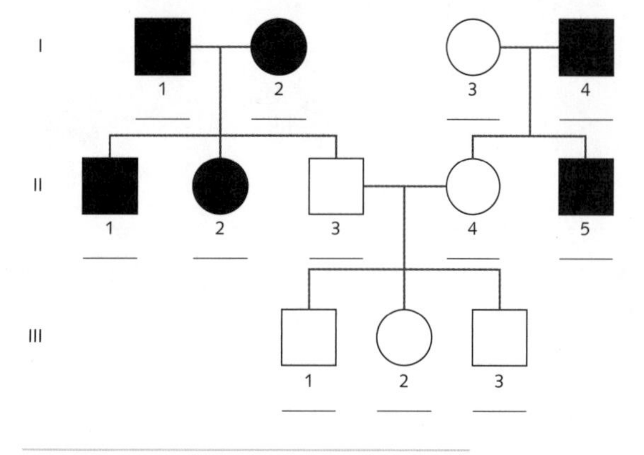

Figure 7.6

3 Determine the type of inheritance shown in Figure 7.7. Provide the reasoning behind your answer by referring to particular individuals in the pedigree chart. Write the genotype of each individual in the spaces provided in the chart.

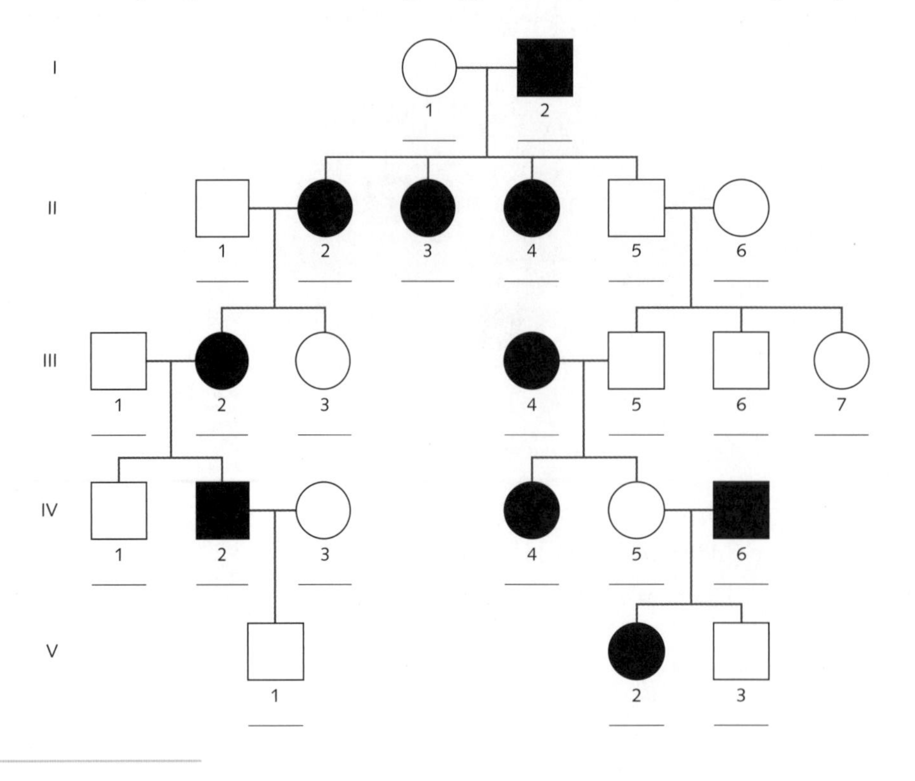

Figure 7.7

4 Determine the type of inheritance shown in Figure 7.8. Provide the reasoning behind your answer by referring to particular individuals in the pedigree chart. Write the genotype of each individual in the spaces provided in the chart.

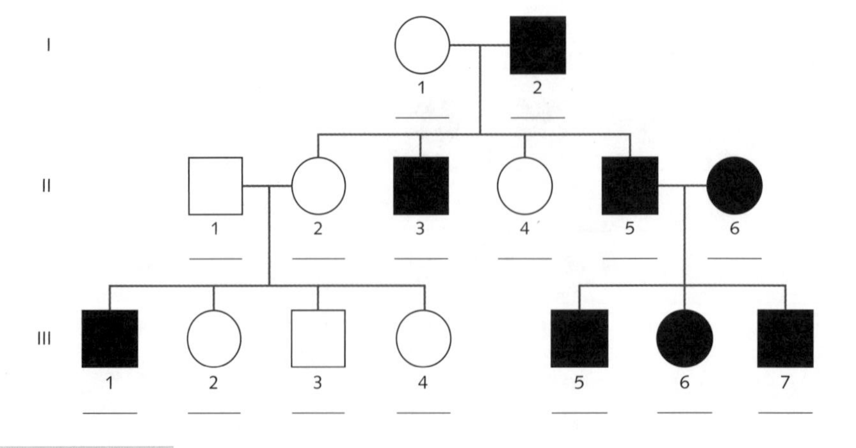

Figure 7.8

7.5 Predicted genetic outcomes for two autosomal genes

Key knowledge

Patterns of inheritance

- predicted genetic outcomes for two genes that are either linked or assort independently

7.5.1 Genes that assort independently

Key science skills

Analyse, evaluate and communicate scientific ideas

- use appropriate biological terminology, representations and conventions, including standard abbreviations, graphing conventions and units of measurement

It is possible to predict the outcome of a cross of two separate genes, each occurring on a separate chromosome. This type of cross is called a dihybrid cross. As these genes occur on separate chromosomes, they independently assort during prophase I of meiosis. Figure 7.9 shows two pairs of homologous chromosomes from a donkey. The chromosome pair on the left contain a gene for ear length (long ears (*E*) and short ears (*e*)) and the chromosome pair on the right contain a gene for coat colour (dark coat (*C*) or light coat (*c*)). Use this information to answer the questions below.

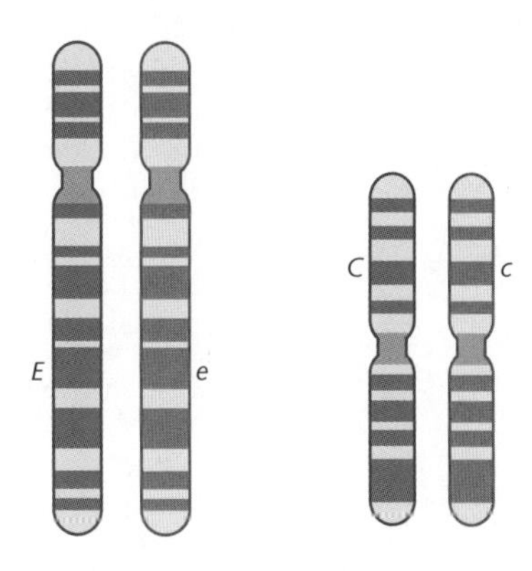

Figure 7.9 Chromosomes that contain the alleles for donkey ear length and coat colour

1 A male donkey that is homozygous dominant for both ear length and coat colour is crossed with a female donkey that is homozygous recessive for both ear length and coat.

a State the genotypes for the male and female donkey in relation to ear length and coat colour.

b State the gametes that each will produce in relation to ear length and coat colour.

c State the genotype of the F_1 generation that will result from this cross.

d Use a Punnett square to show the F_2 generation that will result from a cross of a male and female donkey from the F_1 generation. State the resulting genotypes and their predicted proportions.

2 In pea plants, round seeds are dominant to wrinkled seeds, and purple flowers are dominant to white flowers. Refer to Table 7.2 on page 173 to find out what letters you used to represent each of the alleles that determine these characteristics.

a State the possible genotypes for a pea plant with round seeds and purple flowers.

b What sort of cross would you perform to determine the genotype of this pea plant with round seeds and purple flowers?

c Choose one of the possible genotypes of the pea plant with round seeds and purple flowers and show this cross (from question 2b) using a Punnett square. State the expected ratio of each phenotype in the F_1.

 9780170452632

7.5.2 Linked genes

Key science skills

Analyse, evaluate and communicate scientific ideas

- use appropriate biological terminology, representations and conventions, including standard abbreviations, graphing conventions and units of measurement

TB PAGE 293

It is possible to predict the outcome of a cross of two genes that are situated on the same chromosome (Figure 7.10).

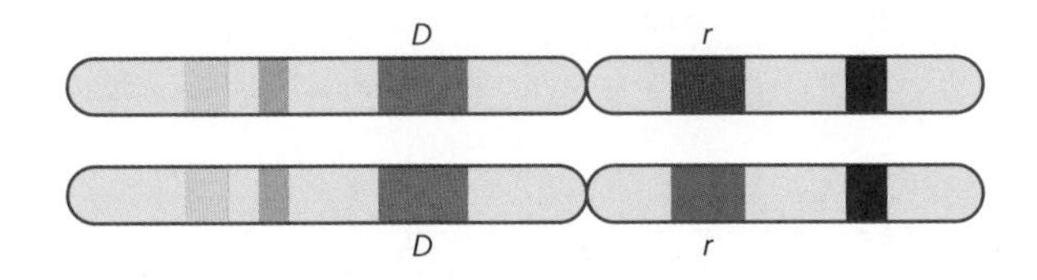

Figure 7.10 Linked genes are those that occur on the same chromosome.

Chromosome 4 in humans has a locus for one of the genes that controls deafness as well as a locus for the gene that controls retinal degeneration. The following notation is assigned to represent the alleles for each gene:

D = deafness

d = normal hearing

R = no retinal degeneration

r = retinal degeneration

The notation is a little bit different from non-linked genes. The notation shows the linked genes and that they travel together during meiosis and fertilisation (as they are attached to the same chromosome). A person's genotype that shows that they are homozygous dominant for deafness and homozygous recessive for retinal degeneration would look like:

$\frac{Dr}{Dr}$ (Figure 7.10). The line in the notation represents a pair of homologous chromosomes, with the alleles above the line appearing on one chromosome and the alleles below the line appearing on the other chromosome. As both genes travel together during meiosis, the gametes produced would be Dr.

Use this information to answer the questions below.

3 **a** Write the possible genotypes of two deaf people heterozygous for both alleles.

b Choose two of the above genotypes and show the resulting cross. Assume there is no crossing over. What is the probability that one of the children will be deaf and also have retinal degeneration? Show your working out below.

7.6 Chapter review

7.6.1 Key terms

TB PAGE 301

Complete this crossword of key terms.

ACROSS

2 Genes located on non-sex chromosomes or autosomes.
3 The characteristics expressed in an organism determined by their genotype and influenced by factors in their environment and epigenetic factors.
6 A chart that uses accepted symbols and shows the inheritance of a particular trait over several generations.
7 The genetic acquisition of characteristics by offspring from their parents.
14 The study of inheritance; the genetic transmission of characteristics from one generation to another.
15 The study of chemical modifications resulting in DNA structural changes that cause changes in the expression of genes but not changes in the DNA sequence.
17 The position a gene occupies on a chromosome.
18 A cross between two organisms for one trait involving the alleles of one gene located at one specific locus on a chromosome.
19 Two organisms that represent the start of a breeding experiment; their offspring are the F_1 generation.
20 A genotype with two different alleles for a single gene locus.
22 A cross using an organism with a recessive phenotype to determine the genotype of an organism displaying a dominant phenotype.
23 The attachment of a methyl group to nucleotides or histone proteins in the DNA molecule, thus altering the structure of the DNA molecule but not the DNA sequence.
24 A genotype with two identical alleles for a single gene locus.

DOWN

1 Both alleles in the genotype are fully expressed in the heterozygote.
4 Offspring with combinations of alleles that are not found in either parent and result from crossing over and recombination of segments of chromatids.
5 A specific combination of alleles for a locus belonging to an individual or cell.
8 The father of genetics.
9 Describes a line of organisms that when crossed with each other always produce offspring with the same phenotype.
10 One of the different forms or alternatives of the same gene found at the same locus on a chromosome.
11 The inheritance pattern shown by a gene located on a sex chromosome.
12 A phenotype that requires only one copy of its allele in an individual for it to be expressed.
13 A cross between two organisms for two different traits that involves two genes located at two different gene loci on the same or different chromosomes.
16 A heritable characteristic; phenotype.
21 A unit of heredity that transmits information from one generation to the next; a segment of DNA that codes for a polypeptide.

Chapter review continued

7.6.2 Practice test questions

PAGE 302

Multiple-choice questions

1 A man who suffers from a rare disease has six children: three boys and three girls. All the girls suffer from the same disease as their father. None of the boys is affected. The most likely mode of inheritance for this disease is

A autosomal dominant.
B autosomal recessive.
C X-linked dominant.
D X-linked recessive.

2 A brown rat is mated with a white female rat (*bb*). The litter produced contains brown pups and white pups only. From this information, which of the following statements is correct?

A The father's phenotype is *BB*.
B The pups include some with the allele combination *BB*.
C The father produces two types of gametes with respect to coat colour.
D The mother produces gametes with allele combinations *b* and *bb*.

3 ©VCAA @2002 VCAA Exam 2 Section A Q6 (adapted) In cattle, the presence or absence of a white saddle-shaped marking across the back of an otherwise coloured coat is under the control of a single autosomal gene.

The gene has the alleles:
S : white saddle present
s : solid-colour coat (for example, black).

With respect to this gene, it is reasonable to predict that the cross

A *SS* × *Ss* results in only solid-coloured cattle.
B *Ss* × *Ss* results in both solid-coloured and white-saddle cattle.
C *SS* × *SS* results in only solid-coloured cattle.
D *Ss* × *ss* results in only white-saddle cattle.

4 The yellow (*c1*) and black (*c2*) pigments in the coats of cats are controlled by a gene located on the X chromosome. The alleles of this gene are codominant.

Females	Males
$X^{c1}X^{c1}$ = yellow	$X^{c1}Y$ = yellow
$X^{c1}X^{c2}$ = tortoiseshell	$X^{c2}Y$ = black
$X^{c2}X^{c2}$ = black	

A cat has a litter of nine kittens with a single father: three black males, two yellow males, two yellow females and two tortoiseshell females. The most likely genotypes of the father and mother are

	Father	Mother
A	$X^{c1}Y$	$X^{c1}X^{c2}$
B	$X^{c1}Y$	$X^{c2}X^{c2}$
C	$X^{c2}Y$	$X^{c2}X^{c2}$
D	$X^{c2}Y$	$X^{c1}X^{c1}$

Use the following information to answer Questions 5–7.

©VCAA @2003 VCAA Exam 2 Section A Q5–7 (adapted) The pedigree in Figure 7.11 shows the inheritance of Tay–Sachs disease, a progressive neurological defect in humans.

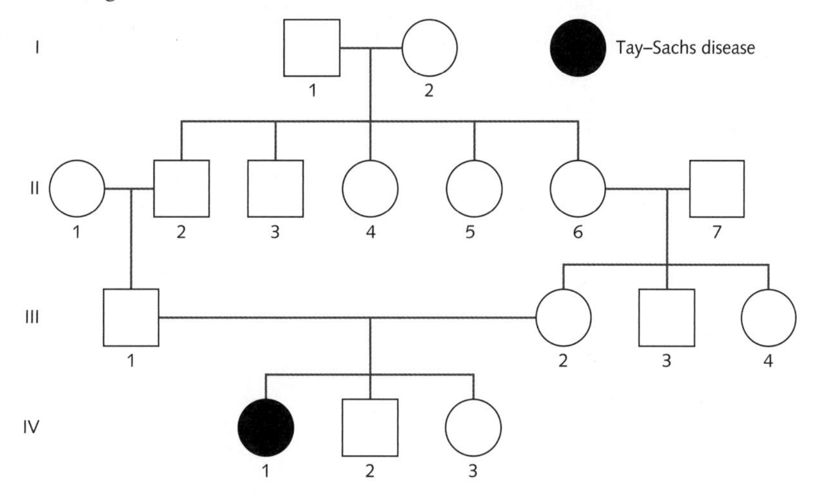

Figure 7.11

5 The individual that is affected by Tay–Sachs disease is

A male.

B female.

C a non-identical twin.

D dead.

6 The chance that a fourth child of III–1 and III–2 is a carrier of Tay–Sachs disease is

A $\frac{2}{3}$

B $\frac{1}{3}$

C $\frac{1}{2}$

D $\frac{3}{4}$

7 The genotype of individual III-1 in relation to Tay–Sachs disease must be

A *TT*.

B *tt*.

C *Tt*.

D unable to be determined from the chart.

Short-answer questions

8 ©VCAA 2003 Exam 2 Section B Q1 (adapted) The back of the leopard frog (*Rana pipiens*) can be either patterned or non-patterned.

Several patterned frogs were allowed to breed and they produced 61 patterned offspring and 24 non-patterned offspring.

a i Which of the phenotypes, patterned or non-patterned, is recessive? 1 + 1 = 2 marks

ii Explain your answer to part **i**.

b Using your own allelic notation, show the genotypes with their respective phenotypes for the parents and offspring of a cross between a heterozygous patterned and a non-patterned frog. 3 marks

Crosses between patterned and non-patterned frogs were performed. Not all crosses produced the same outcome.

	Parents		Offspring
cross A	patterned	patterned	all patterned
cross B	patterned	patterned	$\frac{3}{4}$ patterned; $\frac{1}{4}$ non-patterned

c The parents in crosses A and B have the same phenotypes.

Explain why the outcome of crosses A and B are different. 2 marks

9 In dogs, the allele for black coat colour is dominant over the allele for brown coat colour. On a separate chromosome, the allele for solid coat pattern is dominant over the allele for spotted coat pattern. A homozygous black, spotted dog is crossed with a homozygous brown, solid pattern dog.

a What are the genotypes of the parent dogs? 2 marks

b State the ratio of the phenotypes of the offspring expected in the F_1 generation. 2 marks

c Use a Punnett square to show the F_2 cross. State the ratio of the phenotypes of the offspring expected in the F_2 generation. 4 marks

8 Reproductive strategies

Getty Images/ David Gray

Remember

PAGE 312

Chapter 8 Reproductive strategies will call on content that you have already met in your science studies. Take some time to refresh your knowledge of this content before you enter this chapter. Try to answer the following questions from memory. If you cannot do this, then use a reference to assist you.

1 Where does mitosis occur and what is the outcome?

2 Where does meiosis occur and what is the outcome?

3 Contrast mitosis and meiosis.

4 Contrast haploid cells and diploid cells.

5 What is the name of the process that starts with gametes and ends with a zygote? Why is this process important in terms of chromosome number?

9780170452632

8.1 Asexual reproduction

Key knowledge
Reproductive strategies
- biological advantages and disadvantages of asexual reproduction

8.1.1 Vegetative propagation

Key science skills
Develop aims and questions, formulate hypotheses and make predictions
- identify, research and construct aims and questions for investigation
- identify independent, dependent and controlled variables in controlled experiments
- formulate hypotheses to focus investigation
- predict possible outcomes

Plan and conduct investigations
- determine appropriate investigation methodology: case study; classification and identification; controlled experiment; correlational study; fieldwork; literature review; modelling; product, process or system development; simulation
- design and conduct investigations; select and use methods appropriate to the investigation, including consideration of sampling technique and size, equipment and procedures, taking into account potential sources of error and uncertainty; determine the type and amount of qualitative and/or quantitative data to be generated or collated

Comply with safety and ethical guidelines
- demonstrate safe laboratory practices when planning and conducting investigations by using risk assessments that are informed by safety data sheets (SDS), and accounting for risks

Reinforce

PAGE 316

Maryam and Blake are Year 11 Biology students. They are planning an investigation to find out the best conditions for optimal vegetative propagation in carrots. They know that carrots will regrow if the top of a carrot is placed in a dish of water (Figure 8.1), but they are wondering if the carrot will grow quicker and/or better if different types (liquid/pellet) of fertiliser are added to the water first. Maryam then wondered if the concentration of fertiliser would make a difference. Blake then questioned whether growing the carrots on a material such as cotton wool would also make a difference.

Figure 8.1 Carrots can be propagated vegetatively by growing the top in a dish of water.

1 Choose one aspect of carrot propagation discussed by Maryam and Blake to investigate and write a research question, remembering the rules on how to write a good research question.

2 Write an aim for this investigation.

3 Identify the:

a independent variable

b dependent variable

c controlled variables.

4 Write a hypothesis for this investigation, using the 'If ... then ... ' format.

5 Predict possible outcomes for this investigation if the hypothesis is:

a supported

b refuted.

6 Determine the appropriate methodology for this investigation and design a method to test the hypothesis that you have written. When you are devising the method, you need to take into account:

- sample size
- equipment and material available to you
- sources of error in your data
- the type and amount of data you will be generating and collecting
- how you will record your data
- identifying any risks in your method by completing a risk assessment.

8.1.2 Biological advantages and disadvantages of asexual reproduction

Key science skills

TB PAGE 319

Develop aims and questions, formulate hypotheses and make predictions

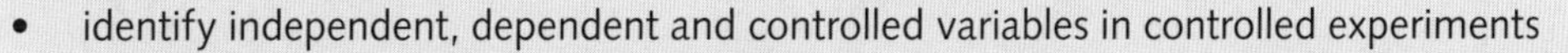

- identify independent, dependent and controlled variables in controlled experiments

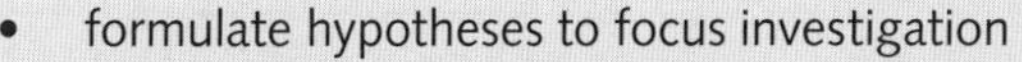

- formulate hypotheses to focus investigation

Generate, collate and record data

- plot graphs involving two variables that show linear and non-linear relationships

Analyse and evaluate data and investigation methods

- process quantitative data using appropriate mathematical relationships and units, including calculations of ratios, percentages, percentage change and mean
- identify outliers, and contradictory or provisional data

Construct evidence-based arguments and draw conclusions

- evaluate data to determine the degree to which the evidence supports the aim of the investigation, and make recommendations, as appropriate, for modifying or extending the investigation
- use reasoning to construct scientific arguments, and to draw and justify conclusions consistent with the evidence and relevant to the question under investigation

Yeast, like all fungi, have a sexual and asexual mode of reproduction. The most common mode is asexual, where the yeast cell grows a small bud to form a new yeast daughter cell. The daughter cell then buds to form another daughter cell and as this continues, chains of yeast cells form (Figure 8.2). Yeast can reproduce in this way every 90 minutes.

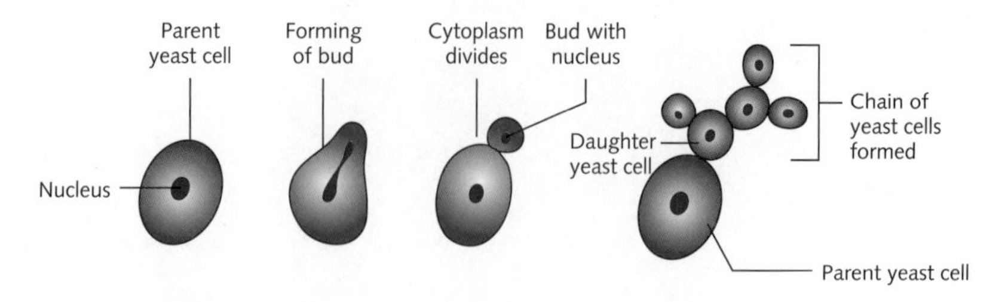

Figure 8.2 Yeast asexual reproduction

Table 8.1 shows the result of an investigation where four test tubes (A1 and A2, and B1 and B2) were each provided with 35 mL of 25% molasses solution and 1 mL of a mixture of yeast and water. Each test tube was stoppered with a holed cork and placed in a water bath at 35°C. Through the holed cork is plastic tubing, which is placed inside an inverted measuring cylinder in a trough of water, as shown in Figure 8.3. The amount of carbon dioxide (mL) produced by anaerobic respiration (respiration in the absence of oxygen) of yeast cells from each test tube was measured every 10 minutes. After 90 minutes, 5 g of table salt was added to test tube B. The results are shown in Table 8.1.

Table 8.1 Results

Time (min)	Test tube A1 (mL)	Test tube A2 (mL)	Test tube A mean (mL)	Test tube B1 (mL)	Test tube B2 (mL)	Test tube B mean (mL)
0	0.8	0.9		1.0	0.8	
10	1.8	1.6		1.9	2.3	
20	3.0	2.8		2.9	2.4	
30	4.8	4.5		5.1	6.1	
40	15.1	5.0		5.3	6.2	
50	5.3	5.0		5.5	6.3	
60	5.4	5.3		5.5	6.6	
70	5.7	5.9		5.8	6.7	
80	5.8	5.9		6.0	6.7	
90	5.8	6.1		0.8	0.6	
100	9.2	7.9		0.9	0.3	
110	9.8	9.9		0.0	0.0	
120	12.9	14.1		0.0	0.0	

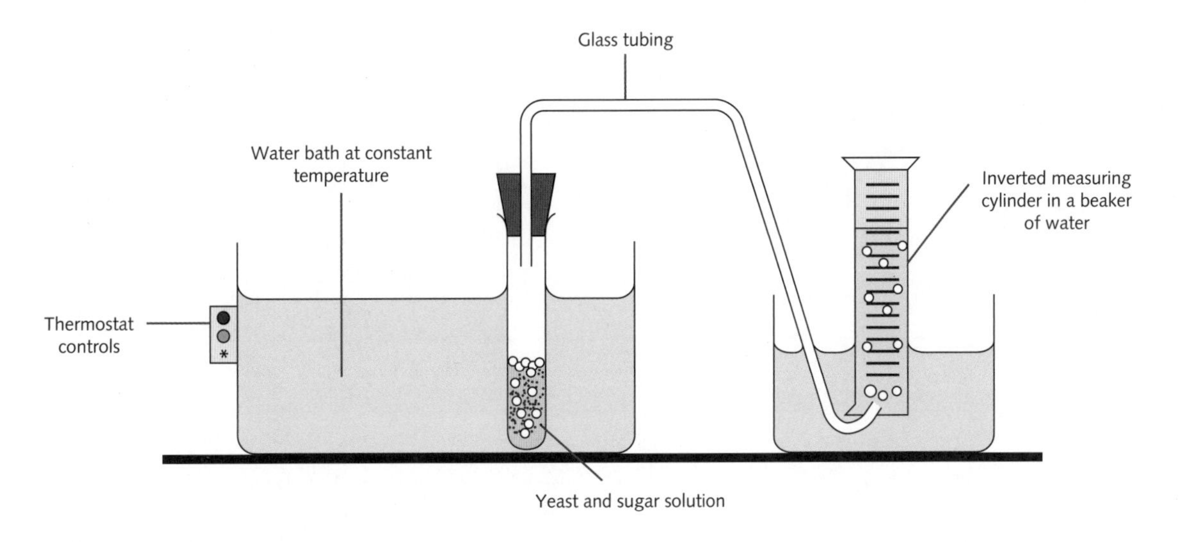

Figure 8.3 Experimental set-up

1 Write a hypothesis for this investigation.

2 Identify the:

a independent variable

b dependent variable

c controlled variables.

3 Identify any outliers in the data.

4 Explain the type of error that could have caused these outliers.

5 What could you do to determine if the outlier was a true value or the result of an error?

6 Calculate the mean mL for test tube A and test tube B and add the results to Table 8.1.

7 Graph the mean results for test tube A and test tube B using the graph paper below.

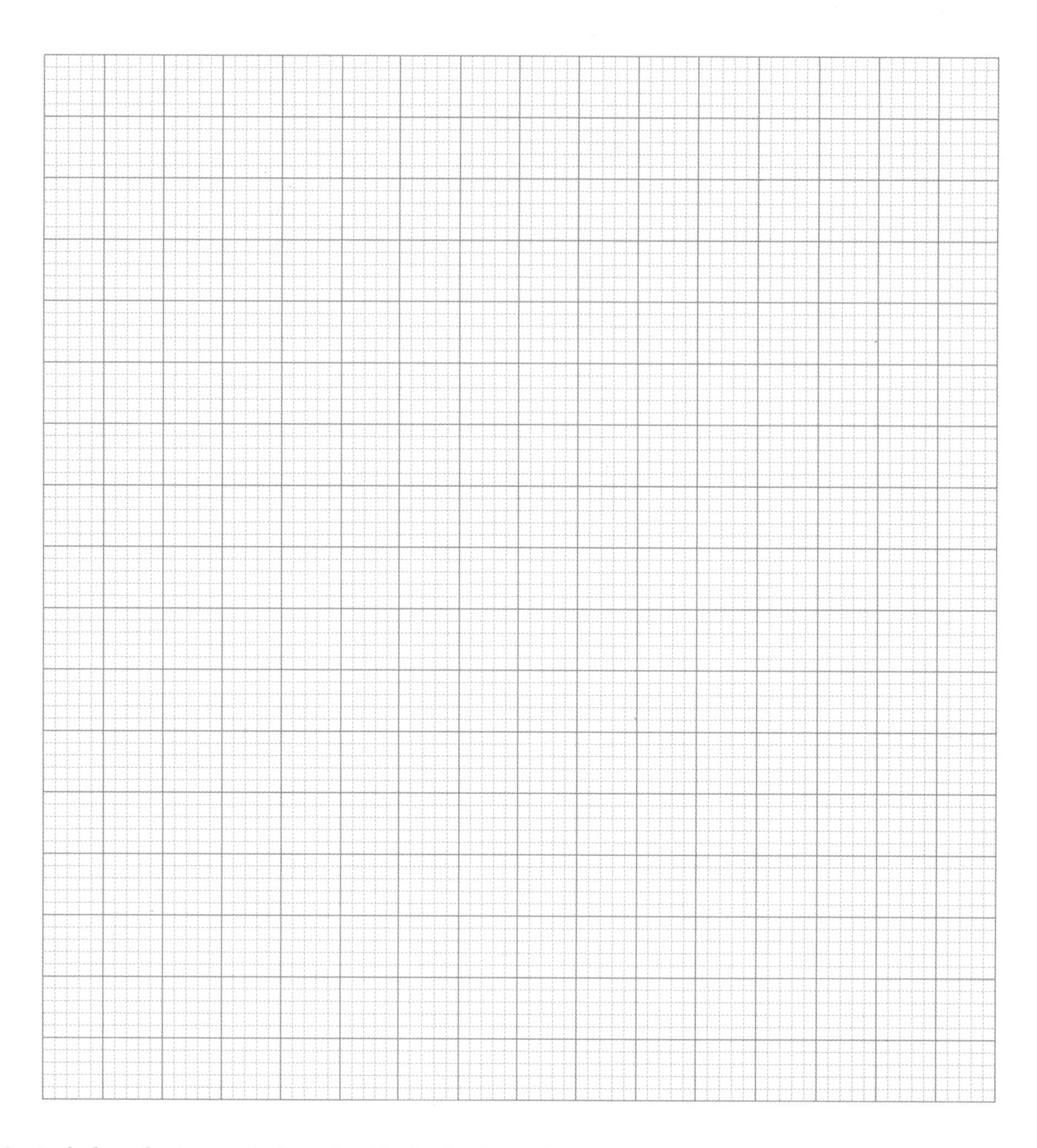

8 Is the hypothesis supported or refuted by the data? Provide your reasoning.

9 Account for the shape of the graph for test tube A.

10 Account for the shape of the graph for test tube B by referring to the genetic diversity of budding yeast.

11 How could you modify or extend this investigation to provide further evidence to support the hypothesis?

12 Write a conclusion of no more than 60 words. In your conclusion, you need to do the following.

- Restate the hypothesis.
- State whether the hypothesis is supported or refuted.
- State the main findings.

8.2 Sexual reproduction

Key knowledge

Reproductive strategies

- biological advantages of sexual reproduction in terms of genetic diversity of offspring

8.2.1 Fertilisation

Key science skills

Analyse, evaluate and communicate scientific ideas

- discuss relevant biological information, ideas, concepts theories and models and the connections between them

Develop

TB PAGE 321

Use a combination of diagrams and words to explain the biological concepts that are mentioned in the statement below.

'Fertilisation is the process where two haploid gametes join together to restore the diploid number in the zygote. The resulting zygote will have a combination of alleles that no other zygote of this species will have, unless they are identical twins.'

8.2.2 Biological advantages and disadvantages of sexual reproduction

PAGE 322

Key science skills

Generate, collate and record data

- organise and present data in useful and meaningful ways, including schematic diagrams, flow charts, tables, bar charts and line graphs

Compare the biological advantages and disadvantages of sexual and asexual reproduction by completing Table 8.2.

Table 8.2

	Sexual reproduction	Asexual reproduction
Advantages		

9780170452632

	Sexual reproduction	Asexual reproduction
Disadvantages		

8.3 Cloning

Key knowledge

Reproductive strategies

- the process and application of reproductive cloning technologies

8.3.1 Cloning

Key science skills

Analyse, evaluate and communicate scientific ideas

- analyse and evaluate bioethical issues using relevant approaches to bioethics and ethical concepts, including the influence of social, economic, legal and political factors relevant to the selected issue

Develop

PAGE 326

Read the following article carefully, taking particular note of any bioethical issues related to the subject.

Mammoth DNA briefly 'woke up' inside mouse eggs

A team of Japanese and Russian scientists have been able to reawaken some 28 000-year-old woolly mammoth cells for a short period of time, but they say that 'cloning the ice age beasts is still a long way off'.

Cells were extracted from a woolly mammoth mummy (*Mammuthus primigenius*) named Yuka that was discovered in the Siberian permafrost in 2011. Scientists were able to extract the least damaged nuclei from cells in the mummy and transplant these into mouse egg cells. Several biological

Figure 8.4 Woolly mammoths have been extinct for 3700 years.

reactions that occur before cell division occurred within the mouse egg cell, but soon stopped.

Egg cells were chosen over other types of cells because egg cells have all the living cellular organelles that are needed to correct and fix damage that has happened within the nuclei. At first, the cellular organelles did try to fix damaged DNA within the chromosomes and piece together the broken bits. The mammoth DNA was far too damaged after spending 28 000 years buried in permafrost and none of the mouse–mammoth cells entered cell division.

As soon as the mammoth died, bacteria from the mammoth's gut and the surrounding environment began breaking down the dead mammoth's cells. Ultraviolet (UV) radiation from the sun broke down more of the genetic material, and those processes continued for aeons. As a result, DNA fragments in the nucleus that survived to today may be only tens to hundreds of bases long, rather than the millions that are found in the DNA of modern elephants.

So, we cannot bring the woolly mammoth back to life just yet, but this is not to say that it will not be achieved some time in the future. Experts continue to scour the ice fields and permafrost of Siberia in search of woolly mammoth specimens that may have been sufficiently well preserved to retain viable cells. If this does become possible, then we need to consider if woolly mammoths would be comfortable living in today's environment, or would it be better to use science to save living species from becoming extinct rather than trying to bring extinct species back to life?

by Laura Geggel 14 March 2019

Adapted from https://www.livescience.com/64998-mammoth-cells-inserted-in-mouse-eggs.html

1 What cloning technique did researchers use to produce the mammoth–mouse cells? Explain this technique.

2 Once the mammoth–mouse cells had been produced, what type of cell division were the researchers hoping to see? Provide a reason for your choice.

3 List three bioethical issues that arise as a result of this research.

4 Choose one of the bioethical issues that you listed in question 3 and discuss using one of the ethical approaches listed on page 49. Consider these points in your discussion.

- Do you think any social, economic, legal and political factors are influencing the decisions of the scientists involved? Provide your reasoning. (Refer to page 137 for a discussion of social, economic, legal and political factors.)
- What course of action do you think needs to be taken to resolve this bioethical issue? Provide your reasons for following this course of action.

8.4 Chapter review

TB PAGE 332

8.4.1 Key terms

Complete the following table by writing the terms or definitions in the spaces provided.

Term	Definition
Asexual reproduction	
	Division of a cell into two, without mitosis; the process by which a prokaryotic cell divides to form two daughter cells
Biodiversity	
	Ethics relating to biological research
Blastocyst	
	Development of a new organism from an outgrowth of the parent organism
Clone	
	Process of producing a cell, tissue or organism genetically identical to its parent
Fragmentation	
	When part of one plant is artificially attached to another plant
Mitosis	
	Agricultural practice of growing a single crop or plant species over a wide area for many consecutive years
Mutation	
	Form of asexual reproduction where growth and development into a complete organism occurs from an unfertilised egg
Spore	
	Cloning of plants

9780170452632

Chapter review continued

8.4.2 Practice test questions

PAGE 333

Multiple-choice questions

1 In eukaryotic cells, asexual reproduction occurs by

A binary fission.
B mitosis.
C meiosis.
D parthenogenesis.

2 An advantage of asexual methods of reproduction is

A using energy to seek a mate.
B long generation times.
C producing a large number of identical offspring.
D ability to adapt quickly to a changing environment.

3 Which of the following is a disadvantage of asexual methods of reproduction?

A There is a large amount of genetic variation within offspring.
B There is a small amount of genetic variation within offspring.
C There is no genetic variation within offspring.
D Offspring are genetically different to their parents.

4 Fertilisation

A is the process in which two diploid cells fuse to produce a zygote.
B results in a zygote that has the same combination of alleles as the parent.
C halves the amount of DNA in the zygote.
D restores a diploid state in the zygote.

5 Which of the following is an advantage of sexual methods of reproduction?

A Offspring each have a unique set of alleles.
B Offspring are suited to a non-changing environment.
C It produces a small number of identical offspring.
D No energy is spent looking for a mate.

6 Reproductive cloning involves making an exact copy of an entire organism. This can be achieved through the following techniques.

A embryo transfer and nuclear splitting
B donor transfer and recipient splitting
C embryo splitting and nuclear transfer
D donor splitting and recipient transfer

7 In an organism with 24 chromosomes in its body cells, how many combinations of chromosomes are possible in its gametes?

A 24^2
B 2^{24}
C 12^2
D 2^{12}

Short-answer questions

8 Roundworm are able to reproduce both sexually and asexually. Figure 8.5 shows the survival rate of roundworm over 30 generations. Three populations of roundworm were grown in separate containers – Populations A, B and C.

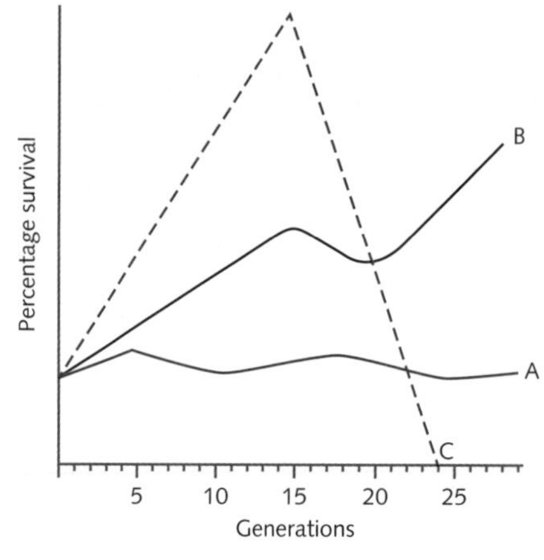

Figure 8.5 **Survival rate of roundworms over 30 generations**

a At one point in the experiment, a bacterium species was introduced into two of the containers. At what generation was the bacterium introduced? 1 mark

b Which population of roundworm was reproducing sexually? Provide evidence from the graph for your choice. 2 marks

c Which population of roundworm was reproducing asexually? Provide evidence from the graph for your choice. 2 marks

d Which population of roundworm represented the control, which was not exposed to the bacteria? Provide evidence from the graph for your choice. 2 marks

e Explain why a control population of roundworm was included in this experiment. 3 marks

Adaptations and diversity 9

Remember

Chapter 9 Adaptations and diversity will call on content that you have already met in your science studies from previous years. Take some time to refresh your knowledge of this content before you enter this chapter. Try to answer the following questions from memory. If you cannot do this, then use a reference to assist you.

TB PAGE 340

1 What is the difference between abiotic and biotic?

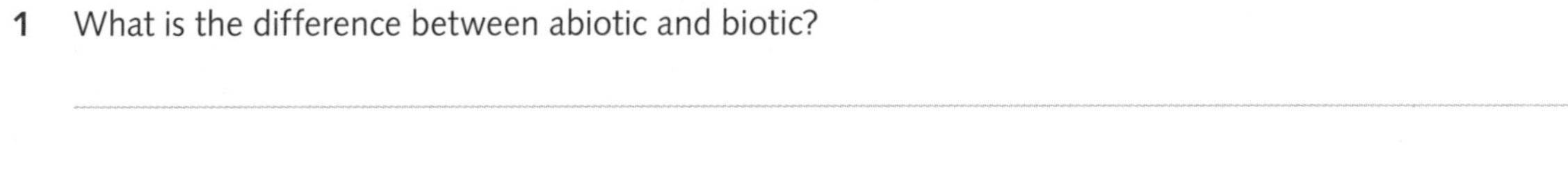

2 What is an ecosystem?

3 What is a community?

4 State two different types of symbiotic relationships between living things. In each case, state which species benefits from the relationship.

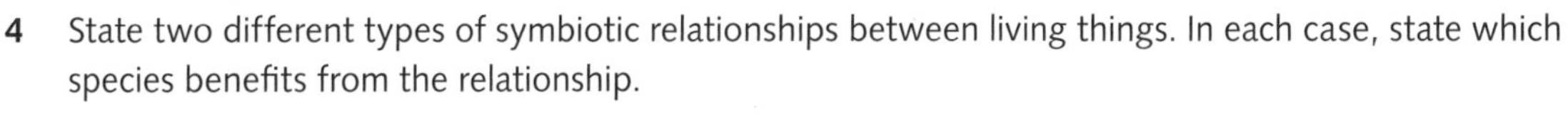

5 What does a food chain show?

6 What is the relationship between a food chain and a food web?

7 What is meant by an 'adaptation' when referring to a living thing?

9.1 Genetic diversity

Key knowledge

Adaptations and diversity

- the biological importance of genetic diversity within a species or population

9.1.1 Genetic diversity

PAGE 341

Refer to Figure 9.1 to answer the questions below.

Figure 9.1

1 List all the biotic factors that you can see.

2 List all the abiotic factors that you can see.

3 Name one population that you can see.

4 What habitat is represented in Figure 9.1?

9780170452632

5 Does Figure 9.1 represent an ecosystem? Support your answer with evidence.

9.1.2 Genetic diversity within a species or population

Key science skills

Develop aims and questions, formulate hypotheses and make predictions
- identify, research and construct aims and questions for investigation

Plan and conduct investigations
- determine appropriate investigation methodology: case study; classification and identification; controlled experiment; correlational study; fieldwork; literature review; modelling; product, process or system development; simulation

Construct evidence-based arguments and draw conclusions
- distinguish between opinion, anecdote and evidence, and scientific and non-scientific ideas

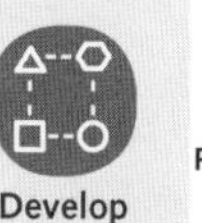

PAGE 341

Read the following extract and answer the questions that follow.

Indian lions at risk of extinction due to lack of genetic diversity

Figure 9.2 An Indian lion

A study of the genomes of extinct and living lions suggests that the near complete absence of genetic diversity in Indian lions is putting the species at risk of extinction.

Researchers from the University of Copenhagen, Barcelona Institute of Science and Technology and others analysed the remains of now-extinct lions, including two Pleistocene cave lions and 12 other historically extinct populations in Africa and the Middle East. They compared the results of this analysis with six modern lions from Africa and India. The report has been published in the journal *Proceedings of the National Academy of Sciences*.

The researchers claim that Indian lions have been living in a comparatively small area for centuries and therefore inbreeding often took place. This resulted in cranial defects, low sperm count and testosterone levels, as well as smaller manes. This lack of genetic diversity may impact the conservation efforts to save the Indian lion population from extinction.

The study found that cave lions and modern lions diverged from the ancestor about 500 000 years ago. It also discovered that within modern lions, two main lineages that used to live in central and western Africa diverged from the ancestors of the subspecies that used to inhabit North Africa about 70 000 years ago.

The descendants of the subspecies from north Africa now inhabit India. This suggests that lions migrated through corridors between sub-Saharan Africa and the Near East that may have existed in the past, for instance, through the Nile basin. It dispels a popular myth about lions being artificially brought to India in the pre-colonial era.

The scientists sequenced the genome, giving them a more complete picture of the evolution of the species through information encoded in the DNA. 'The obtained results demonstrate the power given to us by the era of genome research. We can apply it to discover the secrets of the past by reading the fragments of DNA taken from modern species' ancestors. Apart from that, a troubling reduction in Indian lion genetic material was shown', said one researcher.

Adapted from https://theprint.in/science/indian-lions-at-risk-of-extinction-due-to-lack-of-genetic-diversity-suggests-study/426989/

1 Identify the research question that was being investigated by this study.

2 What do you understand by the phrase '... the near complete absence of genetic diversity in Indian lions is putting the species at risk of extinction'.

3 Describe the methodology of this research.

4 Summarise the researchers' findings.

5 What evidence is provided to support their findings?

6 Science is a global pursuit in which scientists from across the globe share their expertise and knowledge. What evidence is there of this in the extract?

7 Does this extract represent someone's opinion, an anecdote (a story or hearsay) or a scientific idea? Choose one and provide a reason for your choice.

9.2 Adaptations

Key knowledge

Adaptations and diversity

- structural, physiological and behavioural adaptations that enhance an organism's survival and enable life to exist in a wide range of environments

9.2.1 Adaptations of animals: king penguins

TB PAGE 351

Key science skills

Generate, collate and record data

- organise and present data in useful and meaningful ways, including schematic diagrams, flow charts, tables, bar charts and line graphs

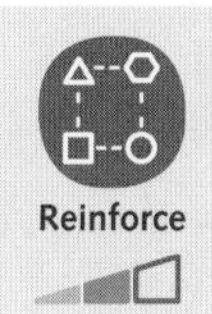

Reinforce

King penguins live on the sub-Antarctic islands such as the Falkland Islands. An adult king penguin stands about 90 cm tall and weighs 10–16 kg. Adult king penguins are covered by four layers of feathers: an outer oily waterproof layer and three inner layers of soft down. Under their skin is a thick layer of blubber: up to 30% of a king penguin's body weight can be blubber. They mainly eat fish, which they catch while diving as deep as 100 m in the Southern Ocean. They happily drink the salty seawater as their stomachs can separate the salt from the water.

King penguins are monogamous, meaning that they mate for life. The male assumes most of the responsibility for egg incubation, keeping the egg warm by nestling it firmly between his feet. Meanwhile, the female goes out to sea to hunt for food. Occasionally they will swap roles. Figure 9.3a shows a king penguin incubating an egg.

King penguins can live in colonies of as deep as 500 000 birds where they huddle together for warmth. The penguin huddle is tightly organised whereby all birds spend some time on the outer perimeter of the huddle as well as on the inside of the huddle. They are able to tilt their bodies so only the smallest part of their flipper is in contact with the cold ice. Figure 9.3b shows the flipper of a king penguin, which is ideally adapted to withstand the cold temperatures of their habitat.

The three types of adaptations that are found in living things are:

» structural – a physical characteristic
» physiological – an organism's function or process
» behavioural – the way in which an organism acts.

King penguins have structural, physiological and behavioural adaptations that enable them to live and breed in their natural Antarctic habitat. Using the information presented here and in Figure 9.3, list these adaptations, and explain how they assist in survival. Record this information in Table 9.1.

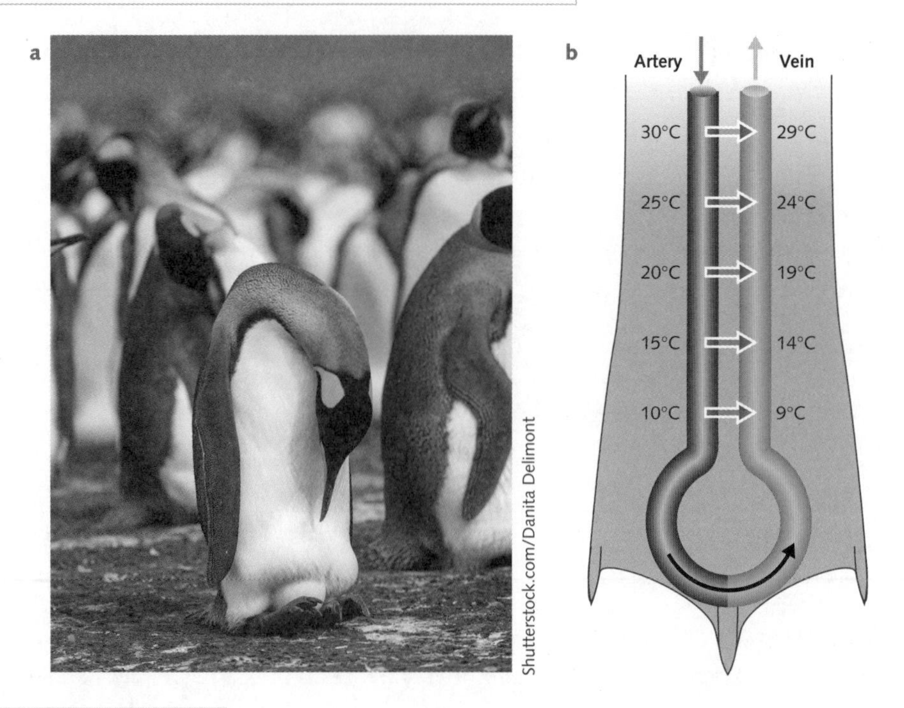

Figure 9.3 a A king penguin incubating an egg. b Countercurrent heat exchange in the legs of penguins helps to reduce heat loss and retain body heat.

Table 9.1

TYPE OF ADAPTATION		
Structural	Physiological	Behavioural

9.2.2 Adaptations of plants

Key science skills
Generate, collate and record data
- organise and present data in useful and meaningful ways, including schematic diagrams, flow charts, tables, bar charts and line graphs

Reinforce

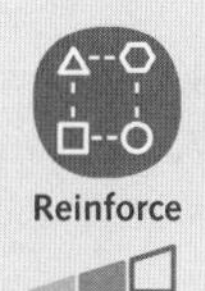

Over thousands of years, Australian plants have adapted to survive extreme conditions such as lack of water, high temperatures and extended exposure to high levels of sunlight.

Complete Table 9.2 by inserting the type of adaptation (structural, physiological or behavioural), the factor in the environment that provides a challenge to survival and how the adaptation assists survival.

Table 9.2

Plant adaptation	Type of adaptation	Environmental factor that provides a challenge for survival	How adaptation assists survival
Sunken stomata in hairy leaves			
Bulbs exposed to cold temperatures to induce flowering			
Woody fruit			
Changes in turgor pressure in the leaf in response to touch causing the leaf to close or fold			
Deciduous trees losing their leaves in winter			
Thick waxy cuticle covering leaves			
Active secretion of salt			

9.3 Survival through interdependencies between species

Key knowledge

Adaptations and diversity

- survival through interdependencies between species, including impact of changes in keystone species and predators and their ecological roles in structuring and maintaining the distribution, density and size of a population of an ecosystem

9.3.1 Interdependencies between species

TB PAGE 355

Key science skills

Develop aims and questions, formulate hypotheses and make predictions

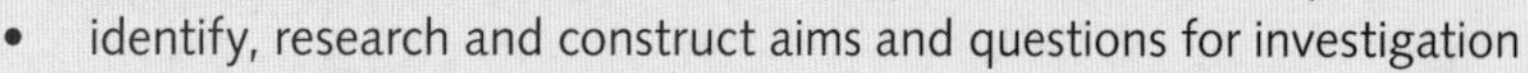

- identify, research and construct aims and questions for investigation

- predict possible outcomes

Generate, collate and record data

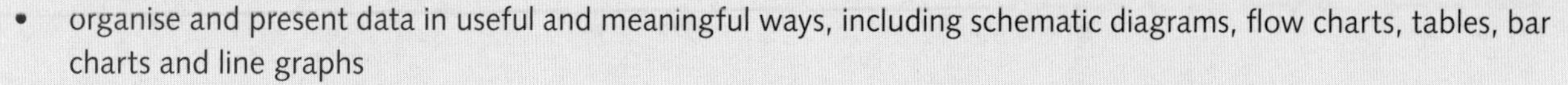

- organise and present data in useful and meaningful ways, including schematic diagrams, flow charts, tables, bar charts and line graphs

Reinforce

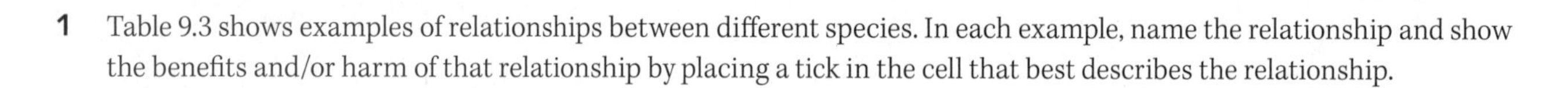

1 Table 9.3 shows examples of relationships between different species. In each example, name the relationship and show the benefits and/or harm of that relationship by placing a tick in the cell that best describes the relationship.

Table 9.3

<table>
<tr><th>Relationship between two species</th><th>Name the relationship between the two species</th><th>Put a cross in the box to show what happens to each species within this relationship, whether it benefits (+), is harmed (–), or neither.</th></tr>
<tr><td>a
Shutterstock.com/Gastev Roman</td><td></td><td><table><tr><td></td><td></td><td colspan="3">Species B</td></tr><tr><td rowspan="4">Species A</td><td></td><td>+</td><td>–</td><td>Neither</td></tr><tr><td>+</td><td></td><td></td><td></td></tr><tr><td>–</td><td></td><td></td><td></td></tr><tr><td>Neither</td><td></td><td></td><td></td></tr></table></td></tr>
<tr><td>b
Shutterstock.com/Andi111</td><td></td><td><table><tr><td></td><td></td><td colspan="3">Species B</td></tr><tr><td rowspan="4">Species A</td><td></td><td>+</td><td>–</td><td>Neither</td></tr><tr><td>+</td><td></td><td></td><td></td></tr><tr><td>–</td><td></td><td></td><td></td></tr><tr><td>Neither</td><td></td><td></td><td></td></tr></table></td></tr>
<tr><td>c
Shutterstock.com/City Escapes Nature Photo</td><td></td><td><table><tr><td></td><td></td><td colspan="3">Species B</td></tr><tr><td rowspan="4">Species A</td><td></td><td>+</td><td>–</td><td>Neither</td></tr><tr><td>+</td><td></td><td></td><td></td></tr><tr><td>–</td><td></td><td></td><td></td></tr><tr><td>Neither</td><td></td><td></td><td></td></tr></table></td></tr>
</table>

<table>
<tr><th>Relationship between two species</th><th>Name the relationship between the two species</th><th>Put a cross in the box to show what happens to each species within this relationship, whether it benefits (+), is harmed (–), or neither.</th></tr>
<tr><td>d
Shutterstock.com/Ihor Hvozdetskyi</td><td></td><td><table><tr><td></td><td colspan="4">Species B</td></tr><tr><td rowspan="4">Species A</td><td></td><td>+</td><td>–</td><td>Neither</td></tr><tr><td>+</td><td></td><td></td><td></td></tr><tr><td>–</td><td></td><td></td><td></td></tr><tr><td>Neither</td><td></td><td></td><td></td></tr></table></td></tr>
<tr><td>e
Shutterstock.com/Elana Erasmus</td><td></td><td><table><tr><td></td><td colspan="4">Species B</td></tr><tr><td rowspan="4">Species A</td><td></td><td>+</td><td>–</td><td>Neither</td></tr><tr><td>+</td><td></td><td></td><td></td></tr><tr><td>–</td><td></td><td></td><td></td></tr><tr><td>Neither</td><td></td><td></td><td></td></tr></table></td></tr>
<tr><td>f
Dreamstime.com/Verdugodario</td><td></td><td><table><tr><td></td><td colspan="4">Species B</td></tr><tr><td rowspan="4">Species A</td><td></td><td>+</td><td>–</td><td>Neither</td></tr><tr><td>+</td><td></td><td></td><td></td></tr><tr><td>–</td><td></td><td></td><td></td></tr><tr><td>Neither</td><td></td><td></td><td></td></tr></table></td></tr>
<tr><td>g
Shutterstock.com/Svetoslav Radkov</td><td></td><td><table><tr><td></td><td colspan="4">Species B</td></tr><tr><td rowspan="4">Species A</td><td></td><td>+</td><td>–</td><td>Neither</td></tr><tr><td>+</td><td></td><td></td><td></td></tr><tr><td>–</td><td></td><td></td><td></td></tr><tr><td>Neither</td><td></td><td></td><td></td></tr></table></td></tr>
</table>

Figure 9.4 shows a tree species that has lost its leaves during winter. The balls of leaves that you can see on some branches are from a different species, which is living in the tree. This species is called mistletoe.

Shutterstock.com/NagyDodo

Figure 9.4 Tree and mistletoe

2 Write four questions that you could ask that would assist you in investigating the type of relationship between the tree and the mistletoe.

9.3.2 More complex interdependencies: competition

PAGE 358

Key science skills

Develop aims and questions, formulate hypotheses and make predictions

- identify, research and construct aims and questions for investigation
- formulate hypotheses to focus investigation

Comply with safety and ethical guidelines

- demonstrate safe laboratory practices when planning and conducting investigations by using risk assessments that are informed by safety data sheets (SDS), and accounting for risks

Generate, collate and record data

- organise and present data in useful and meaningful ways, including schematic diagrams, flow charts, tables, bar charts and line graphs

Analyse and evaluate data and investigation methods

- identify and analyse experimental data qualitatively, handling where appropriate, concepts of: accuracy, precision, repeatability, reproducibility and validity of measurements; errors (random and systematic); and certainty in data, including effects of sample size in obtaining reliable data

Construct evidence-based arguments and draw conclusions

- evaluate data to determine the degree to which the evidence supports or refutes the initial prediction or hypothesis

Gause's competitive exclusion principle states that two species that compete for the same resource cannot coexist. A student wanted to investigate this law by recreating Gause's *Paramecium* experiment. The information below shows the investigative method the student followed and their results. Study this information carefully before answering the questions below.

1 Write an aim for this experiment.

2 Write a hypothesis for this experiment.

Method

1. On nutrient agar, place small samples of *Paramecium* strain 1 and *Paramecium* strain 2.
2. Leave in a warm place to grow for 17 days.
3. Record the number of colonies of each strain every day.

Results

Day	*Paramecium* strain 1 Number of colonies	*Paramecium* strain 2 Number of colonies
0	0	0
1	3	2
2	10	9
3	35	24
4	56	59
5	52	63
6	43	60
7	46	110
8	48	103
9	38	109
10	25	111
11	27	119
12	13	123
13	14	125
14	15	147
15	11	153
16	7	142
17	0	154

1 Graph the data given in the results table using the graph paper below.

2 Does the data support or refute the hypothesis? Explain.

3 Does the data support the competitive exclusion principle? Explain.

4 a Identify two problems with the experimental method that the student followed.

b Choose one problem that you have identified and explain how it affects the validity of the experimental results.

5 Risk is a concern with all experimentation. This student has left out their risk assessment. Create a risk assessment for this experiment.

What are the risks in doing this experiment?	How can these risks be managed to stay safe?

9.3.3 More complex interdependencies: predation

Key science skills

Develop aims and questions, formulate hypotheses and make predictions

- predict possible outcomes

Generate, collate and record data

- organise and present data in useful and meaningful ways, including schematic diagrams, flow charts, tables, bar charts and line graphs

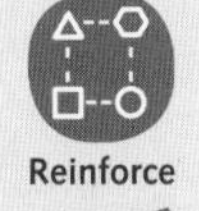

PAGE 359

A researcher counted the number of foxes and rabbits in a certain area of Victoria over 11 years. Their data is shown in Table 9.4. Use the data to plot a graph on the next page to clearly show the relationship between the populations of foxes and rabbits.

Table 9.4 Foxes and rabbit populations from 2010 to 2020 in a certain area of Victoria

Year	Foxes	Rabbits
2020	129	400
2019	103	1700
2018	250	800
2017	96	250
2016	48	2500
2015	46	1856
2014	241	1400
2013	100	270
2012	72	750
2011	219	2000
2010	86	1600

1 Describe the trend shown in your graph. Explain how the predator and prey graphs are related.

2 What is the relationship interdependency between foxes and rabbits?

3 List one benefit of this relationship for the rabbit.

4 Predict how your graph would change if the rabbit haemorrhagic disease caused by calicivirus was introduced into this population in 2016.

9780170452632

9.4 Keystone species

Key knowledge

Adaptations and diversity

- survival through interdependencies between species, including impact of changes in keystone species and predators and their ecological roles in structuring and maintaining the distribution, density and size of a population of an ecosystem

9.4.1 Keystone species in a mangrove ecosystem

Key science skills

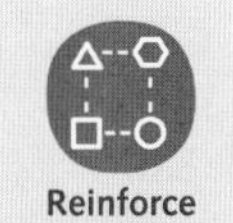

TB PAGE 366

Develop aims and questions, formulate hypotheses and make predictions

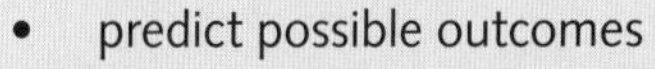

- predict possible outcomes

Generate, collate and record data

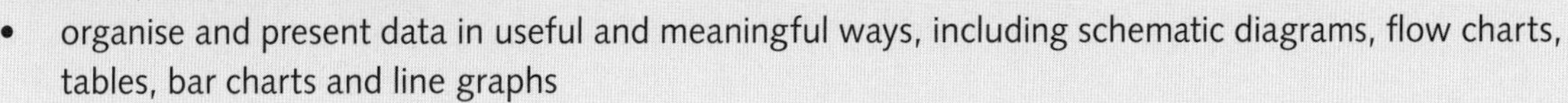

- organise and present data in useful and meaningful ways, including schematic diagrams, flow charts, tables, bar charts and line graphs

Carefully read the following extract, paying particular attention to the relationships between organisms, and answer the questions below.

A mangrove is a woody plant or plant community that lives between the sea and the land, in areas inundated by tides. Thus, a mangrove is a species as well as a community of plants. It is the only tree that is able to survive in saltwater.

Mangrove ecosystems have been called the 'storehouse of materials' as they provide food and shelter for many organisms. Detritus from the mangrove trees falls to the ground and enriches the mud in which the mangroves and algae grow. Mangrove crabs such as the fiddler crab (*Uca lacteal*) undergo the larval stages of their lifecycle in the water surrounding the mangroves. When they mature, they crawl up the mangrove trunks and feed on the leaves. These crabs burrow into the mud beneath the mangroves and aerate the soil, thereby decreasing the sulfide levels in the soil and providing much better growth and reproduction conditions for the mangroves.

Oysters, barnacles, mussels and periwinkles compete with each other for space on the mangrove pneumatophores (roots). Sponges, hermit crabs, shrimps and small fish swim around the mangrove forest during high tide, feeding on the smaller animals. Larger animals such as diamondback turtles and killifishes feed on the smaller animals. Pelicans, herons, egrets and wood ibises feed on the smaller and larger fish.

1 A keystone species is a species that has a disproportionately large effect on other organisms within an ecosystem and is important for maintaining balance within the ecosystem. Identify the keystone species in the mangrove ecosystem above.

2 Provide reasons for your choice of keystone species.

3 Predict the effect on this ecosystem if the keystone species was removed.

4 In the space below, construct a food web for the mangrove ecosystem.

9780170452632

9.5 Adaptations and interdependencies in Australian ecosystems

Key knowledge

Adaptations and diversity

- the contribution of Aboriginal and Torres Strait Islander peoples' knowledge and perspectives in understanding adaptations of, and interdependencies between, species in Australian ecosystems

9.5.1 Aboriginal peoples' knowledge and perspectives

Key science skills

Analyse, evaluate and communicate scientific ideas

- use clear, coherent and concise expression to communicate to specific audiences and for specific purposes in appropriate scientific genres, including scientific reports and posters

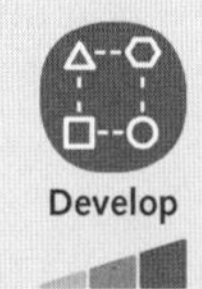

PAGE 369

Read the information below and answer the questions that follow.

Aboriginal fire management and bush fire prevention

Around 23% of mainland Australia is covered in dense grass and scattered trees. This stretches right across the top of Australia from one side to the other. Hot bushfires ravage this area almost annually, not only destroying the plant life but also destroying any animal life that is unfortunate enough to be sheltering there.

One of the recommendations that came out of the Victorian Bushfires Royal Commission that arose as a result of the 2009 Black Saturday bushfires was that 385 000 hectares of land undergo hazard reduction controlled burns each year. In 2018, only 74 000 hectares were burned, and in 2019 only 130 000 hectares were burned. These burns are usually started from the air during the middle of the day.

In contrast, Indigenous Australians have managed this ecosystem through early dry season cultural fire, which can be lit when conditions are right. This could be in the late afternoon or during a morning mist. This produces a 'cool fire'. Its aim is to get rid of invasive species that inhabit the understorey such as bracken fern and African gamba grass, and to encourage native grasses and herbs to grow. The forest canopy remains untouched. This type of burn results in a patchwork of different habitats that provide ideal conditions for the growth of many different plants and animals. It also significantly reduces the chance of a hot fire later in the season.

During a cool burn, bush turkeys can be seen hunting for insects along the edge of the fire and hawks fly overhead, scouring the ground for small animals. Ants and snakes burrow deeper into their underground nests and burrows. Brolgas eat any insects that did not escape the flames. Kangaroos escape the fire to find refuge on rocky outcrops.

After the fire has passed and the ground has cooled, new grass can be seen within weeks. This holds the soil together and provides food for wombats, wallabies and native birds. Wallabies and birds bathe in the cool ash to rid themselves of lice.

The Indigenous cultural fire practice must be adapted to each environment, taking into account the best time to burn, the breeding season of the local animals, and the types of plants in the area.

One recent investigation examined how the Martu people's hunting and fire practices influenced the ecology of the Western Desert in Central Australia. The Martu people were forced to leave the region after the nuclear tests at Maralinga in the late 1950s to early 1960s. Soon after, the animals they hunted by burning the spinifex – bilbies, bettongs and brush-tailed possums – also disappeared. Wild lightning fires raged through the landscape, and invasive predators such as rabbits, foxes and feral cats proliferated.

In the 1980s, mining and exploration threatened their country, so the Martu returned to take it back. As they hunted and burned the spinifex, the animals also came back. The Martu people believed the country became 'sick' without people conducting burns to encourage young grass, bush tomatoes and other plants that desert animals eat. Indigenous Australians believe that cool burning heals their country.

Adapted from https://lens.monash.edu/@politics-society/2020/01/08/1379433/bringing-indigenous-knowledge-into-the-bushfires-conversation
https://www.educationmattersmag.com.au/exploring-aboriginal-histories-and-cultures-through-cool-burning/
https://www.creativespirits.info/aboriginalculture/land/aboriginal-fire-management#Fire,_fauna_and_flora

1 Contrast hazard reduction burns with Aboriginal cool burns.

2 Identify any interdependent relationships mentioned in this extract, state the type of relationship and list the organisms involved in each.

3 Assess the extent that the Aboriginal cool burns assisted in the formation of each of the relationships mentioned in question **2**

Summarise the key points in this extract, using no more than 150 words.

9.6 Chapter review

TB PAGE 381

9.6.1 Key terms

The following sentences contain alternative key terms. Choose the best key term for the context of the sentence and cross out the other one(s). The first one has been done as an example for you to follow.

Example: In a food chain, the ~~first-order consumer~~/producer is always a plant because it can photosynthesise to produce its own food.

1 In an ecosystem, the fungi growing on dead logs is part of the abiotic/biotic factors.

2 An animal that eats both plant and animal material is called a(n) omnivore/carnivore.

3 When a mosquito bites a human, the mosquito is the host/parasite and the human is the host/parasite.

4 Snakes produce poisonous venom to kill their predator/prey. This is an example of a behavioural/physiological/structural adaptation of snakes.

5 Lichen is a combination of green photosynthetic algae and fungi. The algae provide the fungi with food and the fungi provide the algae with protection. This is an example of a commensal/mutual/competitive relationship.

6 Some birds live in the hollows of trees. The birds are provided with a space to shelter but the tree gains no benefit. This is an example of a commensal/mutual/competitive relationship.

7 Species that have large genetic diversity/biodiversity are much more likely to survive an environmental change.

8 If beavers were removed from their natural habitat, then many other organisms in that habitat would be affected. This makes beavers a keyhole/keystone/key species.

9 The geographical range in which a particular species can be found is known as its density/abundance/distribution.

10 The number of organisms in a population within a particular area is known as its density/abundance/distribution.

Chapter review continued

9.6.2 Practice test questions

PAGE 382

Multiple-choice questions

1 When a population of bats is exposed to a bacterial disease, not all the bats will die. This is an example of

A genetic diversity.
B competition.
C commensalism.
D symbiogenesis.

2 What type of adaptation relates to the long loop of Henle in desert-dwelling animals that enables them to excrete concentrated urine?

A behavioural
B structural
C physiological
D systematical

3 The mountain pygmy-possum lives in a cold, wet climate with snow between June and September. During these months the possum hibernates. What type of adaptation is this?

A physical
B physiological
C behavioural
D structural

4 Abiotic factors

A include sunlight, humidity and water pressure.
B do not directly affect living things.
C are not relevant in aquatic ecosystems.
D include all the species that form mutualistic relationships.

5 An example of the competitive exclusion principle is

A a clown fish living in an anemone.
B a killer whale feeding on seals.
C plants that use chemicals to keep other plants from growing near them.
D a worm that lives in the gut of an animal.

6 Consider the predator and prey graphs shown in Figure 9.5.

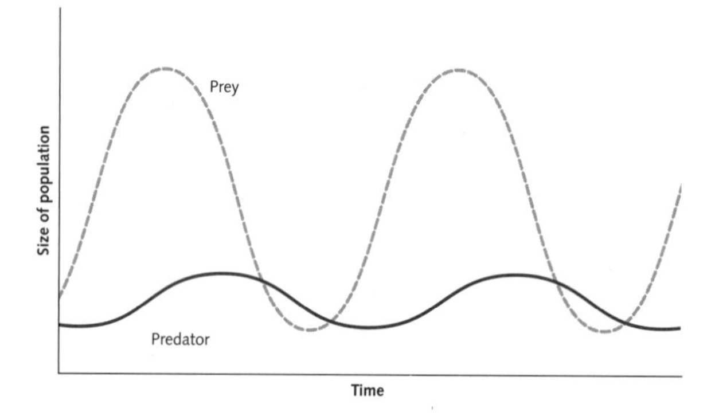

Figure 9.5 Predator–prey relationship

What conclusion can you reach from this data?

A When predator numbers are high, prey numbers are low.
B When predator numbers are high, prey numbers are high.
C Predators control the abundance of prey species.
D There is no relationship between the number of predators and the number of prey.

Short-answer questions

7 The table below summarises the relationships and interactions between living species. Complete the table by filling in the blank cells.
7 marks

Relationship or interaction	Description	Example
	Different species living together and sharing the same resource	
Commensalism		
	Rivalry between species for particular resources	
Mutualism		
	One animal kills another animal for food	
Keystone species		
Amensalism		

8 Elephant dung contains a great deal of fibrous matter, including seeds. Some of their plant food species have evolved to produce seeds that have a coating of rind to protect them from an elephant's digestive juices. Unless the seeds have passed through the elephant's digestive system first, they are unable to germinate.

a Name the type of interaction described above.
1 mark

b State if any of the species benefit from this relationship and if so, what the benefit(s) are.
2 marks

Shutterstock.com/rickyd

Investigating a bioethical issue 10

Remember

Chapter 10 Investigating a bioethical issue will call on content that you have already met in your science studies. Take some time to refresh your knowledge of this content before you enter this chapter. Try to answer the following questions from memory. If you cannot do this, then use a reference to assist you.

PAGE 390

1 Explain all the variables that are used in a controlled experiment.

2 Explain what a control is in a controlled experiment.

3 Explain what is meant by correlation between two variables.

4 When are results of a scientific investigation determined to be valid?

5 Distinguish between a reproducible and a repeatable investigation.

10.1 Beginning an investigation

Key knowledge
Analysis and evaluation of bioethical issues
- ways of identifying bioethical issues

Scientific evidence
- the use of a logbook to authenticate collated secondary data

Scientific communication
- conventions for referencing and acknowledging sources of information

10.1.1 Identifying a bioethical issue

Key science skills
Analyse, evaluate and communicate scientific ideas
- analyse and evaluate bioethical issues using relevant approaches to bioethics and ethical concepts, including the influence of social, economic, legal and political factors relevant to the selected issue

TB PAGE 392

Consider the following three scenarios. For each scenario, identify the bioethical issue and write a problem statement that outlines the dilemma that needs to be resolved. When crafting your problem statements, remember the five Ws: what, who, when, where and why.

Scenario 1

It is early March 2020. COVID-19 is taking hold of the population and doctors are gearing up to cope with the predicted explosion in patient numbers. Doctors fear that there will be an acute shortage of medical resources, including ventilators. Three patients present at the same regional hospital at the same time with severe respiratory failure due to COVID-19. There is only one ventilator available.

Sean is a 25-year-old man who is studying to be a civil engineer. He suffers from insulin-dependent diabetes. Ligaya is 39 years old, employed as a nurse and has three children. She has no other health conditions or disabilities. Ann is 85 years old and has severe dementia. Her husband died 10 years ago. She lives in a nursing home.

1 Identify the issue.

2 Write a problem statement for the issue.

Scenario 2

Sarina is 25 weeks pregnant. She has a regular ultrasound scan as part of her pre-natal care and finds out that her baby has inherited a rare genetic disorder resulting in severe abnormality. The baby will require life-long care. She is offered a late-term abortion.

3 Identify the issue.

4 Write a problem statement for the issue.

Scenario 3

Shortly after 10 a.m. on 10 November 2018, Mindy Barrett took her last breath. Mindy was a 64-year-old Englishwoman who had suffered from several painful diseases for almost three decades. She was forced to give up her highly successful corporate career when she was 39 and had since spent much of her time seeking new medical treatments, to no avail. When the pain became unbearable and it was clear there was no wonder drug on the horizon, she chose voluntary assisted dying as she wanted to die with dignity. She had to travel to Switzerland to access the treatment as it is currently illegal in her home country, the United Kingdom.

5 Identify the issue.

6 Write a problem statement for the issue.

9780170452632

10.1.2 Referencing and acknowledging sources

Key science skills

Analyse, evaluate and communicate scientific ideas

- acknowledge sources of information and assistance, and use standard scientific referencing conventions

As you complete any background research using secondary sources, it is best to record details about each source in your logbook. If you do not record the details of a source, then you might have to retrace your steps and try to locate where you got a particular piece of information from.

There are different methods you can follow to record your secondary sources. The American Psychological Association (APA) citation method has been adopted for referencing in many scientific disciplines, but there are other systems of referencing that you can use, such as the Harvard referencing system. Check with your teacher to find out which system they want you to use.

7 Table 10.1 contains a number of different types of secondary resources. Find each type of secondary resource in your school library, online or elsewhere and complete the table using the APA citation method.

Table 10.1 Types of secondary resources

Source of information	APA citation
Book about genetics	
Book about embryology	
Peer-reviewed science journal	
Popular science magazine	
Website on bioethics	
Television show on current affairs	
Film set in another country	

10.2 Evaluating evidence

Key knowledge

Scientific evidence

- the distinction between primary and secondary data
- the nature of evidence and information: distinction between opinion, anecdote and evidence, and scientific and non-scientific ideas
- the quality of evidence, including validity and authority of data and sources of possible errors or bias
- methods of organising, analysing and evaluating secondary data

10.2.1 Distinction between opinion, anecdote and evidence

PAGE 397

Key science skills

Construct evidence-based arguments and draw conclusions

- distinguish between opinion, anecdote and evidence, and scientific and non-scientific ideas

1 Use three different coloured highlighters to differentiate the following ten statements as one of:

- opinion (a claim or judgement that is formed in the absence of hard evidence)
- anecdote (a claim supported by a personal observation or personal experience)
- evidence (gathered to support an idea or address a claim).

Provide a reason for each of your classifications.

a Seven out of 100 children have been tested and show a severe peanut allergy.

Reason: ______________________________

b I strongly agree with you.

Reason: ______________________________

c I will never forget the time I studied a plant that I had found in the forest, only to find that no one had ever seen that species before.

Reason: ______________________________

d I remember when kangaroos grazed in that field.

Reason: ______________________________

e I would say that brown eggs have more nutritional value than white eggs.

Reason: ______________________________

f I doubt that wearing a mask will stop me from catching COVID-19.

Reason: ______________________________

g My mum says that magpies only warble before a rainstorm.

Reason: ______________________________

h My colleague used to collect mushrooms from the paddock to make mushroom stew, but he doesn't do this anymore because he is scared of eating a poisonous one.

Reason: ______________________________

i DNA was used to place a man at the scene of a crime in 1998.

Reason: ______________________________

j Fifty-four per cent of adults admit to eating chocolate every day.

Reason: ______________________________

10.2.2 Distinction between scientific and non-scientific ideas

Key science skills

Construct evidence-based arguments and draw conclusions

- distinguish between opinion, anecdote and evidence, and scientific and non-scientific ideas

1 Use two different coloured highlighters to differentiate the following six statements as one of:
- scientific idea (supported by evidence that has been systematically collected under controlled conditions to support or refute a claim).
- non-scientific idea (not supported by evidence that has been systematically collected under controlled conditions to support or refute a claim; prejudiced; influenced by external factors).

Provide a reason for each of your classifications.

a Bean plants given a high nitrogen fertiliser will yield twice as many beans as plants that do not receive this fertiliser.

Reason: ______________________________

b Roach Chemicals developed a potion that they claim helps anxious people become peaceful and calm and therefore perform better at their work.

Reason: ______________________________

c Seven out of 10 people who slept on a *Sleeping Goose* mattress said they had a much nicer night's sleep.

Reason: ______________________________

d Better eggs are produced by chickens that listen to classical music, fed corn mash and are allowed to roam in a paddock than chickens who don't experience these things. Anti battery-hen campaigners say that all chickens should be allowed to roam free so they can produce better eggs.

Reason: ______________________________

e Five researchers independently concluded that women over the age of 50 who take a daily dose of vitamin D have significantly stronger bones than similar women who do not take any supplementary vitamin D.

Reason: ______________________________

f Seventy-two people of various age groups completed a survey about the removal of trees from their suburb. The conclusions from the survey showed that over 50 per cent of the respondents were really upset and angry about the removal of the trees.

Reason: ______________________________

10.2.3 Quality of evidence

Key science skills

Generate, collate and record data

- organise and present data in useful and meaningful ways, including schematic diagrams, flow charts, tables, bar charts and line graphs

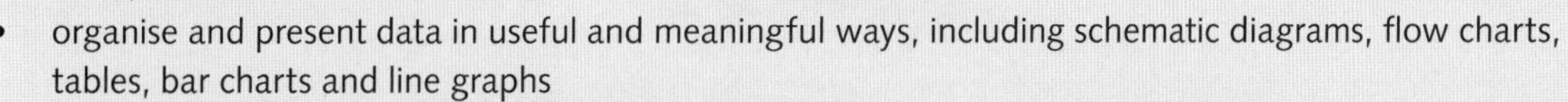

- plot graphs involving two variables that show linear and non-linear relationships

Analyse and evaluate data and investigation methods

- process quantitative data using appropriate mathematical relationships and units, including calculations of ratios, percentages, percentage change and mean
- identify and analyse experimental data qualitatively, handling where appropriate, concepts of: accuracy, precision, repeatability, reproducibility and validity of measurements; errors (random and systematic); and certainty in data, including effects of sample size in obtaining reliable data
- identify outliers, and contradictory or provisional data
- evaluate investigation methods and possible sources of personal errors/mistakes or bias, and suggest improvements to increase accuracy and precision, and to reduce the likelihood of errors

Part A: Assessing the data qualitatively

Read each of the scenarios below and state:

- whether the data is primary or secondary data
- whether the data is free of errors. If you think the data contains errors, state which type of error you think it contains. The three types of errors are personal, random and systematic.

Scenario 1

You measured the height of four of your friends. You measured each friend only once using either a tape measure or a metre ruler. You recorded your results in the table below.

Friend	Height (cm)
Prisha	500
John	578
Zaraan	601
Asim	512

1 a Is this primary or secondary data?

b Is this data free from errors? If yes, which type of error does this data contain?

Scenario 2

Dr Eugene Humphreys was investigating the weight of starling hatchlings. He weighed more than 700 starling hatchings during hatching season. He used the same electronic balance, which he calibrated prior to each weighing. He calculated the frequency of each weight range and his results are shown in Figure 10.1 below. He reported his findings in the *Avian Journal,* March 2020.

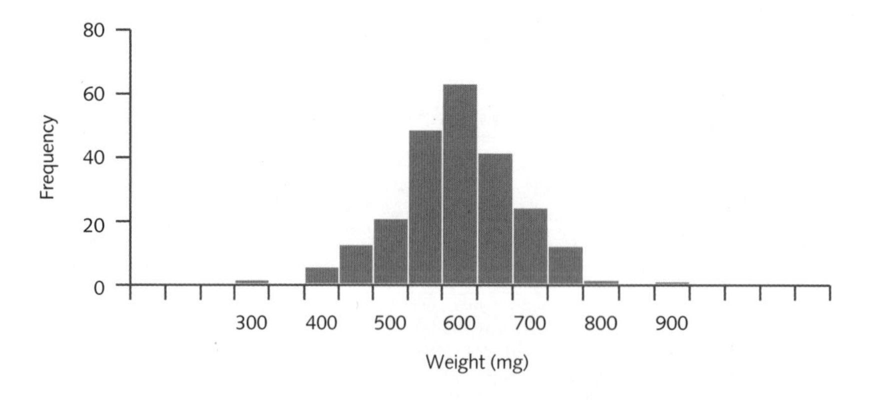

Figure 10.1 Weight frequency of starling hatchlings

2 a Is this primary or secondary data?

b Is this data free from errors? If yes, which type of error does this data contain?

Part B: Organising, analysing and evaluating data

3 Construct a results table in the space below for the following investigation.

- Chu and Oslo want to test the effect of soil pH on the growth of bean seedlings. They decide to test the pH range of 6.2–7.2, testing in intervals of 0.2.
- They decide to have two replicates at every pH.
- They measure the height of the plants in centimetres before the investigation begins, and then once a week for 3 weeks.

4 They record the following data into their logbook. Place this data into the results table you constructed for Chu and Oslo's investigation.

Plant 1:

pH 6.2 Week 1: 3.4; Week 2: 6.9; Week 3: 13.5
pH 6.4 Week 1: 2.8; Week 2: 7.2; Week 3: 16.2
pH 6.6 Week 1: 4.1; Week 2: 5.1; Week 3: 17.8
pH 6.8 Week 1: 3.8; Week 2: 4.4; Week 3: 9.3
pH 7.0 Week 1: 4.2; Week 2: 6.0; Week 3: 10.2
pH 7.2 Week 1: 1.9; Week 2: 4.2; Week 3: 6.7

Plant 2:

pH 6.2 Week 1: 3.2; Week 2: 4.7; Week 3: 12.1
pH 6.4 Week 1: 3.8; Week 2: 6.8; Week 3: 17.2
pH 6.6 Week 1: 2.1; Week 2: 16.0; Week 3: 14.8
pH 6.8 Week 1: 3.6; Week 2: 5.4; Week 3: 10.1
pH 7.0 Week 1: 3.6; Week 2: 6.2; Week 3: 11.2
pH 7.2 Week 1: 2.9; Week 2: 3.9; Week 3: 7.4

5 Identify any outliers in their data and explain how these could have come about.

6 Suggest to Chu and Oslo what they could do to improve the accuracy of their data.

7 Calculate the average height of the bean seedlings at each pH. Add this to your results table.

8 Calculate the percentage change of the bean seedlings at each pH. Add this to your results table.

9 Draw a graph of Chu and Oslo's data using the graph paper below.

10.3 Evaluating a bioethical issue

TB PAGE 400

Key knowledge

Analysis and evaluation of bioethical issues

- characteristics of effective analysis of bioethical issues
- approaches to bioethics and ethical concepts as they apply to the bioethical issue being investigated.

Key science skills

Analyse, evaluate and communicate scientific ideas

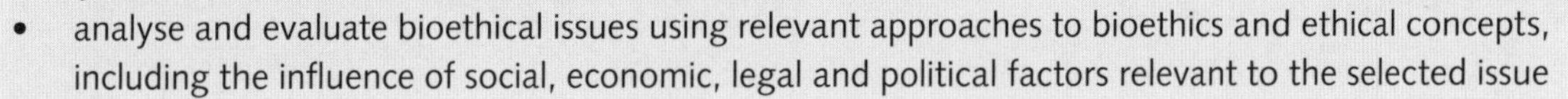

- analyse and evaluate bioethical issues using relevant approaches to bioethics and ethical concepts, including the influence of social, economic, legal and political factors relevant to the selected issue

As part of your assessment for Unit 2 Outcome 3, you will provide a response to an investigation into a bioethical issue relating to genetics or reproductive science or adaptations beneficial to survival. Refer to the Study Design for further information about this.

The activity below provides you with an opportunity to practice some of the skills that you will need to demonstrate when you come to complete this assessment.

Read the information below and answer the questions that follow.

Human challenge trials

Human challenge trials came into existence when Edward Jenner exposed young James Phipp to smallpox in 1796. James did not go on to develop smallpox and his bravery has enabled humankind to eliminate smallpox from the global population through the production of a vaccine. Other vaccines that have been developed using human challenge trials include those for influenza and malaria. These diseases are curable by other methods, but a vaccine ensures the disease is not contracted in the first place. In 2020, the world was overrun by the SARS-CoV-2 virus causing large loss of life, human suffering and vast social and economic shutdowns. Politicians across the world were eager to restart their countries' economies by reopening shops, businesses, schools, tourism and trade.

There is no known cure for SARS-CoV-2. At the time of writing, over 200 potential vaccines were in development, but it remains many months or years before distribution are distributed worldwide. Have human challenge trials been the way to the early development of a vaccine for SARS-CoV-2?

In a human challenge trial, volunteers are placed in randomised, placebo-controlled groups. One group would be vaccinated with the investigational vaccine and the other group would receive a placebo. After getting either, all the volunteers would be exposed to SARS-CoV-2 virus via a nasal spray, quarantined and their immune response and symptoms monitored for anything up to 2 months. This is a quicker process than waiting for the virus to find each volunteer in the community, as is done in normal trials for vaccines. By early July 2020, over 25000 volunteers from 102 countries had signed up to volunteer for the trial. Volunteers were only chosen from young and healthy people. This demographic appears to have the lowest risk of dying from SARS-CoV-2, with data showing that 0.03% of 20–29 year olds died from SARS-CoV-2 and 1.1% were hospitalised.

1 Identify any bioethical issues in the above information.

2 Choose one bioethical issue and identify any social, legal, political or economic considerations that may affect it.

3 Identify two opposing viewpoints about the chosen bioethical issue. Who holds each of those viewpoints?

4 Analyse your chosen bioethical issue in terms of the ethical concepts of integrity, respect, beneficence, non-maleficence and justice. (See page 138.)

10.4 Preparing a report

Key knowledge

Scientific communication

- biological concepts specific to the investigation: definitions of key terms; use of appropriate biological terminology, conventions and representations
- characteristics of effective science communication: accuracy of biological information; clarity of explanation of biological concepts, ideas and models; contextual clarity with reference to importance and implications of findings; conciseness and coherence; and appropriateness for purpose and audience
- the use of data representations, models and theories in organising and explaining observed phenomena and biological concepts, and their limitations
- the influence of social, economic, legal and political factors relevant to the selected research question

Key science skills

Develop

PAGE 408

Analyse, evaluate and communicate scientific ideas

- critically evaluate and interpret a range of scientific and media texts (including journal articles, mass media communications and opinions in the public domain), processes, claims and conclusions related to biology by considering the quality of available evidence
- analyse and evaluate bioethical issues using relevant approaches to bioethics and ethical concepts, including the influence of social, economic, legal and political factors relevant to the selected issue
- use clear, coherent and concise expression to communicate to specific audiences and for specific purposes in appropriate scientific genres, including scientific reports and posters
- acknowledge sources of information and assistance, and use standard scientific referencing conventions

1 Write a 500-word article for a science journal where you investigate, analyse and evaluate the issue of human challenge trials versus conventional clinical development and testing of a vaccine for SARS-CoV-2. (See page 241.)

As part of your article make sure you:

- use clear, coherent and concise expression
- use correct biological terminology
- detail any social, economic, legal or political factors relevant to the selected issue
- provide arguments for your point of view
- construct a reference list of sources used to compile your article.

10.5 Chapter review

TB PAGE 418

10.5.1 Key terms

Select each key term from the list below and match it to its definition.

anecdote
beneficence
dogma
integrity
justice
opinion
paraphrasing
plagiarism
public opinion
quoting
respect
summarising

Key term	Definition
	Principle of a fair and equal distribution of opportunities and responsibilities
	Public understanding, views and attitudes about an issue
	Describes an honest, transparent and professional approach
	Principle of the individual rights of a person or another living organism
	Duty to do more good than harm
	Rewriting a sentence or passage word for word from a source, enclosing the text in quotation marks
	Rewriting a particular passage from a source in your own words and retaining the original meaning
	Claim supported by a personal observation or personal experience
	Act of misrepresenting someone else's work or ideas as your own
	Claim or a judgement that is formed in the absence of evidence
	Putting the main ideas into your own words to capture a broad overview of a source
	Opinion guided by ideological belief rather than by evidence

Chapter review continued

10.5.2 Practice test questions

Short-answer questions

1 Pre-implantation genetic diagnosis (PGD) involves testing human embryos that have been created during the IVF process. Testing identifies any genetic abnormalities and helps select the best possible embryo for transfer to the uterus of the woman. This process is used by couples with no fertility issues but who carry a serious genetic disorder such as Duchenne muscular dystrophy, cystic fibrosis, Huntington disease, Leigh syndrome or fragile X syndrome. These diseases cause severe disability and have no prevention, cure or proven treatments. PGD involves the removal of a few cells from the embryo and testing its DNA before a disease-free embryo is selected for transfer.

A couple who both carry a severe and deadly mutation undergo PGD and request a female embryo be selected for transfer.

Describe one ethical and one social issue that could arise as a result of this request. 2 marks

	Issue/Implication
Ethical	
Social	

2 With the world's population at around 8 billion people, many scientists argue that the best way to feed all these people without creating environmental chaos is to grow genetically modified (GM) crops.

Farmers save seeds from their crops one year to plant the following season. This saves them money. They cannot do this with the GM crop Bt cotton. The seeds of Bt cotton are the legal property of Monsanto and anyone growing Bt cotton must buy the seeds from Monsanto, who manufacture Bt cotton. Bt cotton gets its name from the soil bacterium *Bacillus thuringiensis*. Two genes from *B. thuringiensis* that code for insect-resistant proteins have been inserted into Bt cotton. Studies have shown that farmers who grow Bt cotton are now using 15 per cent of the amount of insecticide than when they grew non-Bt cotton. Australian crops of Bt cotton are picked by machine. In India, Bt cotton is picked by hand and workers have reported skin conditions, which they blame on the proteins in Bt cotton.

Golden rice is not manufactured by Monsanto. Farmers can save the seed from one year's crop to plant the following season. Golden rice has two genes inserted into its genome: one from the bacteria *Erwinia uredovora* and the other from a daffodil, *Narcissus pseudonarcissus*. These two genes produce a rice that contains much higher levels of vitamin A. Vitamin A supports skin, eye and reproductive health, and immune function. Golden rice has been shown to be safe to eat.

Using the above information, describe one economic and one biological implication relevant to the use of Bt cotton and golden rice. Do not use the same implication more than once. 4 marks

	Economic implication	Biological implication
Bt cotton		
Golden rice		

Answers

Chapter 1 Cellular structure and function p. 2

Remember p. 2

1 Cells are the basic building blocks of all living things; they are the basic structural, functional, and biological unit of all organisms. A cell is the smallest unit of life.
2 Organelles are specialised structures or compartments within cells that carry out a specific function.
3 Answers will vary. Examples are nucleus: control centre of cell; mitochondria: produces energy for cell.
4 An organism is a living thing/composed of cells – can be unicellular or multicellular.
5 Unicellular organisms – composed of one cell only. Multicellular organisms – composed of more than one cell
6 Plants, Animals, Fungi, Protists, Archaebacteria, Eubacteria

1.1 Cells are the basic structural unit of life p. 3

1.1.1 The cell theory p. 3

1 Answers will vary. For example: Is the substance made up of cells? Does the substance release carbon dioxide/take in oxygen?
2 Answers will vary. For example: if the cells of the substance contain chloroplasts then it is a plant.
3 There would be maggots on the meat in all three jars.
4 In jar 1, flies were able to enter and lay eggs on the meat. In jar 3, flies were able to lay eggs on the cloth covering; hence, maggots would not be able to get through the cloth or appear on the meat. In jar 2, flies were prevented from depositing eggs in the jar due to the cork in the neck: no maggots would be found on the meat.
5 Jar 1: Flies had contact with meat and were able to lay eggs on meat. Eggs hatched into maggots.
Jar 2: Flies were not able to get contact with meat, and were unable to lay eggs, no maggots appeared.
Jar 3: shows that flies lay eggs but as they were not in contact with meat, could not lay eggs on meat.
6 Answers will vary:
1665 Robert Hooke – observes first cells in cork
1674 Antonie van Leeuwenhoek – observes first living cells in pond water
1833 Robert Brown – discovers nucleus in plant cells
1838 Matthias Jakob Schleiden – proposes that all plant tissues are composed of cells
1839 Theodor Schwann – concludes that animal cells are also composed of cells. States that the cell is the basic unit of all organisms.
1845 Carl Heinrich Braun – states that cells are the basic units of life
1855 Rudolf Virchow – states that cells only come from other living cells
7 A law is supported by the results of many experiments and is accepted as being true, a law is a description of how things behave. A theory is the explanation of an observation that is supported by available evidence.

1.1.2 Prokaryotic and eukaryotic cells p. 5

1

Prokaryotic	Both	Eukaryotic
Smaller	Microscopic	Larger
Simpler	Plasma membrane	Nucleus
No nucleus	Cytosol	More complex
No membrane bound organelles	Ribosomes	Membrane-bound organelles
Circular chromosomes	DNA	Linear chromosomes
No cytoskeleton	May have a cell wall	Cytoskeleton
	May have flagella	

Answers will vary; for example, eukaryotic and prokaryotic cells are microscopic, are surrounded by a plasma membrane and have a cytosol, DNA and ribosomes. Both cells types can have a cell wall and flagella. Eukaryotic cells are larger and more complex than prokaryotic cells. Eukaryotic cells contain membrane-bound organelles such as a nucleus, mitochondria, endoplasmic reticulum. Prokaryotic cells do not contain membrane-bound organelles. Prokaryotic cells contain DNA in the form of a circular chromosome found in the cytosol, whereas eukaryotic cells contain DNA in the form of linear chromosomes found in the nucleus. Prokaryotic cells do not have a cytoskeleton, whereas eukaryotic cells do.

1.2 Size and shape of cells p. 7

1.2.1 Surface area to volume ratio p. 7

Table 1.1 Results

Shape of cell	Surface area	Volume	SA: Vol
Cube	$6 \times 5 \times 5 = 125\ cm^2$	$5 \times 5 \times 5 = 125\ cm^3$	125 : 125 = 1 : 1
Rectangle	$2(5 \times 5 + 5 \times 1 + 5 \times 1) = 70\ cm^2$	$5 \times 5 \times 1 = 25\ cm^3$	70 : 25 = 14 : 5 (2.8 : 1)

1. The thinner, flatter shape had the greater SA:V.
2. The rectangular cell has a greater surface area compared to its volume, meaning it has less volume that needs nutrients/produces waste and a greater surface area over which nutrients can enter and waste can exit the cell.
3. Answers could include: be smaller, have a flatter shape, be more elongated, have folds in the membrane or protusions on the side e.g. root hair cells.
4. Answers may vary; for example, it has shown that different shapes have different surface area to volume ratios, flatter shapes have a larger surface area compared to their volume.
5. Answers will vary; it only compares two shapes, does not show how the SA : Vol changes for different sizes of the same shape.

1.3 What is inside a cell? p. 13

1. Photosynthesis
2. Aerobic cellular respiration
3. Creates a separate compartment from the rest of the cell for processes to occur in, which allows different activities to occur at the same time independent of each other. The membrane also increases the surface area within the cell, allowing more exchange of materials and more cellular functions.
4. Many small discs have a greater surface area than one large disc, this provides a larger surface area for the photosynthetic reactions to occur on.
5. A folded membrane has a greater surface area than a straight circle of membrane, providing a greater surface area for reactions of aerobic respiration to occur on.
6. Each of these functions occurs in its own membrane-bound compartment/organelle, which allows the different functions to occur at the same time without interfering with each other. The folds in the internal membranes of the organelles provide a large surface area for the reactions to occur on.
7. Prokaryotic cells are smaller, so they have a larger SA : Vol compared to larger eukaryotic cells. To compensate for a smaller SA : Vol, eukaryotic cells have compartments, organelles surrounded by membranes. The presence of these membranes increases the surface area allowing for more cellular functions.

1.3.2 Organelles p. 14

Activity: modelling a eukaryotic cell p. 14

Student answers may vary.

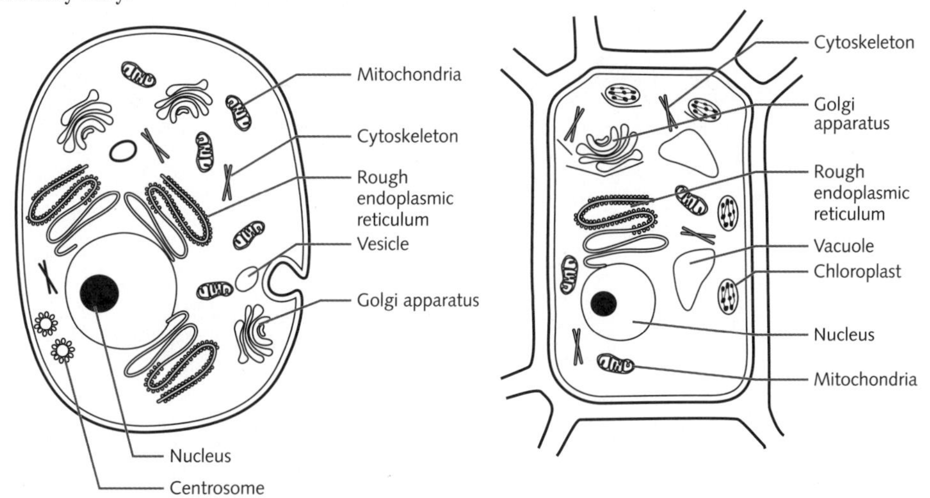

1 The pancreatic cell contains a lot of rough endoplasmic reticulum to allow the production of hormones. It also has mitochondria to provide energy for the creation of the hormone proteins and Golgi bodies to package the hormone to be secreted from the cell.

2 The plant cell contains chloroplasts to allow it to carry out photosynthesis and large vacuoles to keep chloroplasts near the outside of the cell where they will be exposed to more light.

3

Plant cells	Plant and animal cells	Animal cells
Cell wall Large vacuole Chloroplasts Starch granules	Nucleus Nucleolus Plasma membranes Mitochondria Ribosomes Rough and smooth Endoplasmic reticulum Golgi apparatus Lysosome Cytoskeleton	Centrioles Small vacuoles No chloroplasts No cell wall

1.4.1 The structure of plasma membranes p. 21

1

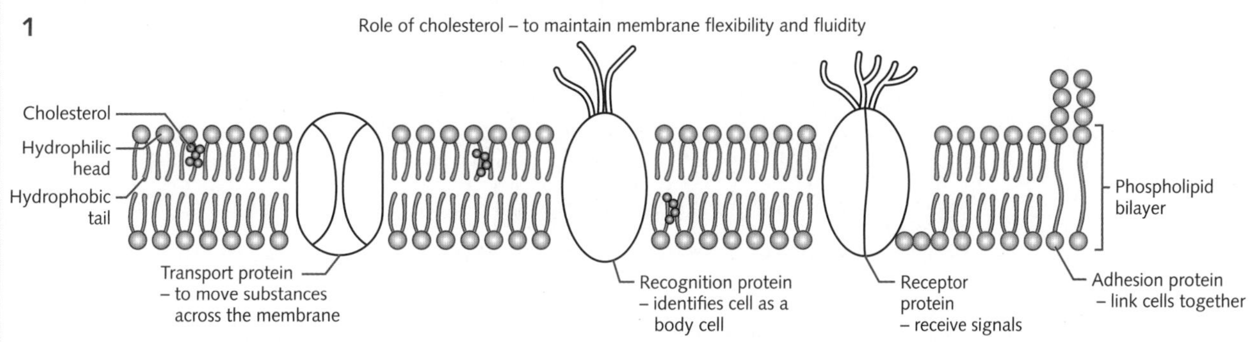

2 The bilateral phospholipid layer describes hydrophilic heads facing the external and internal environments of a cell, while the hydrophobic tails face each other. This structure means that charged molecules find it difficult to move through via diffusion; the layer is also tightly packed and so large molecules cannot move easily either. Proteins, however, are embedded within the membrane and can transport or channel selective molecules and ions from one side to the other. Therefore, the plasma membrane is selectively permeable.

3 The hydrophobic tails are repelled by the watery environment of the inside and outside of the cell, while the hydrophilic heads are attracted to the water and shield the tails from contact with water.

4 This allows cells to be able to change shape, needed for cell movement, to form vesicles needed for endocytosis and exocytosis and to allow cell division to occur.

5 Answers may vary; yes, mosaic describes how the plasma membrane is composed of many different components and fluid describes how all these parts are able to move and are not fixed in place.

1.5 Passive movement across membranes p. 22

1.5.1 Diffusion p. 22

1 a solute, gas, high, low, energy, equilibrium

b concentration, gradient

c equilibrium, solute, gas

d solute, solvent

2

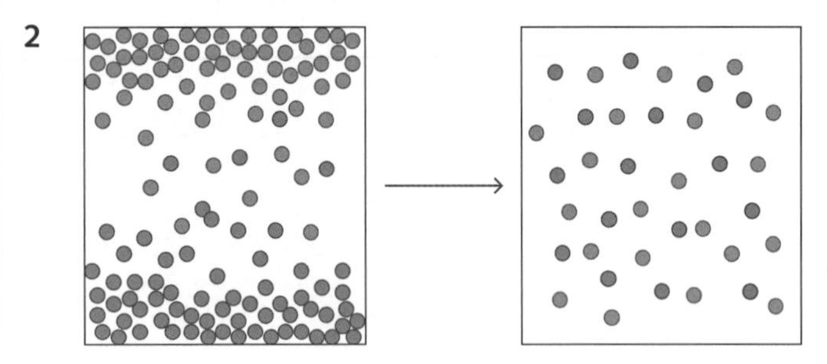

3

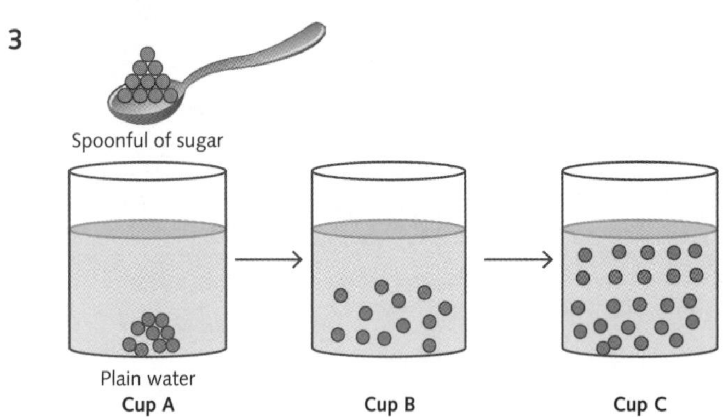

4

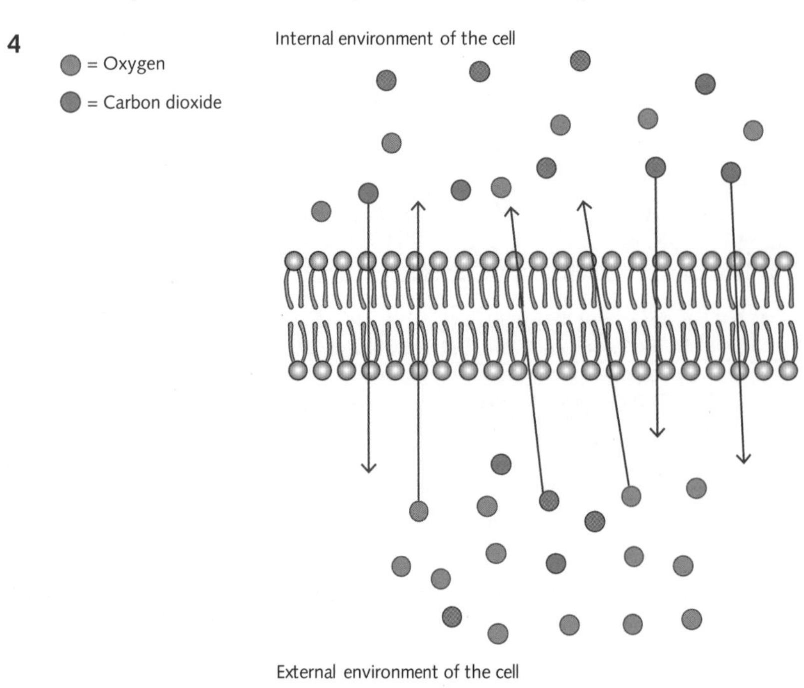

5

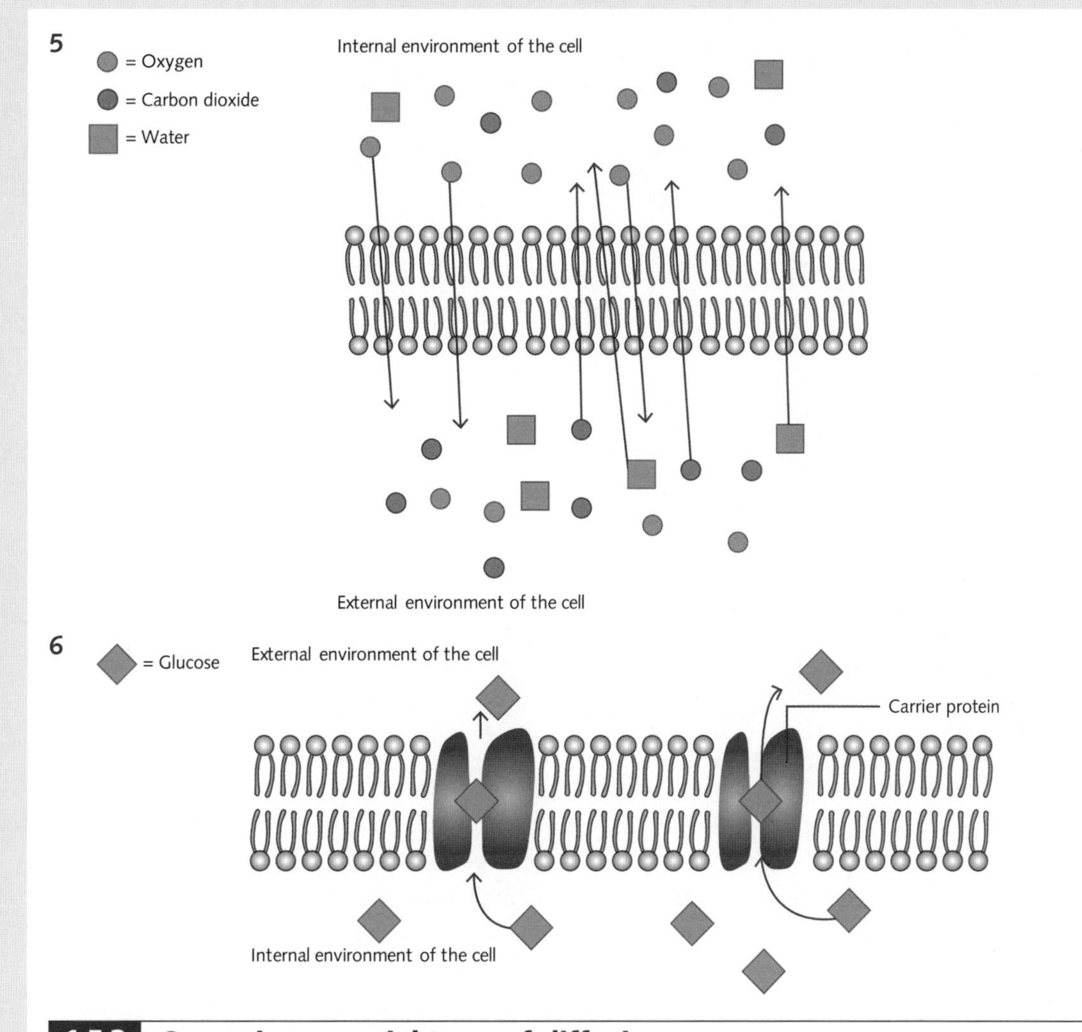

6

1.5.2 Osmosis: a special type of diffusion

p. 24

1 a water, solvent

b solvent, selectively permeable, gradient, energy

2

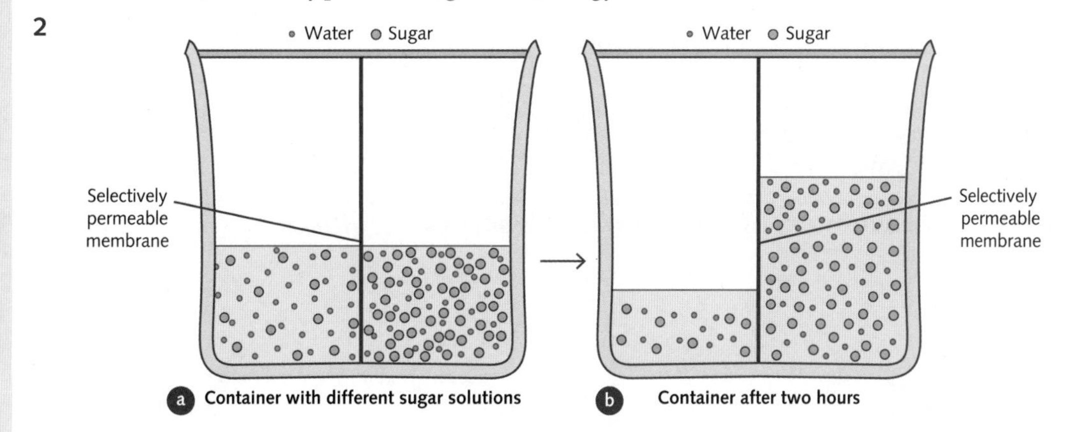

a Container with different sugar solutions

b Container after two hours

3

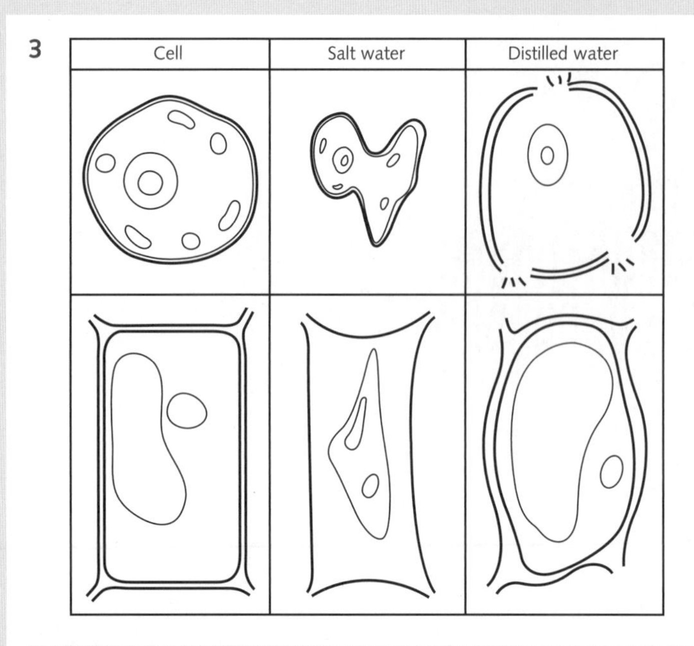

1.6 Movement across membranes using energy p. 26

1.6.1 Active transport p. 26

Student answers will vary. Use of an analogy would be best way to explain to a Year 7 student.
Main points to cover:

- humans are made up of billions of smaller structures called cells (students in year 7 learn about cells)
- each cell has to take in things such as oxygen and sugar and get rid of things such as carbon dioxide to stay healthy
- the diagram is showing part of the barrier of a cell and how things move across that barrier from the outside to the inside of the cell
- imagine you have a balloon full of water, you can't add any more water because the balloon is already full, but you can tip some of the water into an empty cup as it has room to take the water (or some similar analogy). This is called a concentration gradient – the difference in the amount of water in the balloon and the cup. The cell is moving the glucose from where there is not much of it, to where there is lots of it – like trying to put more water into the balloon.
- This needs energy to pump more water in, like an electric pump.
- This energy comes from breaking apart the glucose to release the energy stored within.
- The energy is released from the glucose in the mitochondria and moves to the carrier proteins (the pump) to move the glucose across the cell boundary.

1.6.2 Bulk transport: movement of large molecule across membranes p. 28

Scenario 1

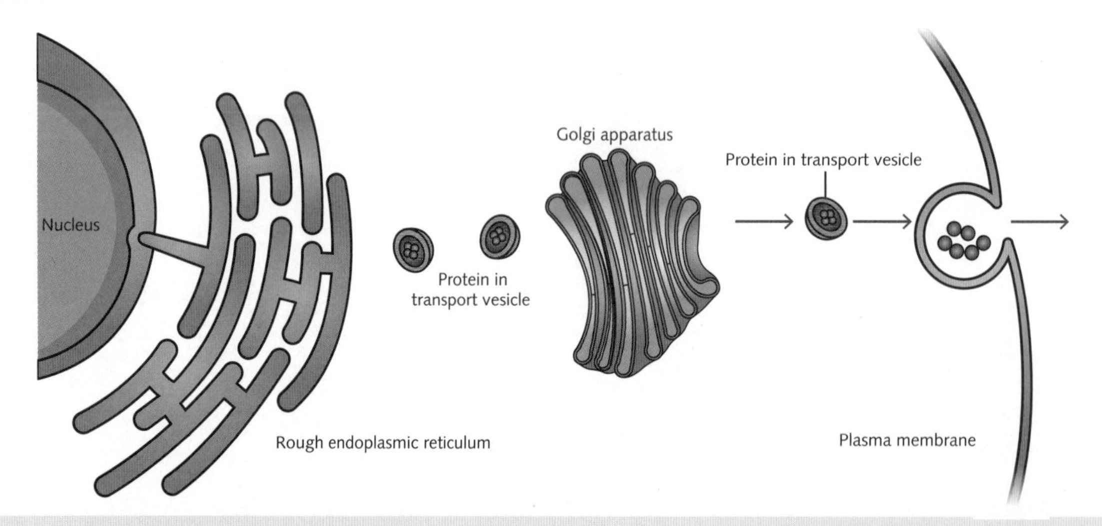

Scenario 2

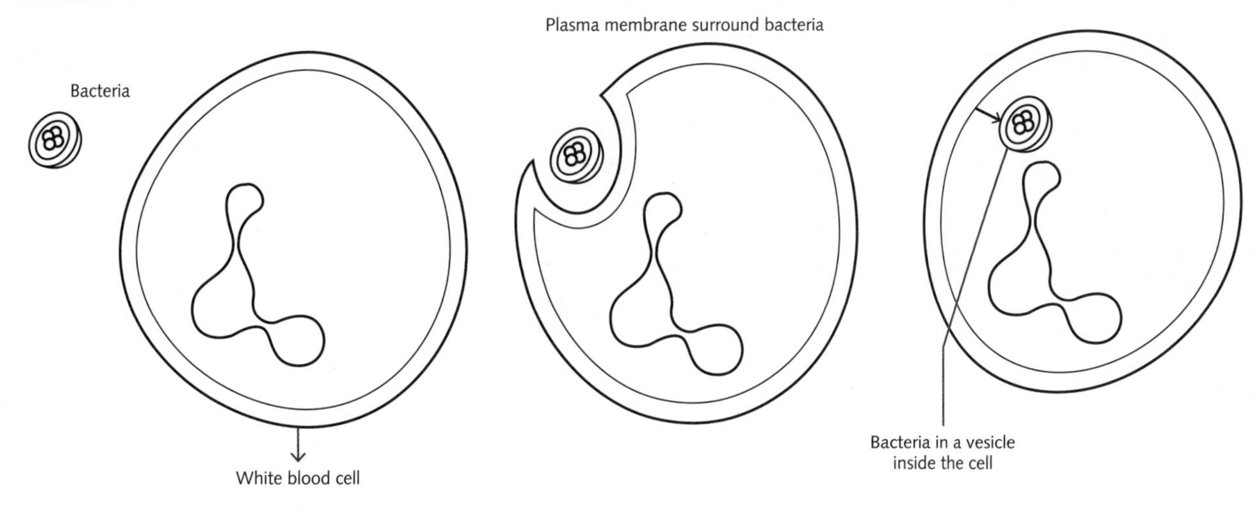

1.6.3 Comparing and contrasting diffusion, osmosis and active transport p. 29

Venn diagrams may vary, for example:

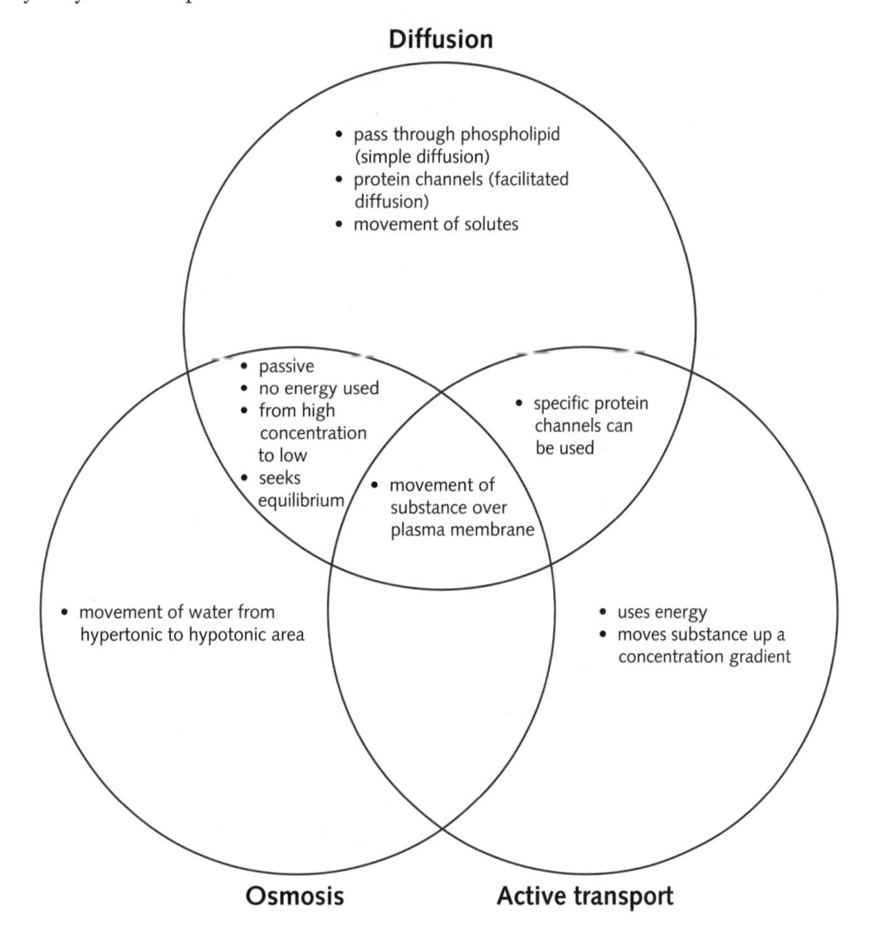

1.7 Chapter review

p. 30

1.7.1 Key terms

p. 30

BEGINNING	What do you get when you mix a solute with a solvent?
Solution	What is the name of the model used to describe the structure of the plasma membrane?
Fluid mosaic model	What is a single-celled organism called?
Unicellular	What is the term for 'water fearing'?
Hydrophobic	What is the name of the organelle in a cell where photosynthesis occurs?
Chloroplast	What is the liquid part of a solution called?
Mitochondria	The movement of water through a selectively permeable membrane is called what?
Diffusion	Where in a eukaryotic cell does respiration occur?
Eukaryotic	What is the movement of a gas or a liquid from an area of high concentration to an area of low concentration called?
Metabolism	What is a cell that has a membrane-bond nucleus and organelles called?
Solvent	What is the name given to all the activities that occur within a cell?
Osmosis	What is the name given to the difference in concentration of a substance from one area to another?
Concentration gradient	What is an organism made up of more than one cell called?
Multicellular	Movement of a substance along a concentration gradient is called ____.
Passive transport	What is a cell with no membrane-bound organelles?
Prokaryotic	What is a form of diffusion that uses carrier molecules embedded in the plasma membrane?
SA:V	A membrane that allows substances to pass through is said to be what?
Solute	What is the relationship between the amount of plasma membrane and cytoplasm in a cell?
Selectively permeable	What is the substance that is dissolved in a solvent to make a solution called?
Active transport	Which type of membrane allows some substances to pass through but not others?
Facilitated diffusion	What is the type of transport molecules that moves them from an area of low concentration to an area of high concentration?
Permeable	What is it called when molecules move equally in all directions?
Equilibrium	END

9780170452632

1.7.2 Practice test questions p. 33

1 D

2 B

3 C

4 A

5 D

6 a A: phospholipid; B: protein

b Concentration gradient (1); the sodium ions would move into the red blood cells (1)

c Osmosis

d There is a concentration gradient between the outside and inside of the cell. There is a higher concentration of water inside the cell, so water leaves the cell to try to equalise the concentrations.

Chapter 2 The cell cycle p. 35

Remember p. 35

1 Cells

2 Prokaryotes and eukaryotes

3 Prokaryote cells are smaller, simpler and do not contain membrane-bound organelles. Eukaryote cells are larger, more complex and contain membrane-bound organelles.

4 Centrioles are rod-like structures only found in animal cells.

5 Chromosomes are made up of DNA, which is organised into genes. Genes carry genetic information between generations.

6 Genes are sections of DNA that code to produce a specific protein.

2.1 Binary fission p. 36

Answers will vary, for example:

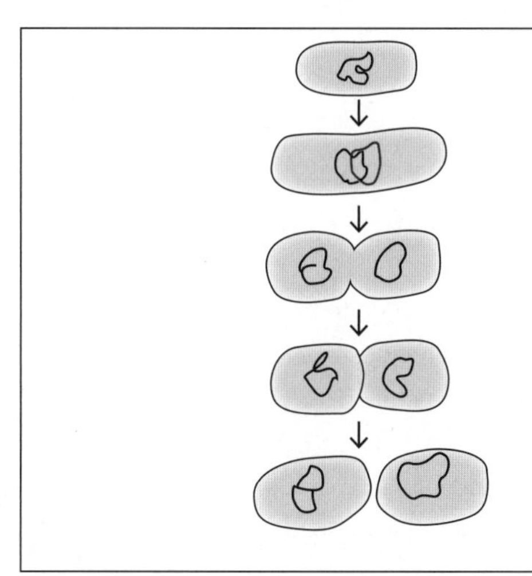

- A prokaryote cell contains one circular chromosome called a nucleoid; this is its DNA.
- When the cell wants to reproduce itself, it must first copy its nucleoid: it now has two identical nucleoids.
- The cell starts to divide in half, the two nucleoids move away from each other, the cytoplasm is split in two and the cell starts to build a new cell wall down the middle between the two cells.
- The cell is divided and now there are two identical cells that each contain identical DNA.

2.1.2 Effect of temperature on binary fission – part A p. 36

1 Answers may vary; for example, to investigate if the rate of binary fission in *E. coli* is different at 4°C compared to 35°C.

2 Answers may vary; for example, will a bacterial colony grow faster due to a higher rate of binary fission at 35°C compared to 4°C?

3 a The temperature of the environment, 4°C or 35°C

b The size of the bacteria colony, diameter of colony or % of plate covered

c Controlled variables should include: same species and strain of bacteria; same culture medium; same amount of culture medium in same sized plate; same amount of bacteria added to plate at the start of the experiment; same amount of time.

4 Answers may vary; for example, if the bacterial colony is incubated at 35°C, then the growth of the bacterial colony will be larger than for a colony incubated at 4°C.

5 Answers may vary; for example, the rate of binary fission will be higher at 35°C and the bacterial colony will have grown to cover more of the plate than the bacterial colony kept at 4°C.

2.1.3 Effect of temperature on binary fission – part B p. 37

1 A controlled experiment – this will show how the independent variable affects the dependent variable; will show a relationship between the I.V. and the D.V.

2 Answers may vary; for example,

1. Collect 10 identical sterile agar plates.
2. Keep the lids on the plates to reduce contamination.
3. Label five plates with 35°C and five plates with 4°C.
4. Use a pipette to measure 1 mL of bacterial solution and place the solution in the middle of the first plate, place the lid on the plate and seal with tape.
5. Repeat step 4 for the following nine plates.
6. Place the 35°C plates in an incubator set to 35°C. Place the other plates in a fridge at 4°C.
7. After 48 hours, remove the plates and record the % of each plate covered by bacterial growth.

3 Answers may vary; for example,

What are the risks in doing this experiment?	How can you manage these to stay safe?
Bacteria may cause disease.	Wear lab coats, safety glasses, gloves, and tie back hair. Keep the lid on the vial of the bacteria colony when not being used. Wash hands thoroughly at the end of the experiment. Disinfect benches and other equipment used before and after the experiment.
The bacteria will grow on the agar plates.	Make sure plates are well sealed, do not open the plates. Autoclave plates to destroy bacteria and dispose of correctly.
Disposable gloves are an allergy risk.	Use low allergenic gloves.

4 Answers may vary depending on how you have decided to measure your results; for example,

Quantitative – measuring diameter of the bacterial colony, counting the number of colonies, determining percentage of agar plate covered.

Qualitative – rating coverage of the plates as high, medium, low etc.

5 a Answers may vary; for example, having more than one bacterial plate at each temperature and averaging results will give more accurate results; using a grid to estimate percentage coverage; using appropriate measuring equipment.

b Answers may vary; for example, using measuring equipment correctly; using a systematic method to ensure all colonies counted once only; using consistent method to estimate percentage coverage.

c Answers may vary; for example, have more than one agar plate at each temperature to ensure the results are repeatable.

d Answers may vary; for example, control all variables except the independent variable, e.g. same number of bacteria; same type of bacteria; same amount, and type of agar in the agar plates; incubate for the same time period.

2.1.4 Effect of temperature on binary fission – part C p. 39

1

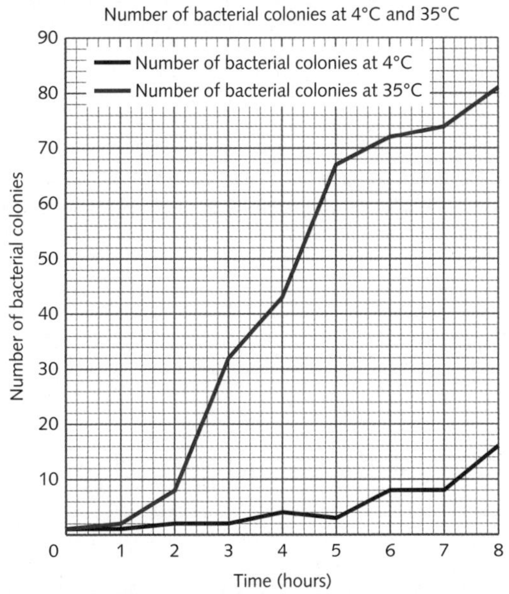

Number of bacterial colonies at 4°C and 35°C

— Number of bacterial colonies at 4°C
— Number of bacterial colonies at 35°C

Number of bacterial colonies

90
80
70
60
50
40
30
20
10
0

1 2 3 4 5 6 7 8

Time (hours)

2 Supports the hypothesis, the number of colonies at 35°C grew exponentially (increased at a much larger rate) compared to at 4°C. After 8 hours there were more than 80 bacterial colonies at 35°C compared to only 16 at 4°C, which shows that the rate of binary fission in the bacteria was much higher at 35°C.

3 Answers may vary; for example, using precise tools to measure the number of colonies, estimate area of plate covered by bacteria, reduce the time to make counting of colonies more accurate/precise.

4 Food should be stored at low temperatures, below 4°C, to reduce food spoilage due to bacteria.

2.2 Eukaryotic cell cycle p. 41

2.2.1 The cell cycle p. 41

1 Yes. Evidence could include: cell volume increases up to time 5, when it halves/returns to the volume of time 0, showing the cell has divided into two. The amount of DNA doubles in the S phase, to prepare for cell division, and then is divided into two new cells in the M phase, mitosis.

2 Times 1 to 2, the cell is growing as the volume is increasing, the amount of DNA is stable

3 Times 3–5, the DNA has been copied in the S phase, the cell continues to grow in volume, the amount of DNA remains stable

4 At the start of the M phase, prophase, time 5

5 Time 5, the graphs show the amount of DNA is the cell halving, providing one set of chromosomes to each daughter cell and the volume of the cell halving as the cell splits into two daughter cells.

6 5 units of time

2.2.2 The phases of mitosis in eukaryotic cells p. 42

Phase	Image	Description
Interphase	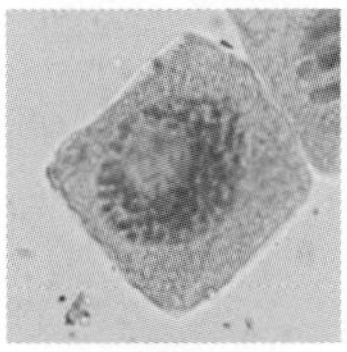	The G_1, S and G_2 stage of the cell cycle. The cell is growing and carrying out its function. Chromosomes are long and unwound, chromatin threads, and cannot be distinguished. The nucleus and nucleolus are visible.
Prophase	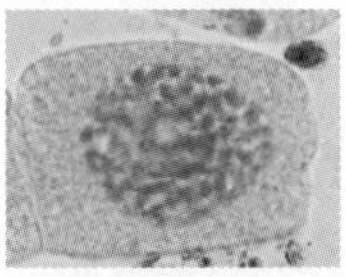	The nuclear membrane has broken down and the chromatin has started to condense and become pairs of chromatids held together by a centromere.
Metaphase	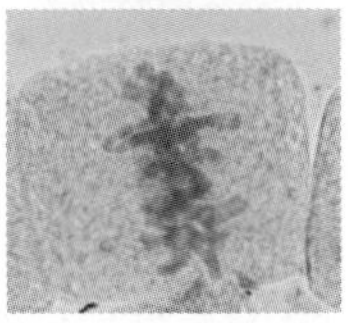	The chromosomes are arranged across the equator (centre) of the cell. The centromere of each chromosome is attached to a spindle fibre.

(continued)

Phase	Image	Description
Anaphase		The spindle fibres contract, pulling the chromatids they are attached to towards the opposite poles of each cell, this separates the sister chromatids of each chromosome, these are each called chromosomes.
Telophase		The DNA in each chromosome begins to unwind and become less visible as chromatin again, a new nuclear envelope forms around each set of chromosomes, nucleoli are visible, and the spindle disassembles.

1 Each cell has an independent cell cycle, therefore at any given time different cells in the same tissue can be in a different stage of the cell cycle.

2 Cell repair, to replace dead cells, growth

3 Cytokines and then G_1, new organelles and proteins are produced, the cell grows and carries out its function

2.2.3 Cell cycle checkpoints p. 47

1 Cyclin E, the concentration of this cyclin increases before the S phase, which is when DNA replication occurs, and declines after the S phase is completed.

2 Cyclin B, the concentration of this cyclin increases just prior to mitosis, the spindle is assembled at the start of mitosis and decreases after mitosis when the spindle is disassembled.

3 Cyclin D, this cyclin is present in all stages, its concentration alters in different stages, increasing in G_1, remaining stable in S and decreasing in G_2 and mitosis.

2.3 Apoptosis p. 48

2.3.1 When apoptosis fails: HeLa cells p. 48

1 Answers may vary. Duty- and/or rules-based approach: the scientists that first established the HeLa cell line should have obtained informed consent from Henrietta or the Lacks family. This is an ethical rule that should have been followed.

2 Answers may vary. Consequences-based approach: there was no financial gain from establishing the HeLa cell line and the cell line is freely available to scientists, leading to benefits to many including the understanding cancer, AIDS virus, impacts of space travel. The many benefits gained from the use of the HeLa cell line could be weighed up against the harm done to the Lacks family from the use of Henrietta's cells without permission.

2.4 Disruption to the regulation of the cell cycle p. 49

2.4.1 Proto-oncogenes and tumour suppressor genes p. 49

Answers may vary:

- Proto-oncogenes and tumour suppressor genes play an important role in regulating the cell cycle, these two groups of genes are involved in promoting and inhibiting cell division.
- Proto-oncogenes stimulate cell division by producing proteins that speed up cell growth and cell division. Proto oncogenes are important for promoting normal growth and development of healthy organs and tissues.
- Tumour suppressor genes produce proteins that inhibit or slow down cell growth, they are involved in repairing DNA damage and can promote apoptosis.
- These two groups of genes therefore act in opposite ways in regulating the cell cycle, with proto-oncogenes accelerating cell growth and division and tumour suppressor genes putting the brakes on cell division. The opposite actions of these two groups of genes keep cells dividing at the correct rate for healthy tissues.

2.4.2 Action of mutagens on the cell cycle p. 51

Answers may vary.

	Types of mutagens		
	Chemical	Biological	Physical
Examples	Chemicals in tobacco smoke Drugs Toxins Teratogenic agents	Virus	UV light X-rays Nuclear radiation
Possible effects	Mutations in tumour suppressor cells Mutations in proto-oncogenes Cancer Physical birth defects	Mutations in proto-oncogenes and tumour suppressor genes Cancer Autoimmune disease	DNA damage Cancer

2.5 Stem cells p. 51

2.5.1 Bioethical issues: embryonic stem cells p. 51

1 Undifferentiated cells that have the potential to replicate and develop into many types of cells

2 Pluripotent. Totipotent stem cells can create all cell types to create an organism and all the cell types needed to develop and nourish the embryo (placenta, membranes needed for embryonic development). Pluripotent stem cells can create all the cell types of an embryo. Multipotent stem cells can differentiate into cells of one tissue type.

3 Embryonic stem cells are harvested from embryos, which are destroyed in the process.

4 Consequences-based approach: embryos used are unwanted and would be destroyed anyway. The stem cells harvested from these embryos can be used for the research and development of treatments for many life-threatening conditions, preventing many deaths and improving the quality of life of many.

5 Answers may vary: will treatments developed from stem cells be affordable for all; stem cell therapies could improve the life quality of many; stem cell therapies could reduce cost of health care for many; there will be more employment in the field of stem cell research; are the treatments 100% safe.

2.6 Chapter review p. 53

2.6.1 Key terms p. 53

1 a A parent cell divides by mitosis to produce two daughter cells.

b Pluripotent stem cells can give rise to all cell types of an organism, multipotent stem cells can only give rise to a limited number of cell types.

c Adult stem cells can be found in the body tissues of organisms and are multipotent; embryonic stem cells come from early embryos and are pluripotent.

d A mutagen can be a chemical, biological, or physical agent that can increase the rate of mutations; a mutation is a change/mistake in the DNA sequence.

e Binary fission is the process by which a prokaryotic cell divides to produce two identical cells; mitosis is nuclear division in eukaryotic cells.

f A chromatid is one strand of a chromosome, a chromatid contains condensed chromatin – long strands of DNA. Chromatids are visible when a cell is dividing, the rest of the time DNA is organised into chromatin.

g In humans, the first 8 weeks of the developing organism after fertilisation is called the embryo, after which it then is called a foetus.

2.6.2 Practice test questions p. 54

1 B

2 C

3 C

4 A

5 A

6 D

7 D

8 a In animal cells, centrioles are involved in forming the spindle fibres; plant cells do not have centrioles, spindle fibres form out of the centrosome (1). Cytokinesis are different in plant cells, they need to form a cell plate to separate the cells – this becomes the new cell wall; due to the absence of cell walls animal cells simply pinch apart with the aid of a contractile ring of microtubules. (1)

b 38 (but each chromosome would now be made up of two chromatids)

9 a Apoptosis or necrosis

b Apoptosis/necrosis

c If apoptosis fails to destroy cells with DNA damage

Chapter 3 Functioning systems p. 56

Remember p. 56

1 Answers may vary.

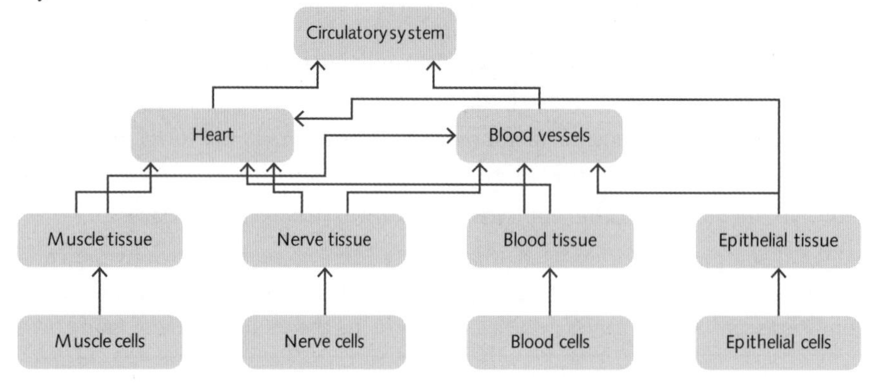

2 Xylem – carries water and minerals. Phloem – carries sugars

3 a Water and carbon dioxide

b Proteins, lipids, carbohydrates

4 Food must be broken down to small molecules to allow it to be absorbed through intestine walls into the bloodstream to be carried to cells.

5 a Carbohydrates, lipids and proteins

b Urea, carbon dioxide

6 Diffusion, osmosis, active transport, facilitated diffusion

3.1 Vascular plants p. 57

3.1.1 Water transport: xylem p. 57

1 Evaporation/transpiration

2 Capillary action

3 Osmosis

4 Water enters root cells via osmosis, this helps to push water up the xylem vessels.

5 Water is drawn up the xylem vessels due to adhesion – water molecules are attracted to the sides of the xylem vessels and this pulls other water molecules up due to cohesion (water molecules are attracted to each other).

6 Water evaporates through open stomata; this helps to pull water molecules up the xylem vessels.

3.1.2 Water loss from the shoot system p. 58

Experiment A: varying intensity of light source

1 Example: To investigate how different light intensity affects the rate of transpiration in plants

2 Does changing light intensity affect the rate of transpiration in plants?

3 a Level of light intensity

b Amount of water lost by transpiration

c Environmental temperature, humidity, surface area of the leaves, type of plant, amount of air movement, time

4 Example: If the light intensity is increased, then the rate of transpiration in the plant will also increase.

5 As the light intensity increases, the amount of water passing through the plant and evaporating will increase.

Experiment B: effect of air movement at varying speeds

6 Example: To investigate how altering the amount of air movement affects the rate of transpiration in plants

7 Does changing the amount of air movement affect the rate of transpiration in plants?

8 a Amount of air movement

b Amount of water lost by transpiration

c Environmental temperature, humidity, surface area of the leaves, type of plant, light intensity, time

9 Example: If the amount of air movement is increased, then the rate of transpiration in the plant will also increase.

10 As the amount of air movement increases, the amount of water passing through the plant and evaporating will increase.

Experiment C: effect of warm air

11 Example: To investigate how changing air temperature affects the rate of transpiration in plants

12 Does changing the temperature of the air around a plant affect the rate of transpiration in plants?

13 a Environmental/air temperature

b Amount of water lost by transpiration

c Light intensity, humidity, surface area of the leaves, type of plant, amount of air movement, time

14 Answers may vary. For example: If the light intensity is increased, then the rate of transpiration in the plant will also increase

15 As the air temperature increases, the amount of water passing through the plant and evaporating will increase.

3.1.3 Nutrient transport: phloem p. 60

1

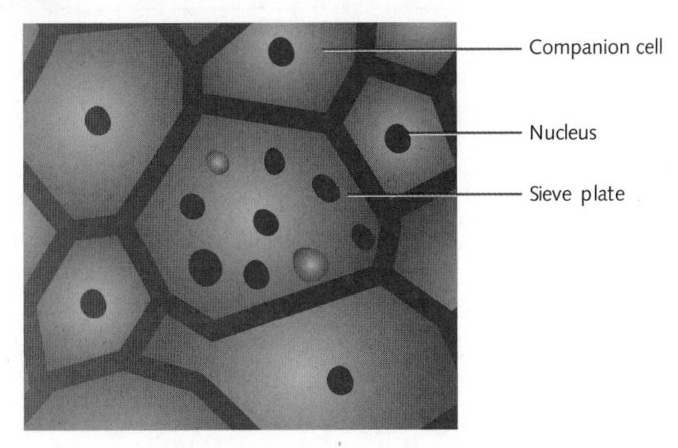

2 Sieve tube cells are elongated and join to form long tubes. They have sieve plates at each end of the cell, which are perforated to allow the cytosol of the cells to flow from one sieve tube cell to another. Sieve tube cells do not contain a nucleus, thus providing more room for sugars. The nuclei of adjacent companion cells control cellular processes in the sieve tube cells.

3 Mesophyll cells

4 Mesophyll cells, source cells move sucrose into the sieve tube cells of the phloem by active transport, this creates pressure that pushes the sucrose through the sieve tube cells, it moves through sieve plates to get to the next sieve tube cell; when the sucrose reaches the sink cells or root cells it moves into them from the phloem.

3.1.4 Translocation in plants p. 63

1 The rate of translocation is constant at 12 arbitrary units (a.u.) until PCMBS is added and then the rate of translocation decreases exponentially.

2 Photosynthetic cells/mesophyll cells (source cells) – no longer able to push sugars into the phloem, this would stop translocation.

Root tissues – if they can no longer take up sucrose from the phloem, this would slow the rate of translocation

Phloem tissues – could affect companion cells or sieve tube cells to slow translocation.

3 Data that has been collected by somebody else

4 Rate of translocation would not decrease much as the graph is showing a negative exponential decrease in the rate of translocation and by 120 minutes the decrease has flattened the rate would be between 3–4 a.u. The plant could die, as the cells would not be receiving enough sucrose to provide energy for them to carry out functions needed to keep them alive; if cells die, the plant will die.

3.1.5 Interaction of transpiration and translocation in plants p. 64

1 a To produce xylem and phloem cells/tissue

b Answers could include: prevent water loss from the tissues; prevent the tubes from collapsing or bursting when under pressure; provide support for the plant.

c

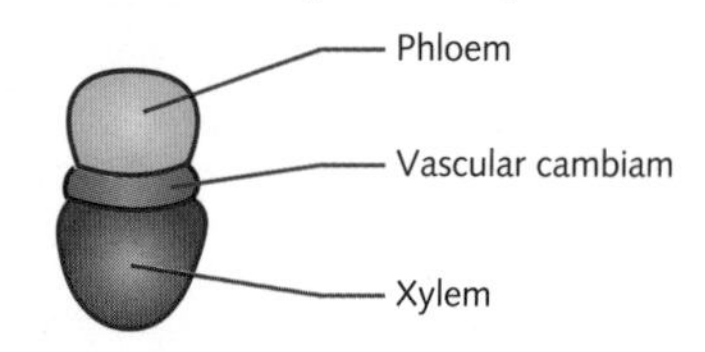

2

	Xylem	Phloem
Which way does the material flow?	Up	Up and down
Are the cells living?	No	Yes
Do the cells contain cytosol?	No	Yes
Do the cells have a nucleus?	No	No
What types of materials are transported?	Water and minerals	Sugars and amino acids
What is the end wall of each cell?	None	Sieve plate

 9780170452632

3.2 Mammalian systems: digestive system p. 66

3.2.1 Mechanical digestion p. 66

1

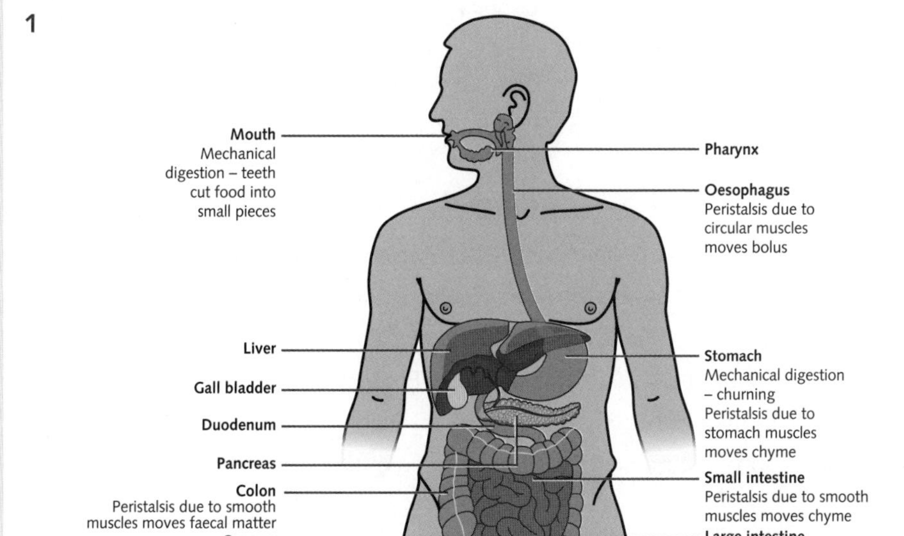

Location in digestive tract	Structures involved	Contents
Mouth	Teeth, tongue, epiglottis	Bolus
Oesophagus	Smooth circular muscles	Bolus
Stomach	Stomach muscles, sphincter muscles	Chyme
Duodenum	Smooth muscles, pyloric sphincter muscle	Chyme
Jejunum	Smooth muscles	Chyme
Ileum	Smooth muscles	Chyme
Large intestine	Smooth muscles	Chyme
Anus	Anal sphincter	Faeces (undigested food, e.g. bacteria, bile, cellulose)

3.2.2 Chemical digestion p. 68

1 Mouth: amylase breaks down carbohydrates

Stomach: pepsin (protease) breaks down protein

Pancreas: trypsin breaks down protein, amylase breaks down carbohydrates, lipase breaks down fats/lipids

Small intestine: trypsin breaks down protein, amylase breaks down carbohydrates, lipase breaks down fats/lipids

2

1	absorb nutrients
2	producing amylase
3	mouth
4	teeth
5	tongue
6	oesophagus
7	longitudinal
8	stomach
9	peristalsis
10	cardiac sphincter
11	chyme

12	small intestine
13	colon
14	faeces
15	anus
16	pancreas
17	liver
18	bile
19	duodenum
20	jejunum
21	ileum
22	absorb water

3.2.3 Digestive enzymes

p. 70

1 a Conclusions could include the following points:

- Starch is broken down to maltose by saliva. Water does not break starch down to maltose, so it is something other than water in the saliva that is breaking down the starch. (Tubes 1 & 2)
- Low pH (acidic conditions) inhibits the ability of saliva to break down starch to maltose. (Tube 3)
- Boiled saliva is no longer able to break down starch to maltose. (Tube 4)

b Conclusions could include the following points:

- Hydrochloric acid on its own does not break down protein (egg white). (Tube 4)
- Pepsin breaks down protein best at low pH (in the present of acid) and at 37°C. (Tube 2)
- If pepsin is in alkaline conditions, protein is not broken down. (Tube 5)
- At 0°C, pepsin will not break down protein, even in acidic conditions. (Tube 3)

2 a Enzyme: pepsin Location: stomach Optimum pH: 2 Substrate: proteins Product: amino acids

b Enzyme: amylase Location: mouth Optimum pH: 6 Substrate: starch Product: maltose

c Enzyme: trypsin Location: small intestine Optimum pH: 9.5 Substrate: proteins Product: amino acids

3.2.4 Absorption in the human gut

p. 72

Answers may include the following points:

- Most digested food is absorbed from the digestive system into the blood system in the small intestine; only alcohol and some drugs will cross the stomach walls to enter the bloodstream.
- The small intestine has thin walls and a large surface area due to its length and the many microscopic finger-like projections called microvilli that are found on the inner lining of the small intestine. Both features make it well suited for absorption.
- The microvilli have a large network of capillaries and lymph vessels, called lacteals that can absorb the nutrients from the small intestine.
- In the small intestine, glycerol and fatty acids are absorbed into the lacteals and end up in the lymphatic system. Water, mineral ions, monosaccharides, peptides, and amino acids are absorbed into the capillaries over the thin walls of the microvilli and are carried into the bloodstream.

3.3 Mammalian systems: endocrine system p. 74

3.3.1 Endocrine system p. 74

1, 2, 3

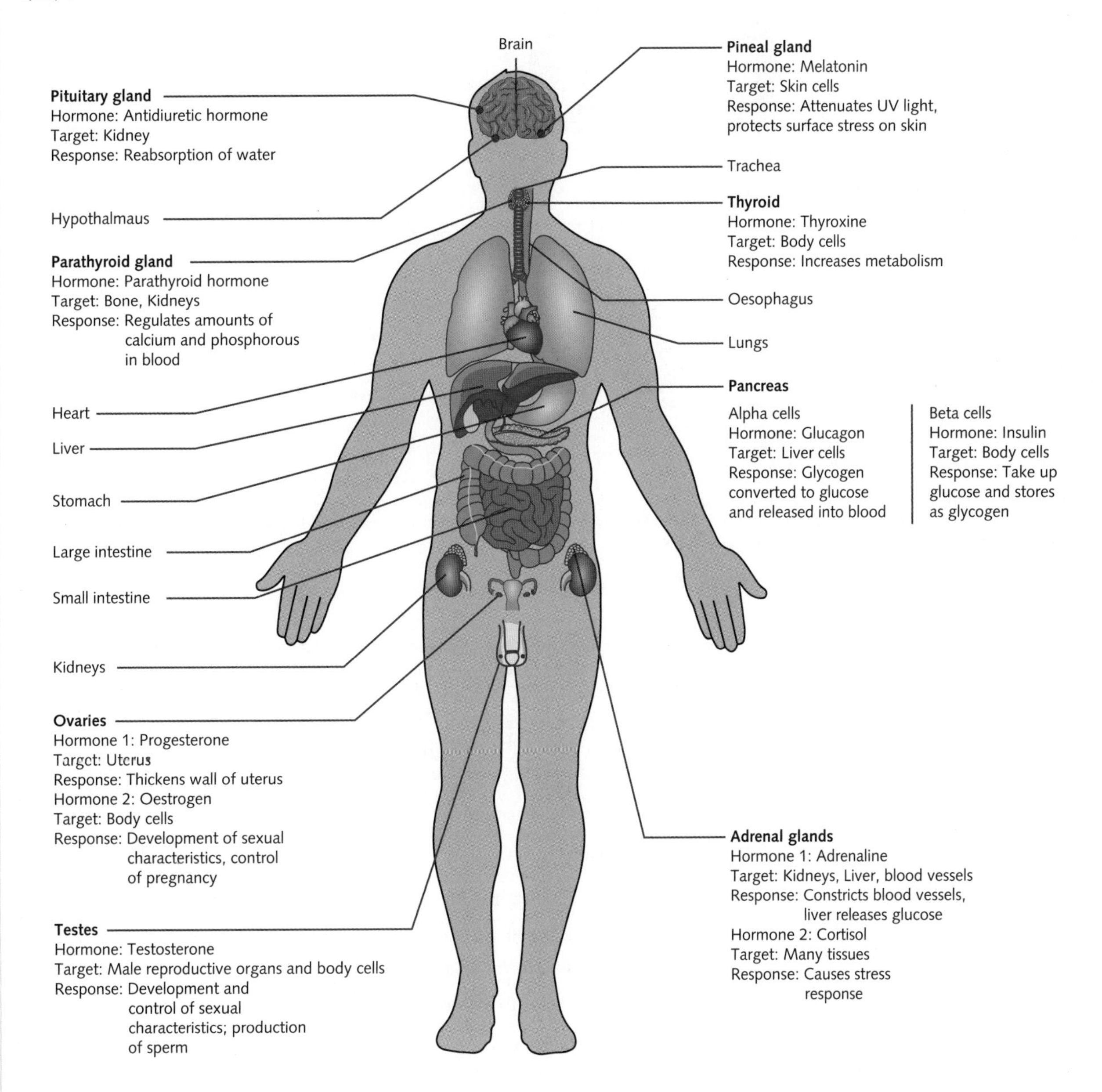

3.4 Mammalian systems: excretory system p. 76

3.4.1 Structure of the excretory system p. 76

1

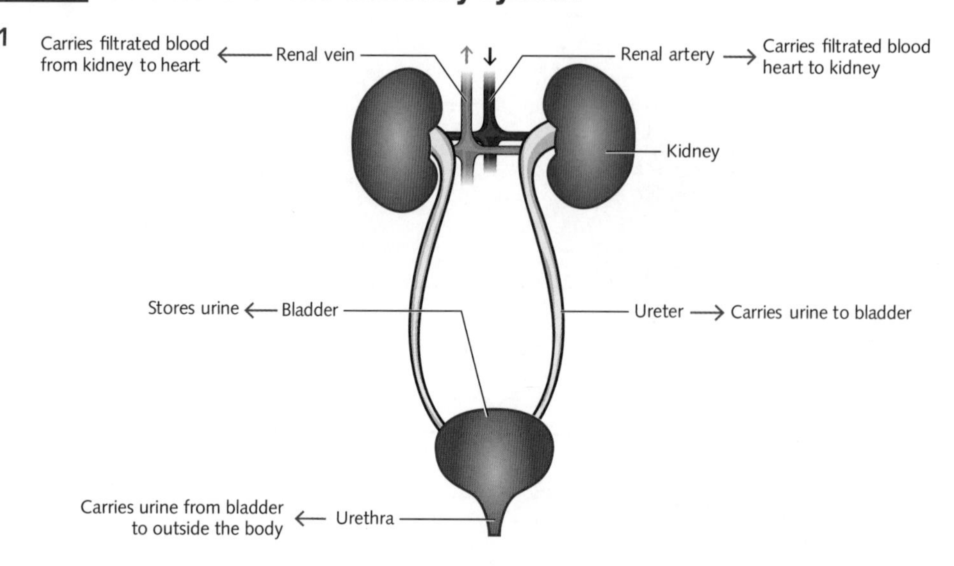

2 Answers may vary, for example:

Filtration of blood – high pressure of blood in glomerulus forces fluids and dissolved substances into the Bowman's capsule.	→	Filtrate moves into proximal tubule where glucose, salt, water, and ions is reabsorbed into capillaries by active and passive transport.
		↓
In the ascending loop, salt, ions and water are reabsorbed by both passive and active transport.	←	Filtrate moves into loop of Henle. In the descending loop, water is reabsorbed into capillaries by osmosis.
↓		
The filtrate moves into the distil tubule, more salt moves back into capillaries by active transport, water is also reabsorbed by osmosis.	→	Filtrate – containing excess water and waste becomes urine in the collecting duct and travels to renal pelvis of the kidney.

3.4.2 Composition of fluids in the kidney p. 78

1 a Proteins are too large to leave the blood vessels and enter the bowman's capsule so are never part of the filtrate.

b All the water is filtered out of the blood but not all the other components, large proteins remain in the blood vessels; this means water will be at a higher concentration in the filtrate.

c As the filtrate moves through the nephron, water is removed and reabsorbed back into the blood stream, meaning the urine becomes more concentrated.

d Other compounds that contain Nitrogen

e The blood plasma and urine contain different concentrations of ions, urine contains a greater concentration of urea, chlorine and N containing compounds than blood vessel leading to a different pH.

f Glucose is needed by the body for cellular respiration, so it is all reabsorbed back into the blood stream, energy is used to reabsorb all the glucose by active transport.

g Ions are reabsorbed in the proximal tubule by passive transport and in the distal tubule of the nephron by active transport, this allows the correct concentration of each ion to be returned to the blood.

3.5 Chapter review p. 80

3.5.1 Key terms p. 80

1 Example

Vascular plants	adhesion, cohesion, cuticle, epidermis, external environment, internal environment, lignin, pH, phloem, parenchyma, root hair cell, root pressure, stomata, terrestrial, tracheid, translocation, transpiration, transpiration stream, transpirational pull, vascular bundle, vascular tissue, xylem, xylem vessel element, vascular plant
Animals	absorption, ammonia, amylase, anus, bile, basal metabolic rate, Bowman's capsule, chemical digestion, chyme, colon, deamination, digestion, digestive system, distal tubule, ductless gland, egestion, endocrine gland, endocrine system, excretion, exocrine gland, filtrate, faeces, gall bladder, gastric juice, gastrointestinal tract, glomerulus, heterotroph, hormones, ingestion, kidney, lacteal, large intestine, loop of Henle, lymph, lymphatic system, mechanical digestion, microvilli, nephron, oesophagus, organ, pancreatic juice, peristalsis, polypeptide, protease, proximal tubule, pyloric sphincter, rectum, renal artery, renal pelvis, small intestine, sphincter, thyroxine, tissue, villi, urea

2

Plants	
Structural terms	**Functional terms**
cuticle, epidermis, lignin, phloem, parenchyma, root hair cell, stomata, tracheid, vascular bundle, vascular tissue, xylem, xylem vessel element, vascular plant	adhesion, cohesion, root pressure, translocation, transpiration, transpiration stream, transpirational pull

Animals	
Structural terms	**Functional terms**
anus, Bowman's capsule, colon, digestive system, distal tubule, ductless gland, endocrine gland, endocrine system, exocrine gland, gall bladder, gastrointestinal tract, glomerulus, kidney, lacteal, large intestine, loop of Henle, lymphatic system, microvilli, nephron, oesophagus, organ, proximal tubule, pyloric sphincter, rectum, renal artery, renal pelvis, small intestine, sphincter, villi, polypeptide	absorption, amylase, basal metabolic rate, bile, chemical digestion, chyme, deamination, egestion, excretion, faeces, filtrate, gastric juice, heterotroph, hormones, ingestion, lymph, mechanical digestion, pancreatic juice, peristalsis, protease, polypeptide, hormone, thyroxine, urea

3 Plant terminology

Structural terms	Functional terms	Diagram
cuticle, epidermis, stomata	transpirational pull, transpiration, transpiration stream	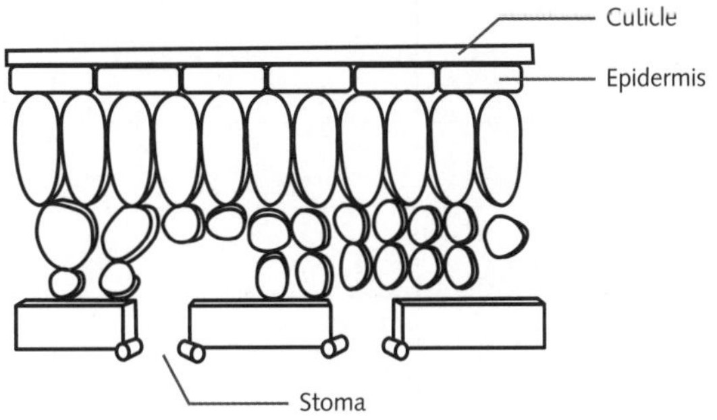
xylem, xylem vessel element, lignin, tracheid, parenchyma, tissue	adhesion, cohesion	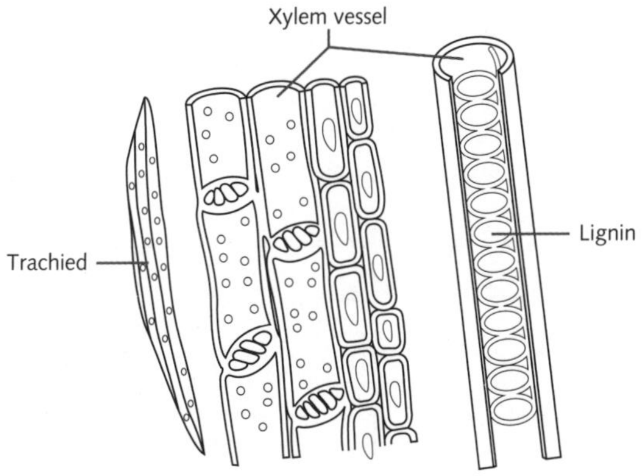

(continued)

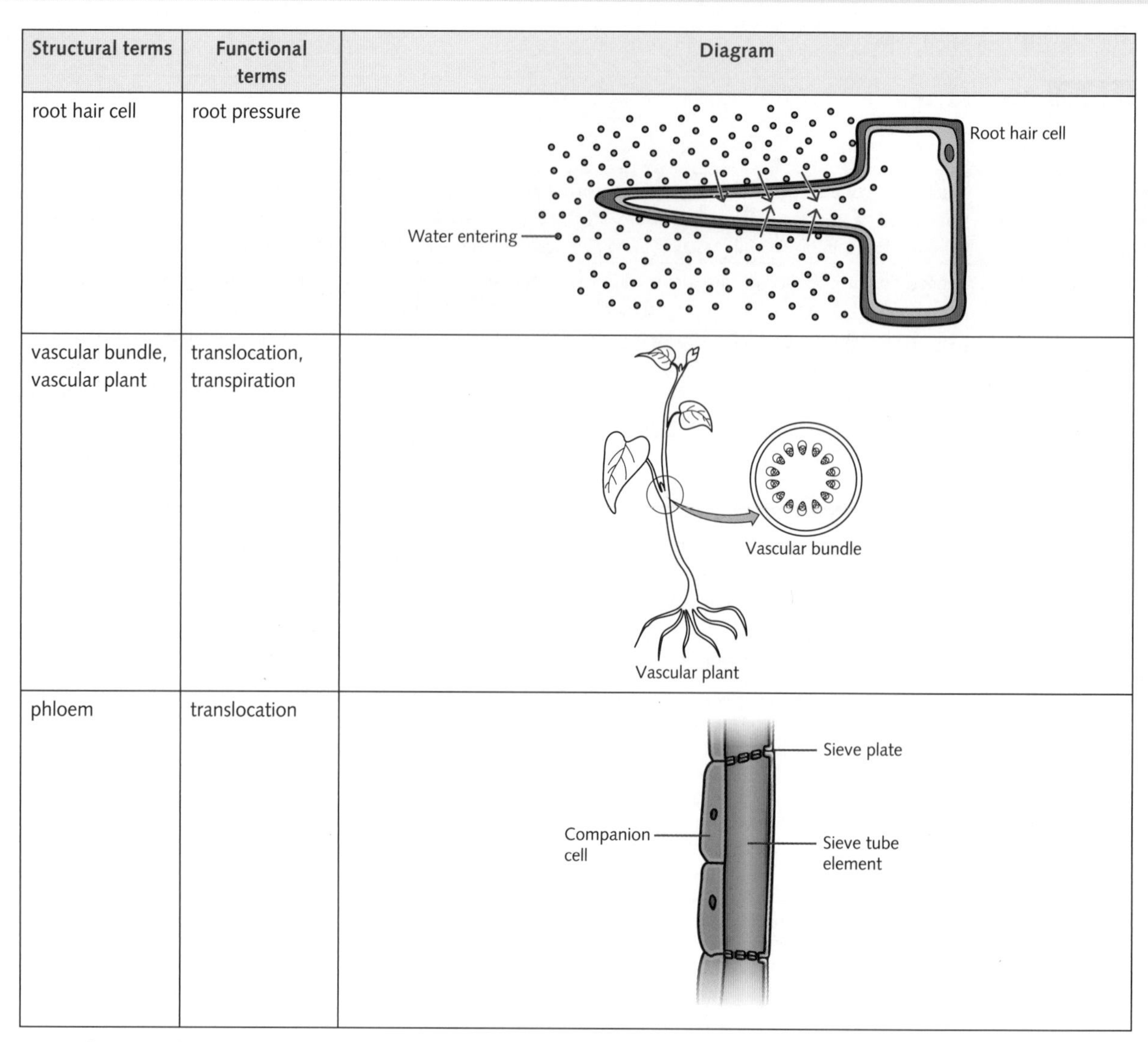

Structural terms	Functional terms	Diagram
root hair cell	root pressure	Water entering; Root hair cell
vascular bundle, vascular plant	translocation, transpiration	Vascular bundle; Vascular plant
phloem	translocation	Sieve plate; Companion cell; Sieve tube element

Animal terminology

Structural terms	Functional terms	Diagram
ductless gland, endocrine gland, endocrine system	hormone, thyroxine, basal metabolic rate	Cell of endocrine gland; Hormones; Target cell responds
kidney, nephron, renal artery, renal pelvis	excretion	Renal pelvis; Renal vein; Renal artery; Kidney; Nephron

(*continued*)

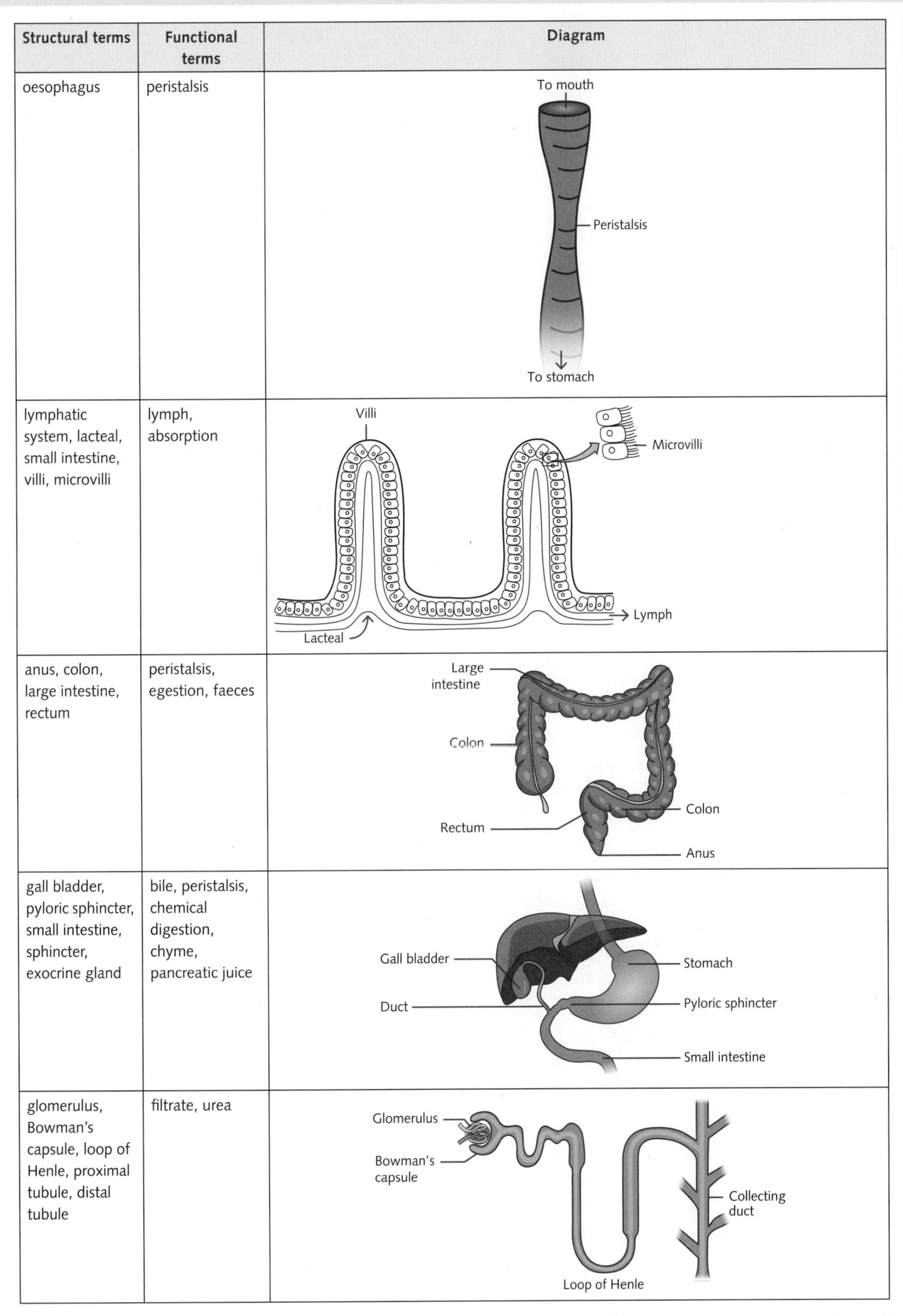

Structural terms	Functional terms	Diagram
oesophagus	peristalsis	
lymphatic system, lacteal, small intestine, villi, microvilli	lymph, absorption	
anus, colon, large intestine, rectum	peristalsis, egestion, faeces	
gall bladder, pyloric sphincter, small intestine, sphincter, exocrine gland	bile, peristalsis, chemical digestion, chyme, pancreatic juice	
glomerulus, Bowman's capsule, loop of Henle, proximal tubule, distal tubule	filtrate, urea	

(continued)

Structural terms	Functional terms	Diagram
digestive system, gastrointestinal tract	amylase, protease, absorption, amylase, bile, chemical digestion, chyme, deamination, digestion, gastric juice, ingestion, heterotroph, mechanical digestion, polypeptide	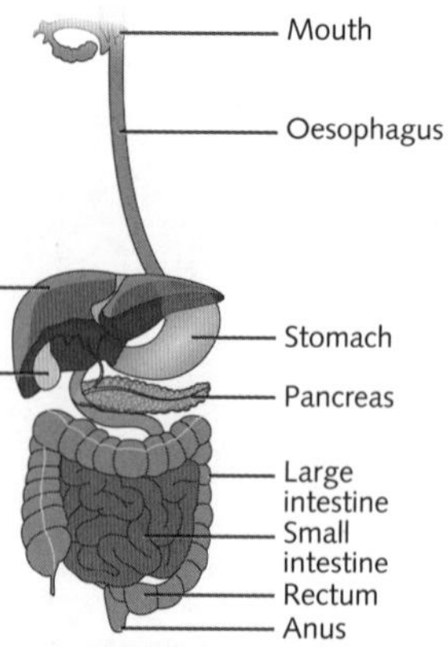 Mouth, Oesophagus, Liver, Gall bladder, Stomach, Pancreas, Large intestine, Small intestine, Rectum, Anus

3.5.2 Practice test questions p. 84

1 B

2 C

3 C

4 C

5 C

6 A

7 **a** Two of the following: temperature, humidity, air movement, light intensity

b The weight change for each leaf would be small: 1 mg is 1/1000 of a gram. By measuring 10 leaves at once the weight change will be larger and more likely to be measured accurately.

c The untreated leaves lost mass due to the evaporation of water through their stomata, as the stomata were not covered with petroleum jelly (1). The leaves lost water rapidly for the first 40 minutes, shown by a rapid loss in mass (1), the loss in mass decreased after 40 minutes and plateaued as the stomata in the leaves would have closed due to water loss (or all water has already evaporated from the leaves) (1).

d The leaves in set Y had their upper surface covered with petroleum jelly, this meant that water could only evaporate from the lower surface of the leaf (1). This meant that less water evaporated from these leaves and they lost less mass than the leaves in set X (1).

8 As the filtrate passes down the descending loop of Henle, water from the filtrate moves back into the bloodstream by osmosis (is reabsorbed back into the bloodstream); (1) thus, the filtrate becomes more concentrated (1). In the ascending loop of Henle, salt is reabsorbed back into the bloodstream by passive and active transport (1). The removal of this salt from the filtrate causes it to become less concentrated or more dilute (1).

Chapter 4 Regulation of systems p. 86

Remember p. 86

1 Water moves in xylem vessels from roots to the leaves. The water is pushed up the xylem by pressure from water entering the roots, adhesion and cohesion of water molecules draw water up the xylem vessels and it is also pulled up the xylem vessels when water evaporates from the leaves.

2 Produce hormones, which are secreted from ductless gland into the blood, they travel in the blood and deliver messages to target cells and tissues leading to a response.

3 Nephrons in kidneys filter blood and remove excess water and waste, which is excreted as urine.

4.1 Regulation of water balance in vascular plants p. 86

4.1.1 Water balance in vascular plants p. 86

1 Answers may vary, for example:

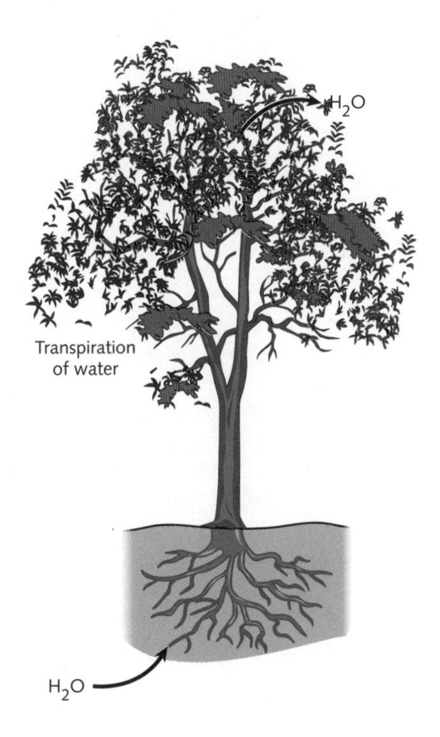

Explain what transpiration is:

- Water enters plants through the roots and moves up the plant stem through special tubes called xylem vessels to the leaves, where water evaporates through tiny openings in the leaf surface called stomata. This movement of water is called the transpiration stream.
- More detail about what happens at each step, explaining how water moves from roots to leaves.
- At the roots – plants have a very large root system that provides a large surface area for water to enter the roots. Water moves from the soil into the roots by a process called osmosis.

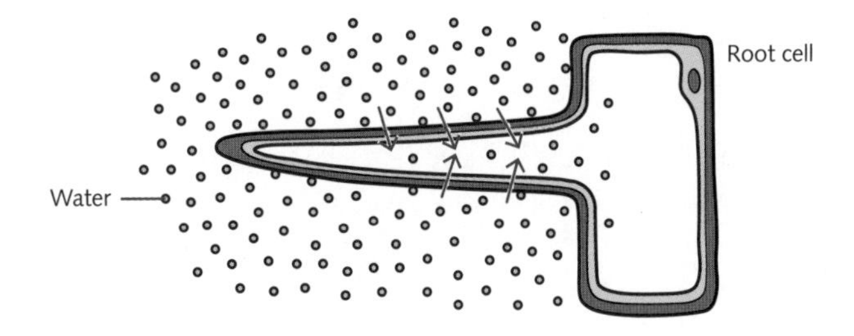

- In the stem: once the water enters the roots it is pushed by the pressure of more water entering the root cells into the xylem vessels.

  Xylem vessels are microscopic tubes formed by long empty dead cells, tracheids are one of these cell types, joined end on end forming a continuous tube that carries water from the roots to the leaves, providing water to all parts of the plant.

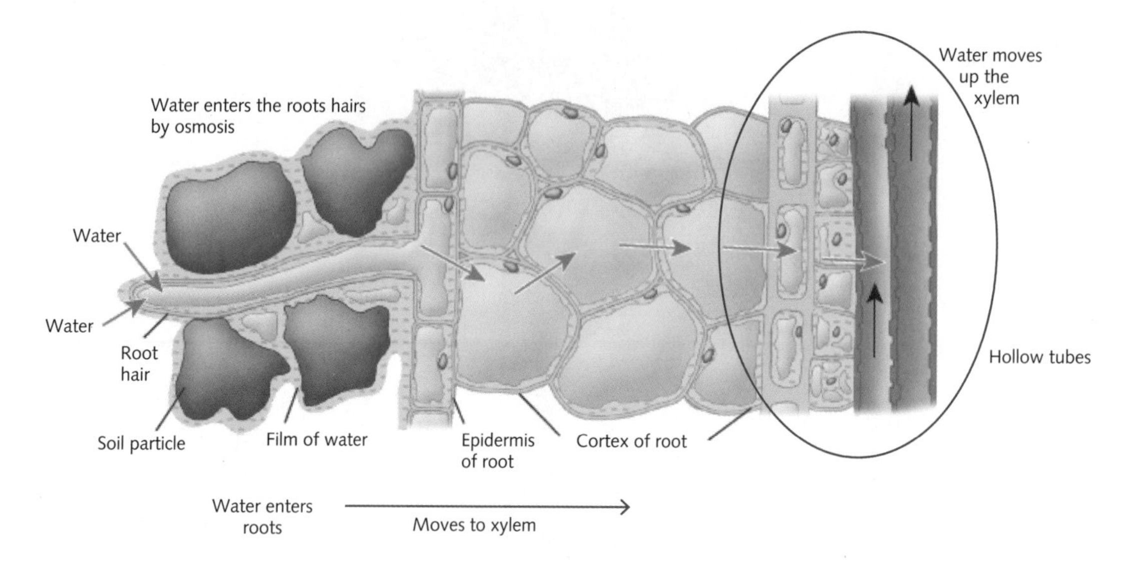

Water is kept moving up the tubes that form the xylem vessels by:

- a push from the roots, more water entering the roots pushes the water up the xylem.

  Water molecules are sticky and form hydrogen bonds with each other and other surfaces. They are attracted to the sides of the xylem vessels and are pulled up the xylem, much like fluids are pulled up a straw – this is called adhesion. Water molecules are also attracted to each other and stick together – this is called cohesion. See below.

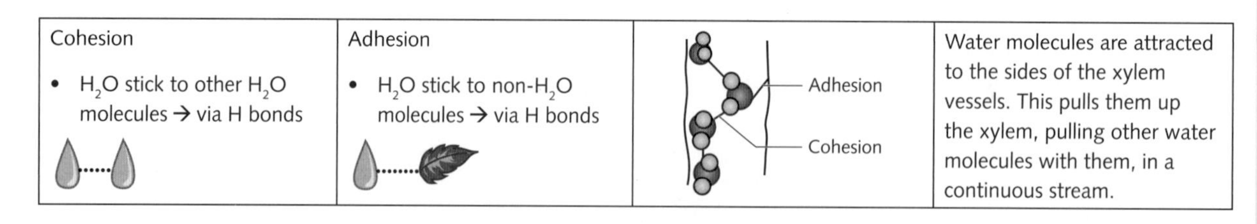

Cohesion	Adhesion		
• H_2O stick to other H_2O molecules → via H bonds	• H_2O stick to non-H_2O molecules → via H bonds	Adhesion Cohesion	Water molecules are attracted to the sides of the xylem vessels. This pulls them up the xylem, pulling other water molecules with them, in a continuous stream.

3 The end of all xylem vessels is the leaves of a plant, leaves have features that prevent water loss, such as a waterproof outer layer but they also have tiny openings called stomata that are needed for gas exchange – oxygen and carbon dioxide. Water can evaporate through these tiny openings – this is called transpiration.

Water evaporating through the stomata also creates a pull that helps to pull water up the xylem vessels.

The amount of water lost from leaves by evaporation is affected by the environmental conditions:

- More water evaporates on dry days compared to humid days. This is because the air can only hold a certain amount of water vapour, if the air is dry around the plant then more water is able to evaporate from the leaf into the air.
- More water evaporates on windy days compared to still days, as the water evaporates from the leaves it makes the air around the plant more humid, if this air is moved away from the plant by wind then the air around the leaves will be able to keep absorbing more water vapour from the plant.

A higher loss of water from the leaves creates a greater water potential in the plant – meaning there is more water in the soil/roots compared to in the leaves – this will cause more water to move through the xylem provided there is enough water in the soil.

If there is no water in the soil, then water will not enter the roots and the flow of water will slow, and plants can wilt as the leaves do not receive enough water. Leaves can close the tiny holes in their surface (stomata) to prevent water loss in these conditions.

4.1.2 Stomatal control and water balance p. 90

1 a A pair of curved cells that surround the stomata and control if the stomata is open or closed

b A plant hormone called abscisic acid, a chemical messenger/signal molecule

c The presence of ABA/abscisic acid causes the closure of stomata.

d Blue light is an input for photosynthesis: the presence of blue light allows photosynthesis to occur, in the chloroplasts inside the guard cells. As the sugar concentration inside the guard cells increases, water moves in due to osmosis. This causes the guard cells to become turgid, which causes the stomatal pore to open.

2 a Water to exit guard cells and stoma to close

b This process prevents water loss from the plant, maintaining water balance in a plant when there is a lack of water. This prevents the plant going into water stress and death of cells and potentially, the plant.

3 They move from an area where there is a high concentration of calcium and potassium ions across the plasma membrane to an area where there is a lower concentration of these ions. The process is passive and does not require the use of energy.

4.2 Human survival in a range of environments p. 92

4.2.1 Internal and external environment of the body p. 92

1

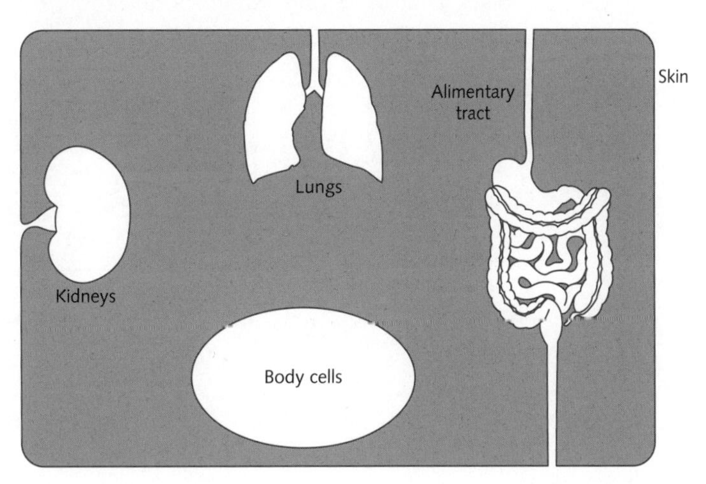

Figure 4.6

2 a–d

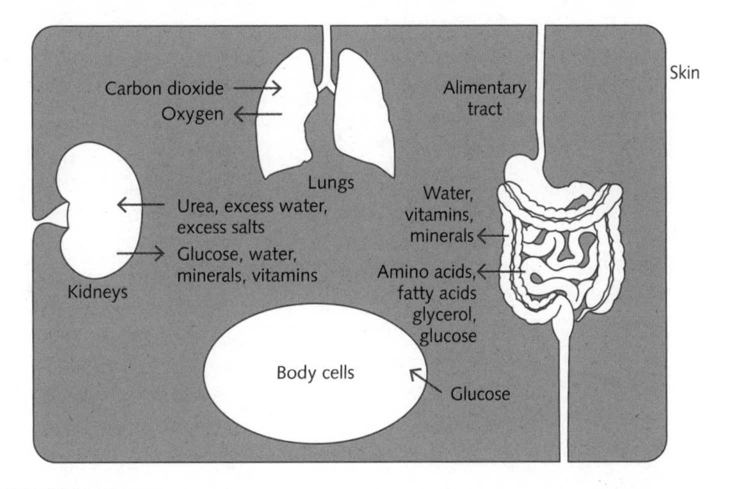

Figure 4.6

4.2.2 Optimum and tolerance ranges p. 93

1 a No, all these organisms would have different temperature tolerance ranges. The optimum temperature range for polar bears, which live in the Arctic, which has lower average temperatures, would be lower than that of humans which would be lower than the optimal range of the desert iguana, which live in a desert environment with higher average temperatures.

b The desert iguana would not survive in the Arctic as the environmental temperature would be much lower than the optimal range for the iguana. The temperature in the Arctic would be in the iguana's zone of intolerance and the iguana would die.

c No, the shark's environment would contain lower oxygen concentrations than were present in a human's environment. The optimum range of oxygen concentration would be lower for the shark compared to the human.

d The shark would die.

2 a The blackstripe topminnow has a larger tolerance range for oxygen concentration than the blacktail shiner.

b The environment of the blackstripe topminnow has a greater variation in oxygen concentration so most likely lives at various water depths in the lake where the amount of oxygen varies. The blacktail shiner most likely inhabits the top levels of the lake where the oxygen is more abundant.

c The blacktail shiner can only survive in an environment that matches its narrow range of optimum oxygen concentration, whereas the blackstripe topminnow would be able to survive in a greater range of environments as it has a greater optimum range for oxygen concentration.

4.2.3 Stimulus–response model p. 95

1 Answers may vary, for example:

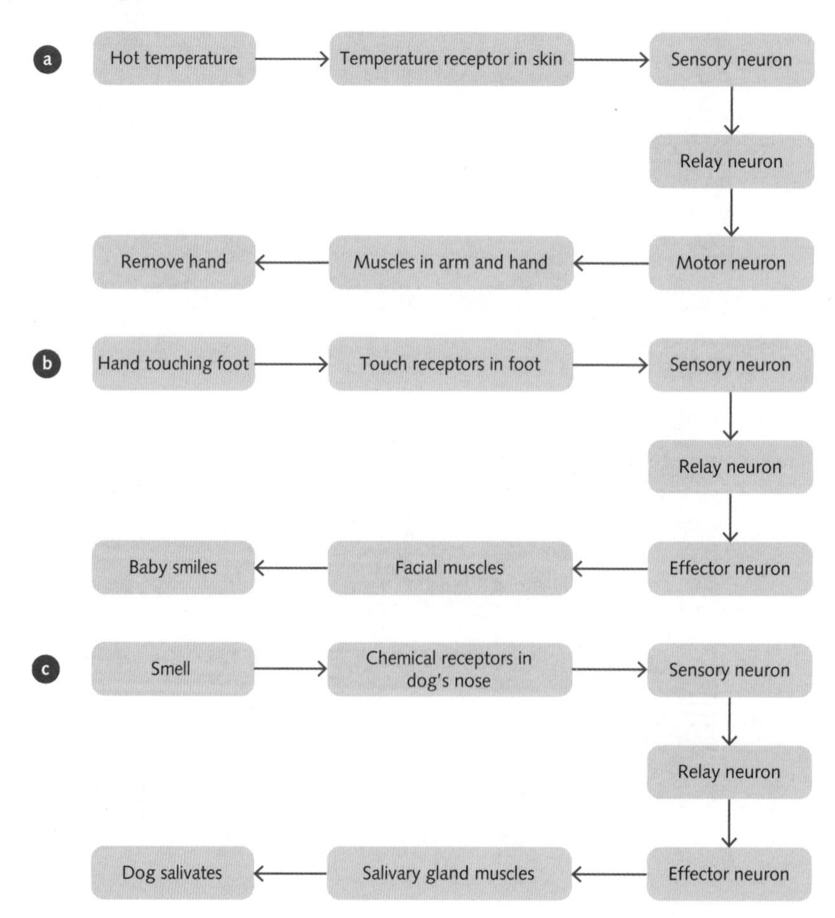

2 Answers may vary: provided a model for how to present the information

3 Answers may vary: working out what the response was

4.2.4 Detection of stimuli from the external and internal environments p. 97

1 a Temperature
  b Light touch
  c Extreme temperature/pain
  d Pressure/touch

2 Allow the organism to sense changes in its environment so it can respond appropriately

3 a Thermoreceptor
  b Mechanoreceptor
  c Mechanoreceptor
  d Mechanoreceptor

4.2.5 Homeostasis using negative feedback mechanisms p. 98

1 a

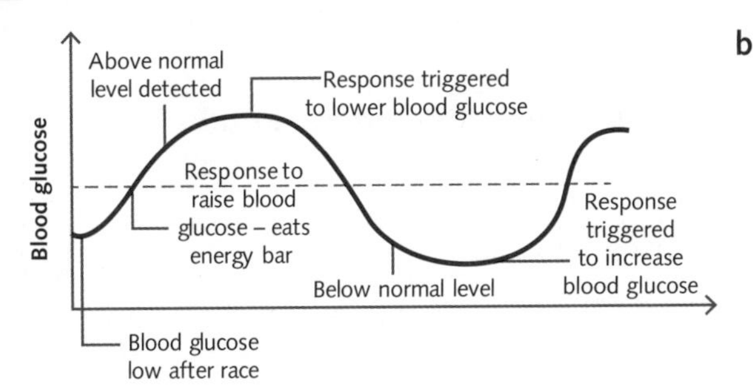

b

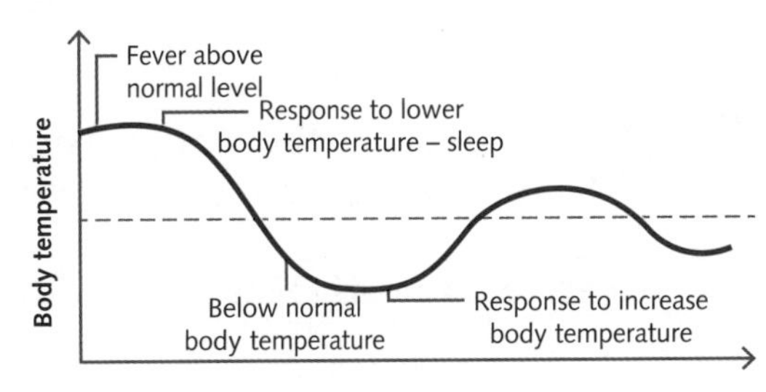

c

[Water] in blood plasma
Drinks 1 L water
Water level increases
Water level
Low water after run

2 a The stimulus of contractions causes the production of prostaglandins, which causes oxytocin to be secreted, this causes more oxytocin to be secreted in the brain, the oxytocin causes more contractions.
  b The response is adding to the original stimulus, the response increases the contractions
  c Oxytocin
  d Muscles in the uterus

4.3 Temperature regulation in the human body p. 100

a

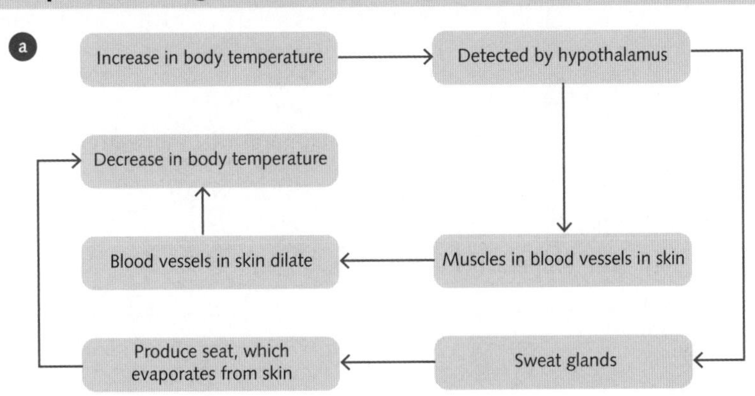

b

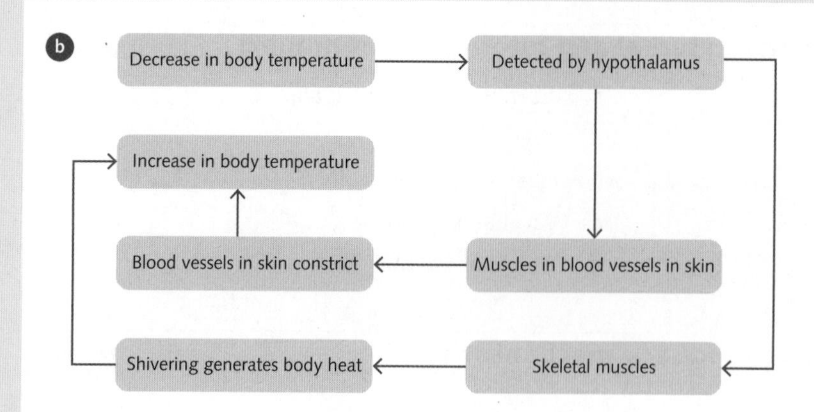

4.3.2 Investigating the body's response to a sudden drop in external temperature

p. 101

Investigation: investigating the body's response to cold

p. 101

Hypothesis may vary, for example: If there is a significant drop in external temperature, then the temperature of the hands will decrease.

I.V.: external temperature

D.V.: temperature of the hands

Controlled variables: same subject, same humidity, same air movement

Apparatus set-up

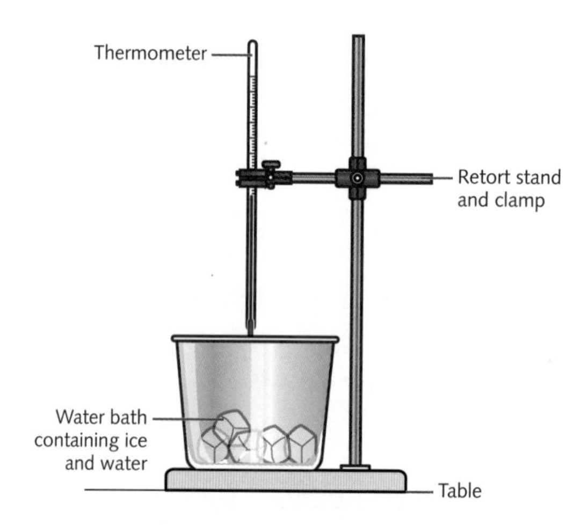

Sample method

1. Set up two water baths, one with water and ice, the other with water at room temperature.
2. Use a retort stand to hold a thermometer in each water bath to monitor the temperature of the water bath.
3. Record the temperature of the subject's right hand.
4. Immerse the subject's right hand in the room temperature water bath for 20 seconds.
5. Dry the subject's hand and immediately record skin temperature of the hand.
6. Repeat steps 4 & 5 for 10 minutes.
7. Give the subject a 3-minute recovery period.
8. Record the temperature of the subject's left hand.
9. Immerse the subject's left hand in the ice bath for 20 seconds.
10. Remove and dry the hand and record the temperature of the hand.
11. Repeat steps 9 & 10 for 10 minutes
12. Give the subject a 3-minute recovery period.
13. Repeat steps 1–12 using the water at room temperature.

Results

1

Table 1 Observations of subject after immersion of hand in ice water

Time after immersion (mins)	Observations	
	Right hand	Left hand
1	Hand pink and swollen, nails are blue	Hand pink and swollen, nails are blue
2	Hand looks worse	Subject reports hand hurting, subject is left-handed
3	Hand very swollen, white/red in colour, looks cold	
4		No change from 3 minutes
5		Hand is redder, subject is distressed
6		No change from 5 minutes
7		No change, subject cannot feel hand
9–10		No change
Recovery	Swelling reduces, hand regains some colour	

Other observations: subject distressed and crying during the recovery time

2

Table 2 Drop in temperature on body surface after immersion in ice water

Time after immersion (mins)	Right hand		Left hand	
	Temperature (°C)	% Temperature change	Temperature (°C)	% Temperature change
0	31.3	0	31.1	0
1	30.8	1.6	30.5	1.9
2	30.1	3.8	29.8	4.2
3	28.5	8.9	28.2	9.3
4	26.5	15.3	25.9	16.7
5	23.1	26.2	24.3	21.9
6	22.0	29.7	22.1	28.9
7	21.4	31.6	20.8	33.1
8	20.6	34.2	20.2	35
9	19.5	37.7	19.7	36.6
10	19.2	38.6	18.8	39.5
After recovery	25.4	18.8	24.9	21.5

3 Sample graph

Graph 1 Drop in temperature on body surface after immersion in ice water

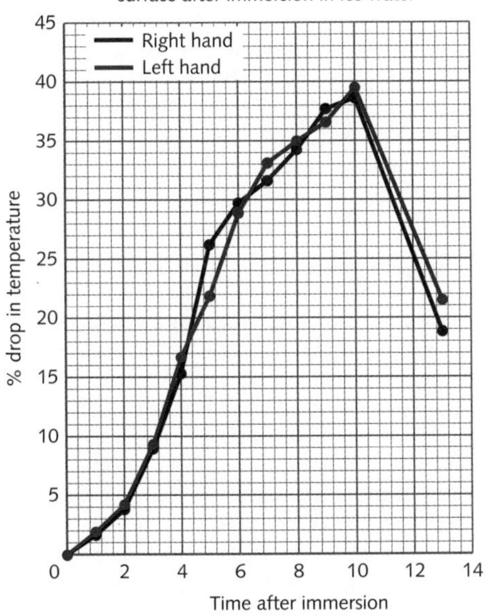

Discussion

1 They were close but not identical, the 0.2°C difference may have been due to inaccuracy in reading the thermometer.

2 Immersing the subject's hand in room temperature water to observe the effect on the temperature. A control is important to check that the observed results are only being caused by the independent variable – the ice water.

3 No, the temperature of the hands was still lower than the baseline temperature, more time is needed for the hands to return to the baseline.

4 The maximum % temperature change in the right hand was 38.6% compared to 39.5% in the left hand. The left hand had a 0.9% higher maximum temperature loss.

5 The drop in external temperature led to a decrease in the temperature of the hands, both hands showed a drop in temperature after being immersed in ice water, with the % decrease in surface temperature of the hands increasing over time as shown by graph 1. Need to compare to control data which is not given.

6 Observations of the subject show a change in the colour of the hand after immersion in ice, this would suggest there was a change in blood flow to the hand.

Blood flow to the hand in the ice bath would have been restricted due to constriction of the blood vessels going to the hand, leading to a drop in temperature of the skin of the hand.

7 No, the data for both hands match suggesting there were no errors or uncontrolled variables

8 Answers may vary, for example: What is the impact of hot water on the temperature of the hand? How long does it take for skin temperature to return to the base temperature? Do feet respond in the same way to exposure to a drop in temperature?

9 On their extremities, nose, ears, fingers, toes, hands, feet. As was observed in the experiment the body restricts blood flow to extremities if they are exposed to the cold, this enables the body to maintain its core temperature.

10 Motor nerves carry a message to the muscles around the walls of blood vessels, the motor nerves cause these muscles to contract, this constricts/narrows the blood vessels. Blood carries heat around the body, when it flows near the skin, heat can be lost to the surroundings if the environmental temperature is lower than the temperature of the blood. This lowers the temperature of the blood and would lower the body temperature of the organism. By constricting blood flow to the skin, less body heat is lost to the cold environment, which stops heat being lost from the blood and prevents the body core temperature from dropping.

4.3.3 Temperature regulation in other animals. p. 109

1 Answers may vary, for example:

Organism	Physiological	Structural	Behavioural
Penguin	Counter current blood flow in feet	Thick feathers	Huddling
Kangaroo	Vasodilation of vessels in forearms to allow temperature loss when kangaroos lick them	Areas of their body with little fur to allow heat loss	Crepuscular – most active at dusk and early morning when cooler
Lyrebird	High metabolic rate to generate body heat	Body shape, thick feathers	Roosting in high trees at night to escape cold of the ground

2 External temperature

3 Body temperature of ectotherm

4 Temperature of underground burrow

5 The graph shows that ectotherm's body temperature is gained from the environment. This is shown between 10am–2pm and 5pm–9pm when the animal is out of its burrow their body temperature rises due to the high environmental temperature. Otherwise its temperature matches that of its burrow.

6 Points to make:

Prior to 6am and up to 9am animal is asleep in its burrow.

10am the animal leaves its burrow to feed, lie on a rock, mate, find water and returns to its burrow at 2pm.

2pm–4pm the animal sleeps.

5pm the animal again leaves its burrow to feed, lie on a rock, mate, find water and returns to its burrow at 9pm.

10pm the animal returns to its burrow to sleep over night.

4.4 Regulation of blood glucose levels p. 111

4.4.1 Homeostasis for the regulation of blood glucose p. 111

1 a After a meal, receptors detect an increase in blood glucose.

The pancreas secretes insulin.

The liver removes excess glucose.

The excess glucose is converted into glycogen and stored in the liver.

Receptors detect a decrease in blood glucose.

The pancreas secretes glucagon.

The liver converts stored glycogen into glucose.

The liver adds glucose to the blood.

b

Increase in blood glucose	→	High blood glucose detected by receptors	→	Pancreas secretes insulin
↑				↓
Liver converts glycogen to glucose and adds it to the blood				Liver cells take up glucose and store it as glycogen
↑				↓
Pancreas secretes glucagon	←	Low blood glucose levels detected by receptors	←	Blood glucose drops

2 a Sample graph

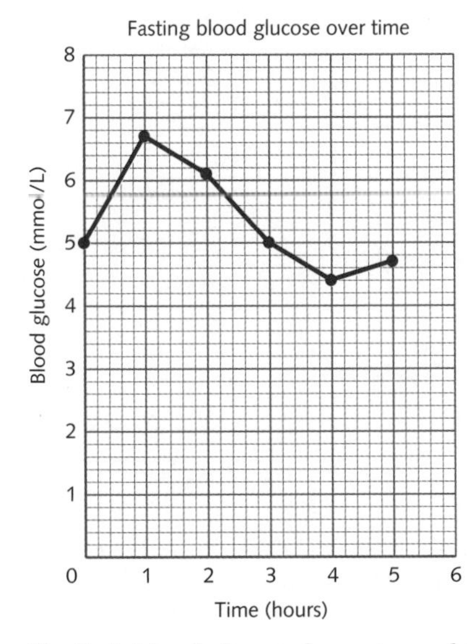

b Negative feedback, after an increase in blood glucose at time 1 hour, the response leads to a decrease in blood glucose

c Blood glucose level

d Time after consuming a sugar drink

e 5.32 mmol/L

f No, 1 hour after the sugar drink the blood glucose rises to 6.7 mmol/L and is still at 6.1 mmol/L at 2 hours after the sugar drink.

g No, their blood glucose is lower than 7 mmol/L 2 hours after the sugar drink.

h Hypoglyceamia – is when the blood glucose level is too low

Hyperglyceamia – is when the blood glucose level is too high

i No, their blood glucose has returned to the normal level of between 4 and 5.4 mmol/L after consuming the sugar drink.

j The blood glucose level drops back down after the sugar drink to the normal level of between 4 and 5.4 mmol/L.

k At 5 hours, the blood glucose has started to increase slightly in response to a drop below the starting level, this means stored glycogen has been converted to glucose and returned to the blood.

4.4.2 When blood glucose homeostasis system malfunctions p. 114

1

Glucose
Glucagon
Insulin
Relative concentration in blood
Meal eaten
Time

2 a If there is excess glucose in the blood it will be excreted in the urine, if there was sugar in the urine it would taste sweet.

b They use urine strips, that carry an indicator that changes colour to show the glucose content of urine.

4.5 Water balance regulation p. 115

4.5.1 Osmoregulation in humans. p. 115

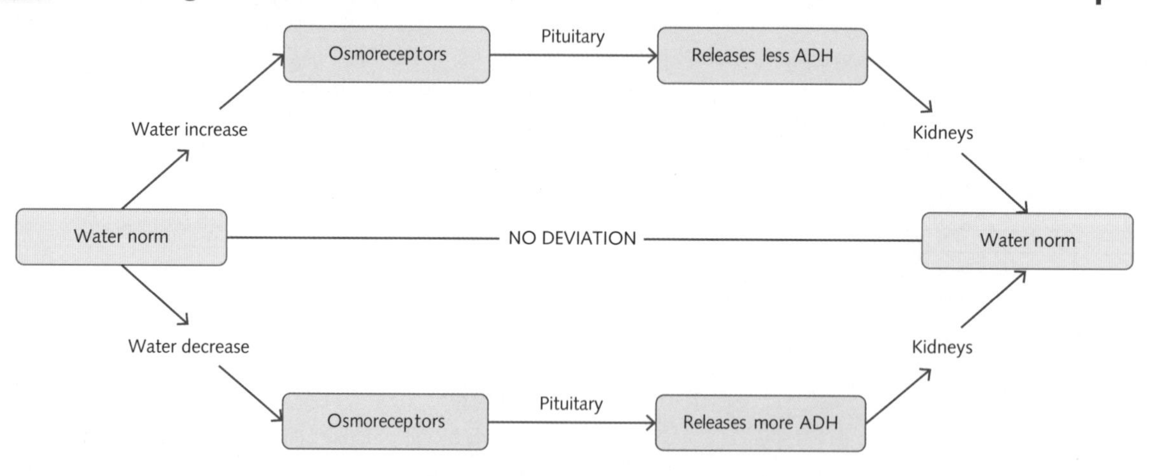

4.5.2 The mechanism of osmoregulation p. 116

1 Order of fill the gap terms: dilute, osmoreceptors, brain, pituitary, antidiuretic hormone, nephron, impervious, water, dilute urine, pelvis, concentrated, hypothalamus, pituitary gland, ADH, renal, distal tubules, more, water, water, concentrated

9780170452632

2 a

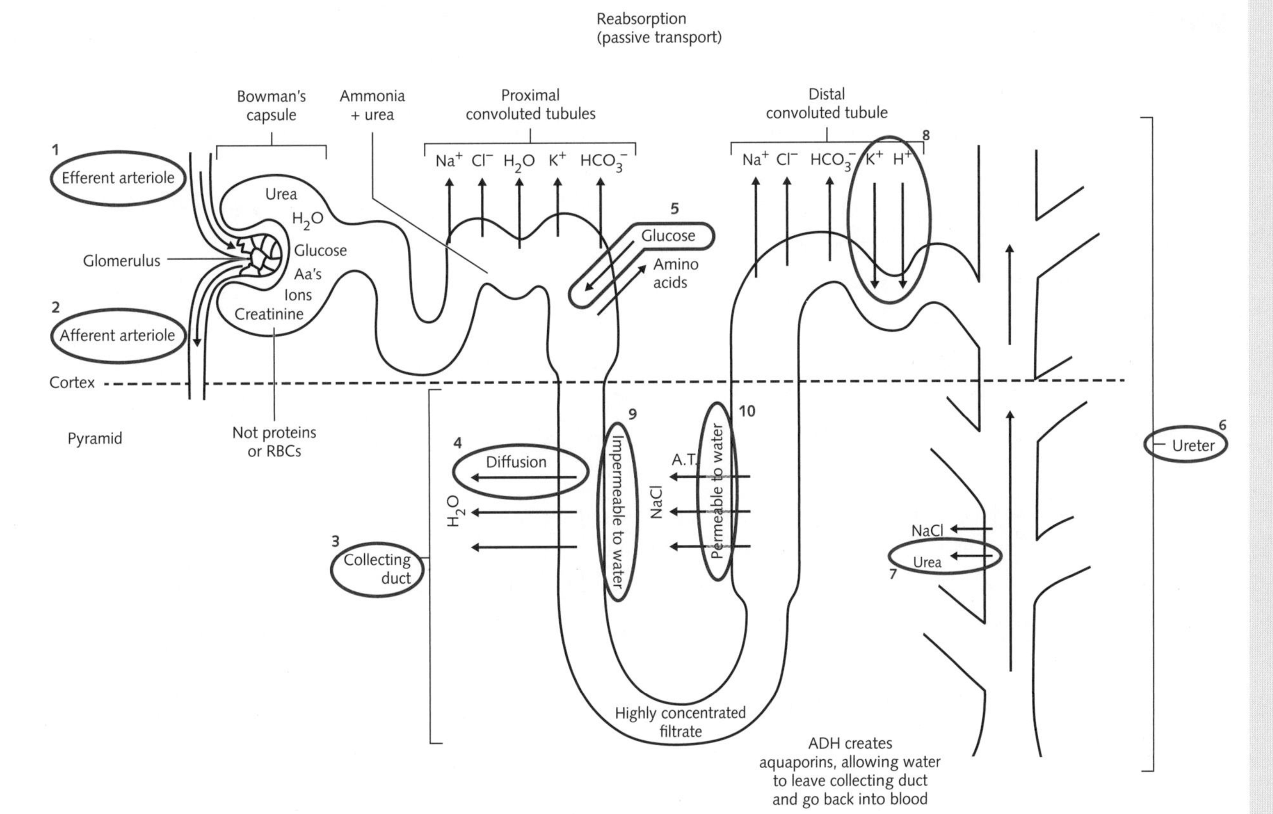

b
1 Afferent arteriole
2 Efferent arteriole
3 Descending loop of Henle
4 Osmosis
5 Arrow should be in the opposite direction
6 Collecting duct
7 Urea should be water
8 Arrows should be in the opposite direction
9 Permeable to water
10 Permeable to salts

4.5.3 Water balance in other animals p. 118

1 Terrestrial animals constantly lose water to the dry air; they must have adaptations to lower water loss and increase uptake. They also must have adaptations to allow them to nitrogenous waste and excrete excess salt to maintain salt balance.

2 a A saltwater animal transitioning to a terrestrial environment, would no longer be able to excrete ammonia into their surroundings as they are no longer surrounded by water to dilute it, they would have to develop an alternative method for eliminating nitrogenous waste. They would have issues with maintaining salt balance, as they have adaptations to eliminate salt, not gain salt.

b A freshwater animal transitioning to a terrestrial environment would have the issue of losing too much water to the environment, they would have to develop adaptations to reduce water loss to the environment. They would also have to develop an alternative method to eliminate nitrogenous waste as they will not have enough water to excrete ammonia.

c A freshwater fish is adapted to increase water loss and increase salt intake, their problem on land would be excess water loss and maintaining water balance, whereas marine fish are adapted to reduce water loss and maximise salt loss – they would have issues maintaining their salt balance.

d Answers will vary. Fresh water fish – land environment only presents an issue with lack of water for these fish, they could have first occupied areas that limited water loss such as humid areas, mud flats, burrows etc. They would not also have to cope with the change in salt content of their environment.

Saltwater fish – they already have adaptations to limit water loss and this would have enabled them to survive on land.

3 Ability to produce concentrated urine, that contains urea instead of ammonia to enable them to conserve water. A method of gaining or eliminating salts from their body that does not rely on being in a watery environment.

4.6 Regulation and control of basal metabolic rate and growth by thyroid hormones p. 119

4.6.1 Analysing a science journal p. 119

1 Sample sentences:

Humans who live longer have a high production of a hormone that stimulates their thyroid but does not change the rate that their cells use energy.

People that have good health into old age have not been studied much by scientists to try and work out what factors have allowed them to keep their good health.

Three organs – the hypothalamus, pituitary and the thyroid all work together to keep the levels of thyroid stimulating hormone high when the level of thyroid hormone is low and vice versa.

Living longer is thought to be linked to having high levels of thyroid stimulating hormone and low levels of thyroid hormone, but scientists do not understand how these levels are maintained.

The hypothalamus, pituitary gland and thyroid are all involved in growth, development and the rate that energy is used by cells.

It has been found that people with one parent still alive in their 90s and one of this parent's brothers/sisters still living into their 90s have produced a greater amount of thyroid stimulating hormone but still have the same cellular activity and similar levels of thyroid hormone compared to people without a parent in their 90s.

These healthy offspring were found to have a similar rate of cellular energy use and core temperatures as their spouses.

2 See answer options to question 1 above.

3 Answers may vary – hopefully, yes.

4.7 Chapter review p. 121

4.7.1 Key terms p. 121

1 Headings could be: water balance, temperature regulation, blood glucose regulation, survival in different environments, regulation of metabolic rate

2 Survival in different environments. One grouping could be:

Homeostasis: external environment, physiological stress, tolerance range, internal environment, interstitial fluid, optimum range, negative feedback

Stimulus–response model: effector, receptor, response, stimulus, set point, feedback mechanism, negative feedback

Receptors: hypothalamus, exteroreceptor, interoreceptor

4.7.2 Practice test questions p. 122

1 D
2 A
3 C
4 A
5 B

6 a (1/2 mark for each statement)

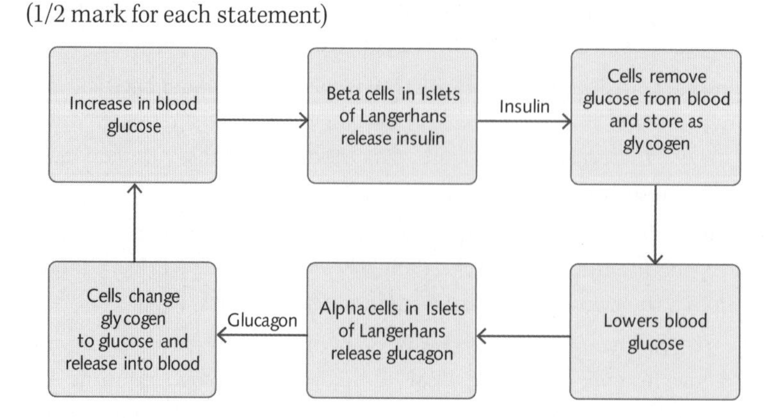

b Homeostasis (1)

c chemoreceptors (1)

d Negative feedback is when a response to a stimulus acts to reduce the original stimulus (1) reversing the direction of the change in a variable (1)

e Consume something that contains a lot of sugar e.g. jellybeans, honey, sweet drink (1). This increases the amount of glucose in the blood counteracting the hypoglyceamia (1).

Chapter 5 Scientific investigations p. 125

Remember p. 126

1 Ask a question/observation, research question, hypothesis, experiment, collect data, analysis, conclusion
2 A prediction based on theory or knowledge
3 Any three of the following: classification and identification; controlled experiment; correlational study; fieldwork; modelling; product, process or system development; or simulation
4 A system of moral principles to consider when undertaking scientific investigation. The principles are based on what is good and bad for society.
5 a An experimental condition set up to compare to the condition receiving the independent variable
b The variable that is altered or manipulated in a scientific investigation
c The variable that changes because of changes to the independent variable; the variable that is measured
6 Details of practical activities starting with the observations and ideas for an investigation up to the conclusions that are drawn.

Part A: Investigation p. 126

5.1 Choosing your topic p. 126

5.1.1 Getting started: organisation p. 126

Students should complete their own checklist.

5.1.2 Getting started: assessment p. 126

Answers will vary. Students should complete the rubric in Table 5.2 and tally scores and write a comment in Table 5.3.

5.2 Types of scientific methodology p. 131

5.2.1 Identifying the different types of scientific methodology p. 131

Scientific methodology	Definition	Research question
Fieldwork	Scientific investigation that is undertaken outside the laboratory	In what areas of my local parks can blue-tongued lizards be found?
Correlational study	When two factors are studied to see if one affects the other	Will the compulsory wearing of facemasks during the 2020 pandemic mean there will also be a drop in the number of cases of lung cancer due to second-hand smoke?
Product, process or system development	To design an object or product, process or system that will assist a human in meeting their biological demands and requirements	How can I best help people who suffer from arthritis and cannot pick up anything?
Controlled experiment	Investigation where all variables that could affect the results of an experiment are kept constant except the one under independent variable	Does the amount of light affect how much fruit my tomato plants produce?
Simulation	Use a model to simulate real life or part or whole of a system to gain knowledge of its functioning	How does DNA replicate itself?
Modelling	Making a representation of something usually on a smaller or larger scale so it can be seen and manipulated	What does DNA look like?
Classification and identification	Investigation of a larger group of living things and to classify them by placing them into like groupings and identify them by naming them	How many different spider species live in the schoolyard?

5.2.2 Designing a scientific investigation p. 132

Answers will vary depending on question selected – check your plan with your teacher.

5.3 Quantitative and qualitative data p. 134

5.3.1 Primary data p. 134

1 Sample question: In which area of the school can the most bacteria be found?

2 Sample hypothesis: If an agar plate is exposed to the air in the boys' toilets, then a larger number of bacteria will grow on it compare to other areas of the school.

3 Controlled experiment

4 Sample

Risk	How it will be managed
Exposure to pathogenic bacteria	Petri plates will be sealed with tape and not opened. At end of the experiment all agar plates will be autoclaved to kill all bacteria.

5 Answers will vary. Samples could be:

Qualitative data – description of the growth on the plate (colour, profile etc.)

Quantitative data – be number of colonies growing, % of plate covered with bacterial growth

6 Sample results table:

Table 1 Bacterial growth on agar plates after exposure to different school areas

School area	Bacterial growth after 2 days at 30°C		
	Description	Number of colonies	% plate covered with bacteria
Staff room sink	1 large colony, 5 smaller and a film covering 1/5 of plate	6	20
Year11 locker room	4 large colonies and 1 smaller, 2 colonies are a dark colour, 2 are white	5	20
Year 7 classroom	3 large and 12 small colonies, there is a dark area of growth in the centre of the largest colony	15	40
Boys' toilets	4 large and 3 small colonies, dark and light areas of in one colony	6	50
Girls' toilets	1 large and 3 small colonies	4	50

7 A control would be an agar plate that is not opened/exposed to the air, this is to show that bacterial growth that would occur if the dish was not exposed to check that it was the exposure to air in different locations that was causing the bacterial growth and that bacteria did not naturally grow on agar without any exposure.

5.3.2 Quality of primary data p. 135

1 New results table

Table 5.5 Salt concentration and light transmission

Salt concentration (%)	Transmission of light (%)							
	Trial 1	Trial 2	Trial 3	Trial 4	Trial 5	Trial 6	Mean	Mean without trial 1
0	78.57	75.27	77.23	78.4	64.88	124.66	83.12	84.09
3	92.82	69.71	85.23	79.54	88.91	57.96	79.03	76.27
6	100.05	66.51	88.39	78.29	73.66	61.54	78.07	73.68
9	110.05	64.91	80.71	109.43	68.29	52.96	81.05	75.26
12	117.18	59.91	75.66	81.96	65.01	49.95	74.94	66.5
15	115.46	66.03	72.55	81.06	65.72	55.37	76.03	68.14
18	120.67	60.48	69.31	74.63	58.43	54.51	73.00	63.47

2 No, these results are not precise. There is a range of different values for light transmission at each salt concentration, some differ by as much as 46%.

3, 4 See the table.

5 Answers may vary. Sample answer: The measurement from the light meter may have been affected by other light sources in the room, which would make the experiment invalid as other sources of light could have caused the results. The light intensity from the torch may have reduced during the period of the experiment and this would make the results invalid.

6 Trial 1

7 Random error or personal error

8 See table above. All the averages have changed significantly except for salt concentration 0. All other means have dropped significantly.

5.4 Health, safety, and ethics p. 137

5.4.1 Ethical guidelines p. 137

Scenario 1

1 Answers may vary slightly: should Dr Witt be refusing to reveal the results of his study despite his views on development in national parks.

2 Integrity, information is being withheld that is needed to make the best decision in the interest of all stakeholders.

3 Answers will vary. Examples:

Social factors: access of the public to the national park could be increased by building on the tourist centre, education of the public could be increased due to the tourist centre, enjoyment of the natural environment may be reduced by the development of the tourist centre.

Economic factors: increased income to the national park authority from the tourist centre, increased income for people employed at the tourist centre, more jobs may be created in the toursist centre.

Scenario 2

4 Who should have access to DNA data of individuals?

5 Respect – the rights of individuals to not share their DNA data are not being considered

6 Sample answers. Legal: sharing of personal information of individuals, accuracy of test results, what the DNA data can be used for

Political: relationship with other countries, how should personal DNA sequencing be regulated, what sort of licence should these companies have.

Part B: Scientific evidence p. 139

5.5 Collecting and analysing data p. 139

5.5.1 Analysing your data p. 139

Dataset 1

Step 1

Type of water	Trial number	Percentage change
Tap water	1	+317.39
	2	+380.95
Deionised water	1	+320.00
	2	+258.88

Step 2: Answers will vary. Sample: If plants are watered with deionised water, then they will grow taller when compared to plants watered with tap water.

Step 3: Sample answer: The results do not support the hypothesis. The plants watered with tap water increased their height by 317.39% and 380.95%. Plants watered with deionised water increased their height by 320% and 258.88%.

Dataset 2

Step 1 and step 2

Concentration of sucrose solution (mol dm^{-3})	% change in length of the potato strip	Mean % change in length of potato strip
0	+4.00	+5.57
	+7.14	
0.2	+2.00	+1.99
	+1.98	
0.4	+2.00	+3.02
	+4.04	
0.6	–2.0	–2.47
	–2.94	
0.8	–3.0	–2.98
	–2.97	
1.0	–3.03	–4.48
	–5.94	

Step 3: Sample hypothesis

If potato strips are placed in increasing concentrations of glucose solution then those in the more concentrated glucose solution will lost weight compared to those in more dilute solutions.

Step 4: The hypothesis was supported. Potato strips in a glucose concentration less of 0.4, 0,2 and 0 mol dm^{-3} all showed a % weight gain, + 3.02, +1.99 and +5.57 respectively. Potato strips in higher glucose concentrations of 0.6, 0.8, 1.0 mol dm^{-3} all showed a % weight loss: –2.5, –2.98 and –4.48 respectively.

Part C: Scientific communication p. 141

5.6 Communicating your results p. 141

5.6.1 Presenting your work as an article p. 141

Answers will vary – check your work with your teacher.

5.7 Chapter review p. 143

5.7.1 Key terms p. 143

Accuracy is a measure of how close your data is to the true value whereas precision is a measure of how close repeated data measurements are to each other. Data can be precise without being accurate.

Quantitative data has measurements with a numerical value, qualitative data is descriptive and does not include numbers.

The independent variable is the one that the experimenter changes, it causes changes in the dependant variable, which is what the experimenter measures for the results.

The research question is the specific question that you are trying to answer, an aim gives the reason for carrying out a particular investigation.

Ethics are a system of moral principles that can be used to determine what is good or bad for society, bioethics is applying ethical principles to biological research.

Random errors are caused by unknown factors, personal errors are caused by mistakes made by the experimenter.

Methodology is a broad description of the approach taken in an investigation; method is a step-by-step description of the process followed.

Modelling is the act of building a model. A simulation is the process of using a model to study the behaviour of an actual or theoretical system.

An observation is what you make with your senses, a hypothesis is a prediction to try and explain the observation.

Integrity relates to honestly sharing all available information and non-maleficence relates to doing no harm to others.

Social refers to issues that have an impact on many members of a population/society, whereas political relates to government policies and regulations.

Repeatable is when the same experimenter repeats an experiment and gets similar results each time, reproducible is when another experimenter repeats the experiment and gets similar results.

5.7.2 Practice test questions p. 145

1 C

2 C

3 D

4 A

5 D

6 **a** Type of cooking: microwave or stove top

b Time it takes for the potatoes to cook

c Size of the potato pieces

d If potatoes are cooked in a microwave, then the time needed to cook them will be less than potatoes cooked on a stove top. (1 mark for including IV and DV, 1 mark for including how the IV will affect the DV)

e One of the following: 1 mark for limitation, 1 mark for explanation example limitations

There is no explanation of how it was determined if the potatoes were cooked – this may have made the cooking time longer or shorter if there was not an objective way to determine when the amount of cooking was the same for both methods.

The potato pieces in each method may have been different sizes, this could have made the cooking time longer or shorter.

The method of measuring the cooking time in the microwave can only give time in 30 seconds increments – this would not give an accurate cooking time.

The experiment includes only one trial for each cooking method, this would impact on the accuracy of the results.

7 Marks: 1 for title, 1 for correct type of graph, 1 for correct scale, 1 for axes labels

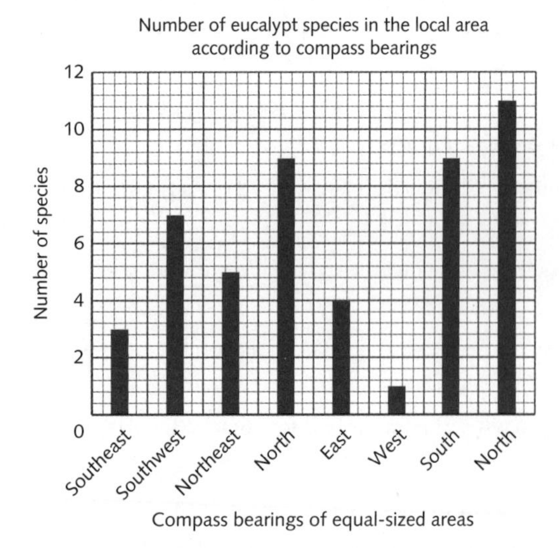

Chapter 6 Chromosomes to genomes p. 148

Remember p. 148

1 Because it is passed from parent to offspring

2 In the plasmid in the nucleoid

3 In the nucleus mitochondria and chloroplasts

4 Fertilisation

5 Mitosis

6.1 Distinction between genes, alleles and a genome p. 148

6.1.1 Structure of DNA: making a model p. 148

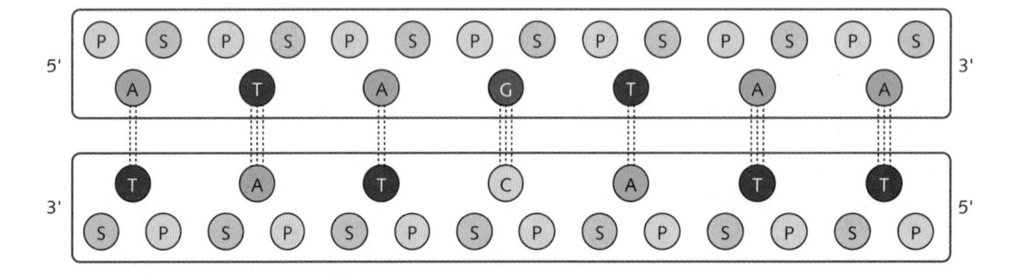

6.1.2 The ABC of DNA p. 152

1 a An information molecule that is the universal basis of an organism's genetic material; it contains instructions, written in a chemical code, which is used to produce proteins

b One zone of each strand is made up of identical repeating units, the sugar and phosphates, while the other zone is made up of differing units, one of four different nitrogenous bases.

c Hydrogen bonding between complementary base pairs of each strand

2 a Before cell division: mitosis, or meiosis

b In the nucleus

c Each new DNA molecule retains one of the strands for the original DNA molecule and one newly made strand.

d This rule refers to the nitrogenous bases. Adenine always pairs with thymine, cytosine always pairs with guanine

3 a Instruction to make a protein

b A component of a nucleotide, for DNA there are four different nitrogenous bases: adenine, guanine, thymine, and cytosine

c Alternating sugar and phosphates

d 3′ is the 3rd carbon in the deoxyribose sugar, the 3′ end of a DNA molecule ends with a sugar. 5′ is the 5th carbon in the deoxyribose sugar, the 5′ end of a DNA molecule ends with a phosphate group.

5 The monomer or subunit of nucleic acids – DNA. It is composed of a sugar, deoxyribose for DNA, a phosphate group and 1 of four nitrogenous bases, adenine, thymine, guanine, or cytosine. A molecule of DNA is composed of many nucleotides joined.

6 a A nucleotide

b Deoxyribonucleic acid

7 36% thymine and 14% cytosine

8 a There are two hydrogen bonds between adenine and thymine and 3 hydrogen bonds between cytosine and guanine.

b 32% thymine (18% C and 18% G – leaves 64% for A and T)

6.1.3 Genes, alleles, and genomes p. 153

1 Temperament gene from the Labrador and coat genes from the poodle

2 Docile and easily trained alleles from the Labrador and non-shedding and hypoallergenic coat alleles from the poodle

3 To be used as assistance dogs by the Royal Guide Dogs

4 They have worked out the base sequence for 150 000 different areas of the Australian labradoodle's DNA.

5 The genome of the Australian labradoodle is largely poodle with only a few genes from the Labrador. Their evidence was based on comparing 150 000 areas of the labradoodle's genome to the genomes of the Labrador and the poodle, this seems to be substantial evidence.

6 Answers may vary. For example: Yes, it should be considered a different breed because its genome contains a combination of genes from two different breeds, meaning that it is genetically different from both the Labrador and the poodle.

Or

No, the genome of the labradoodle is very similar to that of the poodle, there is not enough difference for it to be considered a new breed.

6.1.4 Genomics p. 155

Person A

Person B

```
1   GGATGCGATG GCTGCGGCGT CCTGGGGCGA GGCGCTGACG TGAGCTCGGC GCACCTGGGC
61  TGGGCAGGTA AGGGCTGGTG CGGAACGGGG AGAGGAACCT GCAGTCCCTA CTTGGGTAGA
121 GCCAGGCGCC CCTTGGCTAA GACGTCGAGG AGCGTGGTAG CGACGGGTGA TCTTCGCTGC
181 GGACTTGGTT CGGAGGGACG TCCGCTTCTG GTGGACACAT TGAGCAAAGG CCTGGGCTGT
241 AGAGACAGGG TTGTACCAGG AAGGGGTGGA TGACCCTGAC CCAGCTAAAT GGAAGGCCCA
301 TCTTATACTC ATGAAATCAA CAGAGGCTTG CATGTATCTA TCTGTCTGTC TATCTATCTA
361 TCTATCTATC TATCTATCTA --------------------- TGAGACAGGG TCTTGCTCTG
    TCACCCAGAT TGGACTGCAG
421 TGGGGGAATC A
```

6.1.5 Human genomics research p. 156

1 Answers may vary, for example:

- If you find out you carry an allele for a genetic disorder, should you have children?
- If you find out you carry an allele for a genetic disorder, do you tell the rest of your family in case they also carry it? What if they do not want to know?
- Who are you required to inform about any discoveries of genetic disorders?

- Should companies be able to profit from any discoveries made in this way?
- Should this information be available to law enforcement to be used in identifying suspects for crimes?
- Does the company provide you with enough information to give informed consent?

2 Answers will vary depending on the issue, examples for the above issues – these are not definitive.

- Non maleficence – the right to parent a child that you know could have a disorder that would reduce its quality of life
- Respect – every individual has the right to decide for themselves if they want to know about their genome or not
- Integrity – is it right to reveal information?
- Justice – if there is a profit to be made from a discovery in your genome, should the company be the only ones to benefit?
- Integrity – were participants informed that their data could be used in this way?
- Integrity – were participants informed of all the possible uses of their data?

3 **Sample answers given to Point 1 above.**

Factor	How this factor may influence your issue
Social	Does everyone have a right to have children? Is it right to deny your parents any grandchildren? Could I use IVF to bypass the faulty gene? Could I adopt a child instead?
Economic	If you have a child with a genetic disease, how much will it cost you or the government to support the needs of that child? Will you need to take time off work/leave work to support that child?
Legal	If I adopt a child, what are the legal hurdles that I have to jump? If I have a child with a disability are there laws in place to assist them/support them?
Political	Is there any government assistance that I can access to support a disabled child? Is there any carer support for me if I have to give up my job to care for a disabled child?

6.2 Chromosomes p. 157

6.2.1 Chromosomes p. 158

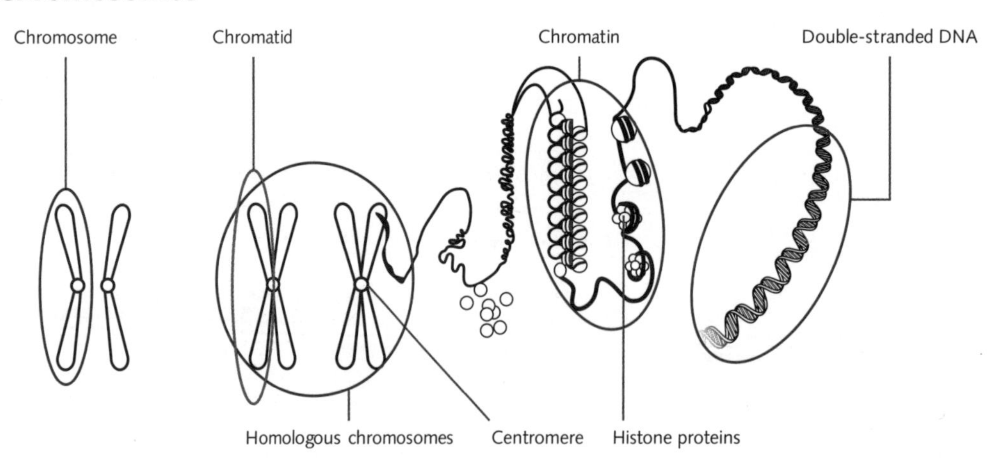

6.2.2 Autosomes and sex chromosomes p. 158

1 46

2 a Chromosome number 1 → 22

b Sex chromosomes are X and Y

3 X and Y

4 XX

5 The mother

6 The father

7 One from the mother and one from the father

6.2.3 Variations in nuclear chromosomes of eukaryotes p. 159

1 a The number of complete sets of chromosomes in a cell

b Having two complete sets of chromosomes in a cell

c A cell containing only one set of chromosomes

d Having a single set of chromosomes

e Having more than two complete sets of chromosomes in a cell

f Having an abnormal number of chromosomes in a cell, e.g. one extra or one less

2

Species	Number of chromosomes in haploid cells (*n*)	Number of chromosomes in diploid cells (2*n*)
Human	23	46
Fruit fly	4	8
Chimpanzee	24	48
Bat	22	44
Koala	8	16
Kangaroo	8	16
Tasmanian devil	7	14
Rice	12	24
Platypus	26	52
Chicken	39	78
Spinach	6	12
Cucumber	7	14

6.2.4 The endosymbiotic theory p. 160

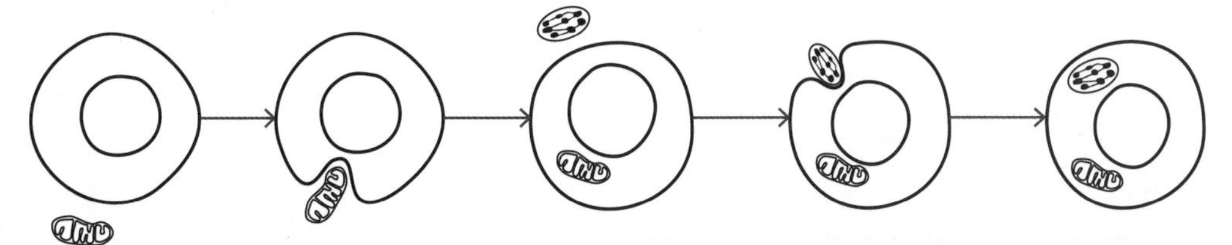

The mitochondria were once a free-living prokaryote.

A larger prokaryotic cell engulfed the 'mitochondrion' prokaryote by endocytosis and brought it inside the cell enclosed by a membrane.

Chloroplasts were also once free-living prokaryotes.

A larger prokaryotic cell also engulfed the 'chloroplast' prokaryote by endocytosis and brought it inside the cell enclosed in a membrane.

The cell now contains a mitochondrion and chloroplast that:
- have their own circular chromosome of DNA like prokaryotes
- are enclosed in a double membrane, inner membrane is the plasma membrane of the prokaryote and the outer membrane was formed when they were engulfed
- can divide by binary fission like other prokaryotic cells.

6.3 Karyotypes for identifying chromosomal abnormalities p. 161

6.3.1 Chromosomal abnormality p. 161

1 47

2 3

3 Klinefelter syndrome

4 Symptoms include: taller than average; long legs, short torso, broad hips; reduced facial and body hair; reduced muscle strength; breast development; lethargy.

6.4 Production of gametes and sexual reproduction p. 163

6.4.1 Introducing variation through meiosis p. 163

1 a Two gametes each with an extra chromosome, two copies of the chromosome that failed to disjoin and two gametes missing copies of the chromosome that did not disjoin.

b Two gametes with the correct number of chromosomes, one gamete missing the chromosome that failed to disjoin and one gamete with two copies of this chromosome.

2 & 4 Sample graph

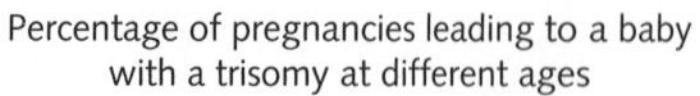

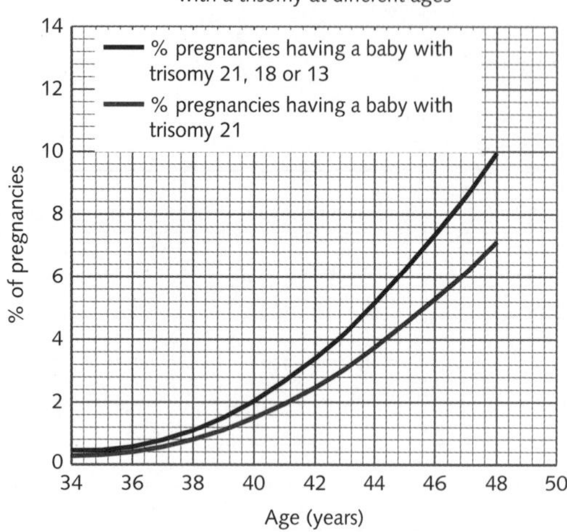

3 As maternal age increases, the chance that a baby will be born with trisomy 21, 18 or 13 increases.

4 See above for graph.

Maternal age (years)	% Pregnancies having a baby with trisomy 21, 18 or 13	% Pregnancies having a baby with trisomy 21	% Pregnancies having a baby with trisomy 18 or 13
34	0.45	0.26	0.19
35	0.46	0.29	0.17
36	0.57	0.39	0.18
37	0.78	0.55	0.23
38	1.1	0.79	0.31
39	1.52	1.1	0.42
40	2.04	1.49	0.55
41	2.67	1.94	0.73
42	3.4	2.46	0.94
43	4.23	3.06	1.17
44	5.17	3.73	1.44
45	6.21	4.47	1.74
46	7.36	5.28	2.08
47	8.61	6.16	2.45
48	9.96	7.11	2.85

 9780170452632

5 Answers may vary; for example:

Trisomy 21 is the more common in babies than trisomy 12 and 18 at all ages, it accounts for more than half of all the babies born with one of the three trisomy.

The chance of a pregnancy of a 48-year-old woman producing a baby with trisomy 21 increases by approximately 27 times to 7.11% compared to only 0.26% at 34 years of age. In comparison, the chance of having a child with trisomy 18 or 13 only increases by 2.6% for the same ages. The chance of having a trisomy 21 child increases more with age than the chance of having a trisomy 18 or 13 baby.

6.4.2 Meiosis, fertilisation and mitosis p. 165

Step 1:

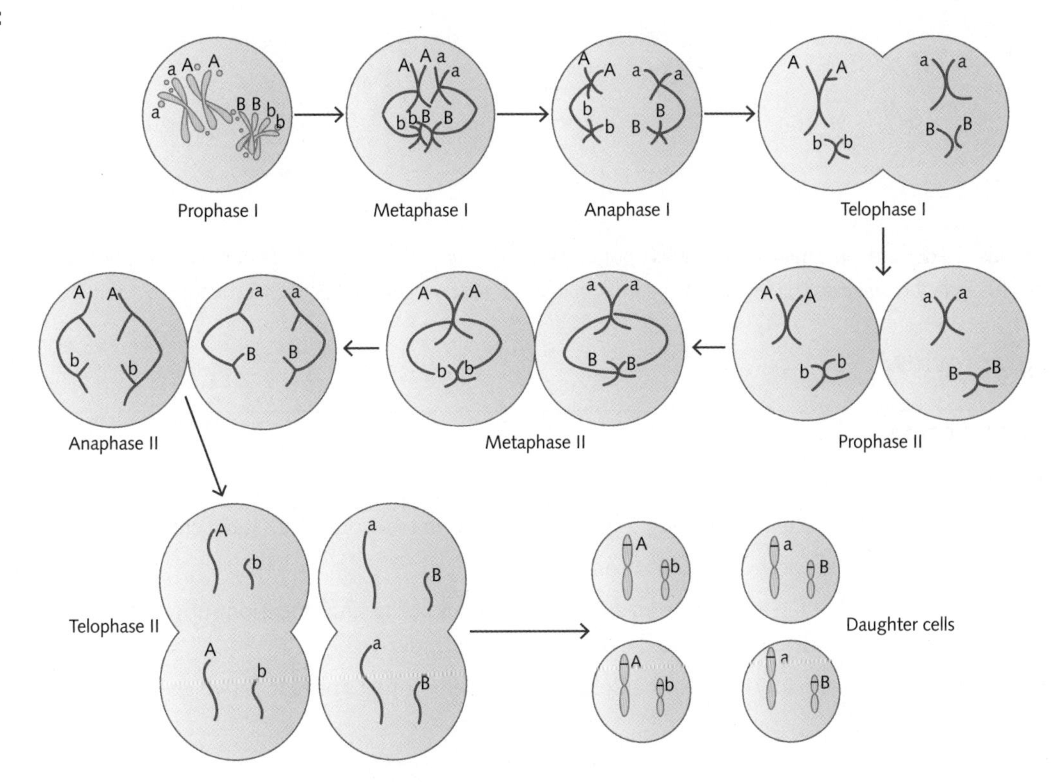

Step 2:

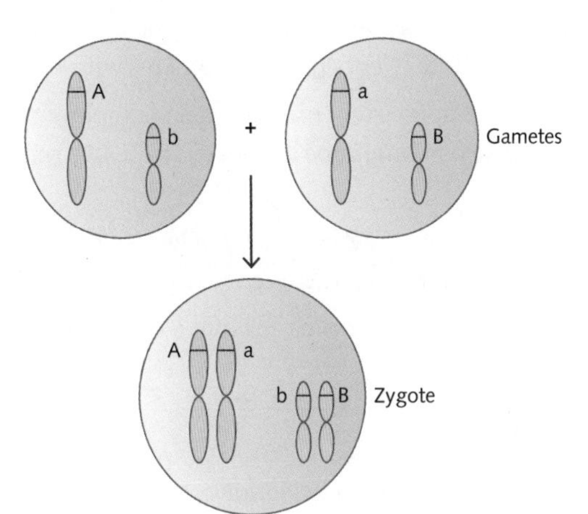

Step 3:

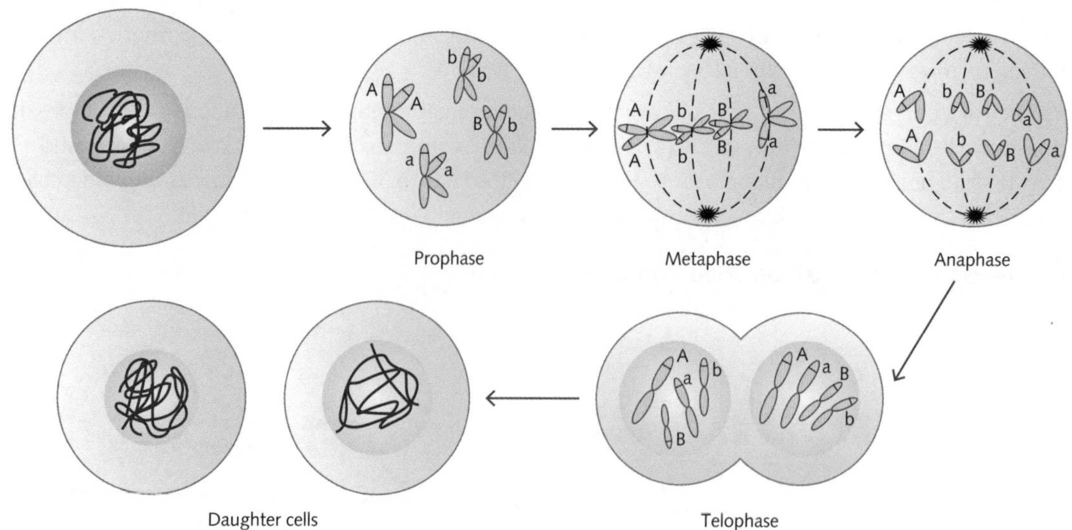

All somatic cells are diploid, but meiosis produces haploid gametes – containing only one set of each chromosome, when two gametes fuse via fertilisation the resulting zygote will again have two completes of chromosomes – will be a diploid cell.

6.5 Chapter review p. 168

6.5.1 Key terms p. 168

1 Sexual and asexual
2 A body cell, compose the tissues of the body – are diploid (contain two complete sets of chromosomes)
3 One of two identical strands of a replicated chromosome.
4 A pair of chromosomes that have the same size, shape and genes at the same locations one chromosome in a homologous pair is maternally derived and the other is paternally derived.
5 Meiosis produces four genetically different haploid daughter cells; mitosis produces two genetically identical diploid daughter cells.
6 9
7 When two homologous chromosomes fail to separate during anaphase I of meiosis or two sister chromatids fail to separate during anaphase II of meiosis. Both events produce gametes with missing or extra chromosomes.
8 Crossing over is the exchange of parts of non-sister chromatids on homologous chromosomes during meiosis I. It creates chromosomes with new allele combinations and leads to greater genetic variety in gametes, leading to more variety in offspring.
9 DNA is composed of many nucleotides joined together, each nucleotide has a phosphate group, a deoxyribose sugar and one of four nitrogenous bases (adenine, thymine, cytosine, or guanine).
10 An autosome is a chromosome that does not determine sex (not a sex chromosome), they exist in homologous pairs in diploid cells.

6.5.2 Practice test questions p. 169

1 Disagree, the building blocks of RNA and DNA are nucleotides.
2 Agree, the bases are different sizes, if you change the base pairing it will alter the distance between the strands, wider at some points and narrower at others.
3 Disagree, DNA replication is semi conservative and 50% of the parent DNA would be in the new DNA molecules.
4 Agree, a gene is a section of DNA.
5 1 mark per mistake

Segment 1	Segment 2	Segment 3	Segment 4
A-C-G-G-C T-T-C-C-G	G-G-T-G-A C-C-A-C-T	G-A-T-T-A C-C-A-A-T	C-A-A-T-T G-T-T-A-C

9780170452632

6 a Non-disjunction when the chromosome has not separated properly (1) during anaphase II as the centromere has not been broken and the two chromatids have not separated (1).

b 45

c Female

Chapter 7 Inheritance p. 171

Remember p. 171

1 A gene is a section of DNA that codes to produce a protein.

2 Allele

3 Autosomes

4 X and Y

5 A pair of chromosomes that have the same size, shape and genes at the same locations.

6 Diploid cells contain two complete sets of chromosomes and are body cells that form all the tissues of the body. Haploid cells contain only one set of chromosomes, they are gametes and are found in the sex organs: ovaries and testes.

7.1 Patterns of inheritance p. 172

1

Characteristic	Dominant form	Recessive form
Stem length	T = tall	t = short
Flower position	A = axial	a = terminal
Seed (endosperm)	Y = yellow	y = green
Seed shape	R = round	r = wrinkled
Flower colour	P = purple	p = white
Pod colour	G = green	g = yellow
Pod shape	I = inflated	i = constricted

2

Characteristic	Ratio
Stem length	3 : 1
Flower position	3 : 1
Seed (endosperm)	3 : 1
Seed shape	3 : 1
Flower colour	3 : 1
Pod colour	3 : 1
Pod shape	3 : 1

3 All the ratios are the same 3 dominant : 1 recessive

7.1.2 The relationship between genes, alleles, and traits p. 174

1 Answers based on allele symbols used in question 1, you will have to substitute your own symbols, but follow the pattern of two upper case, two lower case or one of each.

a RR

b ii

c Pp

d aa

e GG

2 a Parents: RR × rr

	r	r
R	Rr	Rr
R	Rr	Rr

100% chance offspring are Rr
100% chance of offspring have round seedpods

b Parents: Rr × Rr

	R	r
R	RR	Rr
r	Rr	rr

25% chance offspring are either RR or rr.
50% chance are Rr

75% chance they will have round pods and 25% chance they will have wrinkled pods.

3 Parents: Pp × Pp

gametes	P	p
P	PP	Pp
p	Pp	pp

25% chance PP. 50% chance of being Pp and 25% chance pp
75% chance purple flowers and 25% chance white flowers

4 Alleles: R – red, r – silver grey

Parents: Rr × rr

	r	r
R	Rr	Rr
r	rr	rr

50% chance offspring are Rr, 50% chance rr
50% chance offspring have red fur, 50% chance offspring will have silver grey fur

5 a Parents: Cc × Cc

	C	c
C	CC	Cc
c	Cc	cc

25% chance offspring are CC. 50% chance are Cc and 25% chance they are cc
75% chance they will have not have cystic fibrosis and
25% chance they will have cystic fibrosis

b Parents: Cc × CC

	C	C
C	CC	CC
c	Cc	Cc

50% chance offspring are CC. 50% chance are Cc
100% chance they will have not have cystic fibrosis

7.1.3 Dominance is not always clear-cut p. 176

Incomplete dominance

1 a frizzled feathers × straight feathers (Hint: Parents will be $T^FT^F \times T^ST^S$)

	T^S	T^S
T^F	$T^F T^S$	$T^F T^S$
T^F	$T^F T^S$	$T^F T^S$

100% chance of being $T^F T^S$
100% chance of having slightly frizzled chicken feathers

b slightly frizzled × slightly frizzled: $T^F T^S \times T^F T^S$

	T^F	T^S
T^F	$T^F T^F$	$T^F T^S$
T^S	$T^F T^S$	$T^S T^S$

25% chance $T^F T^F$, 50% chance $T^F T^S$, 25% chance $T^S T^S$
25% chance frizzled feathers, 50% chance of slightly frizzled feathers,
25% chance straight feathers

c frizzled feathers × slightly frizzled: $T^F T^S \times T^F T^S$

	T^F	T^S
T^F	$T^F T^F$	$T^F T^S$
T^F	$T^F T^F$	$T^F T^S$

50% chance $T^F T^F$, 50% chance $T^F T^S$
50% chance frizzled feathers, 50 % chance of slightly frizzled feathers

Codominance

2 white feathered turkey × black and white feathered turkey: $F^B F^B \times F^B F^B$

	F^B	F^W
F^B	$F^B F^B$	$F^B F^W$
F^B	$F^B F^B$	$F^B F^W$

50% chance $F^B F^B$, 50% chance $F^B F^W$
50% chance black feathers, 50% chance black and white feathers
therefore you would expect a genotypic and phenotypic ratio of 1:1 in the F_1

3 $F^R F^W$

4 a red camelia × white camelia: $F^R F^R \times F^W F^W$

	F^W	F^W
F^R	$F^R F^W$	$F^R F^W$
F^R	$F^R F^W$	$F^R F^W$

100% chance $F^R F^W$
100% chance flowers with red and white blotches
Therefore you would expect a genotypic and phenotypic ratio of 1 in the F_1

b white × red/white camelia: $F^W F^W \times F^R F^W$

	F^W	F^W
F^R	$F^R F^W$	$F^R F^W$
F^W	$F^W F^W$	$F^W F^W$

50% chance $F^R F^W$, 50% chance $F^W F^W$
50% chance white flowers, 50% chance red/white flowers
therefore you would expect a genotypic and phenotypic ratio of 1:1 in the F_1

c red/white camelia × red/white camelia: $F^R F^W \times F^R F^W$

	F^R	F^W
F^R	$F^R F^R$	$F^R F^W$
F^W	$F^R F^W$	$F^W F^W$

25% chance $F^R F^R$, 50% chance $F^R F^W$, 25% $F^W F^W$
25% chance red flowers, 50% chance red/white flowers,
25% chance white flowers therefore you would expect a genotypic and phenotypic ratio of 1:2:1 in the F_1

7.2 Genetic material, environmental factors, and epigenetic factors p. 179

7.2.1 Epigenetics p. 179

1 Sample aim: To investigate the link between DNA methylation patterns and obesity in women
2 Correlation study
3 Women within the normal weight range had 11% less DNA methylation than obese women. After losing weight it obese women had different methylation patterns on their DNA.
4 Yes, the data showed a link between the amount of DNA methylation and obesity, with obese women having 11% more methylation than normal weight women. Also, weight loss by obese women led to a change in the methylation of their DNA.
5 Extend the study to include a greater sample size and diversity, including women from different populations.
6 Being obese has altered the epigenome of these women, causing more methylated tags to be added to their DNA. When the women lost weight, their epigenome was altered by the removal of some of these tags.

7.3 Patterns of inheritance p. 180

7.3.1 Monohybrid crosses p. 180

1 Bb
2 Parents: Bb × Bb

	B	b
B	BB	Bb
b	Bb	bb

a 25%
b The cross is investigating the inheritance pattern of one trait controlled by one gene at one locus.

3 a The black guinea pigs could have the genotype BB or Bb, cross the black guinea pigs with white guinea pigs

If homozygous black

BB × bb

	b	b
B	Bb	Bb
B	Bb	Bb

All offspring will be black

If heterozygous black

Bb × bb

	b	b
B	Bb	Bb
b	bb	bb

50% of offspring will be expected to be black and 50% will be expected to be white

b Test cross

c If all the offspring are black then the black guinea pig is most likely homozygous dominant, if there are any white offspring then the black guinea pig is heterozygous.

4 a Alleles: L – long wing l – vestigial wings

Parental: LL × ll

There is 0% chance of having a vestigial wing offspring in the F1 as all the flies will be heterozygous and have long wings.

In the F2 there is 25% chance the offspring will have vestigial wings.

b R – red eyes, r – non red eyes

Parents: Rr × Rr

	R	r
R	RR	Rr
r	Rr	rr

25% chance of RR, 50% chance of Rr, 25% chance rr

75% chance red eyes, 25% chance non red eyes

Therefore, the expected ratio of red eyes to white eyes is 3:1.

7.3.2 Sex-linked inheritance p. 182

1 a X^RX^R **b** X^RX^r **c** X^rX^r **d** X^RY **e** X^rY

2 a $X^RX^r \times X^rY$

	X^r	Y
X^R	X^RX^r	X^RY
X^r	X^rX^r	X^rY

50% chance of producing a child with red-green colour blindness

b $X^R X^R \times X^rY$

	X^r	Y
X^R	X^RX^R	X^RY
X^R	X^RX^r	X^RY

0% chance of producing a child with red-green colour blindness

3 a 50% of the sons would be expected to suffer from Duchene muscular dystrophy

b $X^D X^D \times X^dY$

	X^d	Y
X^D	X^DX^d	X^DY
X^D	X^DX^d	X^DY

i. 100% chance their daughter will be a carrier

ii. 0% chance their next son will suffer from the disease

4 a Not a reason, only males carry a Y chromosome and show these traits. It would not be possible for a daughter to inherit the trait if the gene were located on the Y chromosome.

b Not possible, if the trait were autosomal dominant, one of the parents would need to have the trait to be able to pass on an allele for a dominant trait to their daughter. Neither of the parents have the trait so it cannot be a dominant trait.

c Possible, the trait could be autosomal recessive – neither parent has the trait, but they could both be carriers/heterozygous, enabling them both to pass on an allele for an autosomal recessive trait to their daughter.

d This is not possible, males only have one X chromosome and cannot be a carrier of a recessive trait, if the trait were X-linked recessive, the father would need to have this allele and would have the trait to be able to pass the allele on to his daughter, as the father does not have PKU this is not possible.

7.4 Pedigree charts for autosomal and sex-linked inheritance p. 184

7.4.1 Pedigree charts p. 184

1 a,b,e,f

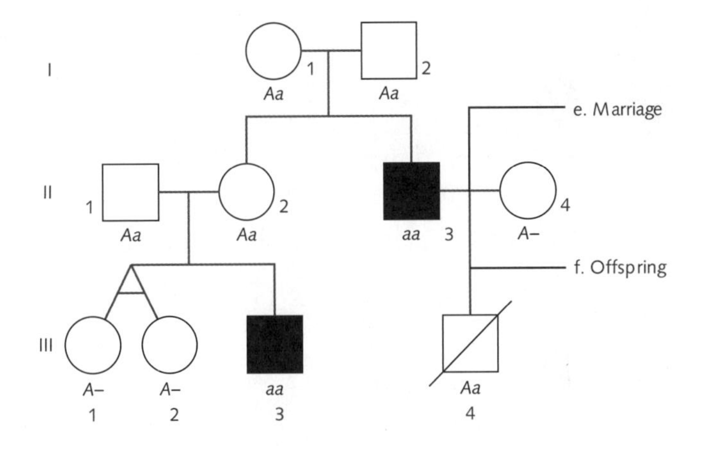

c I-2, II-1, II-3, III-3 or III-4

d I-1, II-2, II-4, III-1, or III-2

g III-1 and III-2

h III-4

i II-3 or III-3

2 Autosomal dominant, individuals I-1 and I-2 both have the trait but have had a child II-3 that does not have the trait, this is only possible if the parents are both heterozygous for the trait and have each passed an allele for the recessive trait to their son II-3. If it was X-linked all daughters of an affected male must inherit the trait, I-4 has the trait but has not passed it to his daughter II-4.

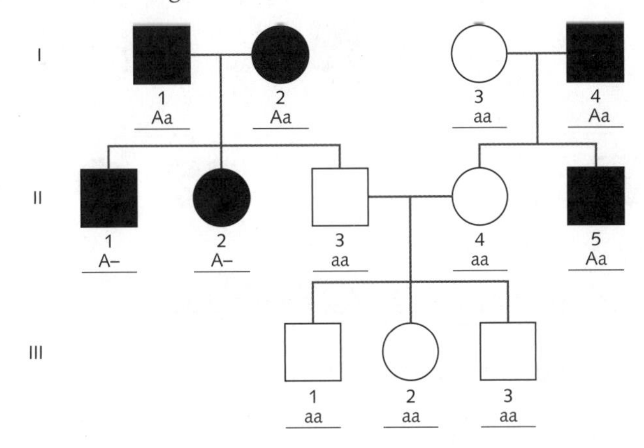

3 X- linked dominant, affected father I-2 has passed the trait to all his daughters, II-2, II-3, II-4 and not his son II-5. Daughters must inherit one X chromosome from their father so they will all get the allele for the dominant trait from their father, sons inherit their X chromosome from their mother.

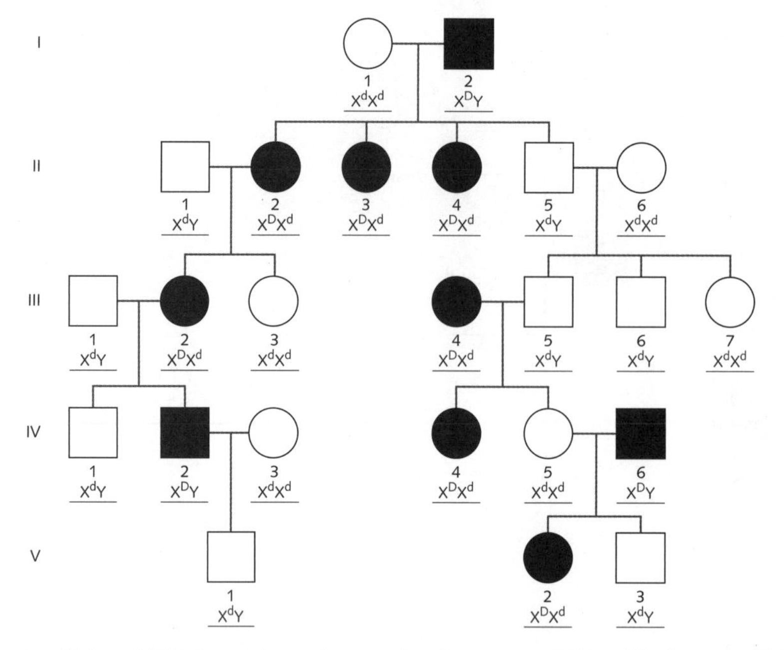

4 Autosomal recessive, II-1 and II-2 do not have the trait but have a son III-1 with the trait, II-1 and II-2 must be heterozygous/carriers to be able to pass on one allele each for the recessive trait. The trait is present in equal numbers of males and females so is not X-linked.

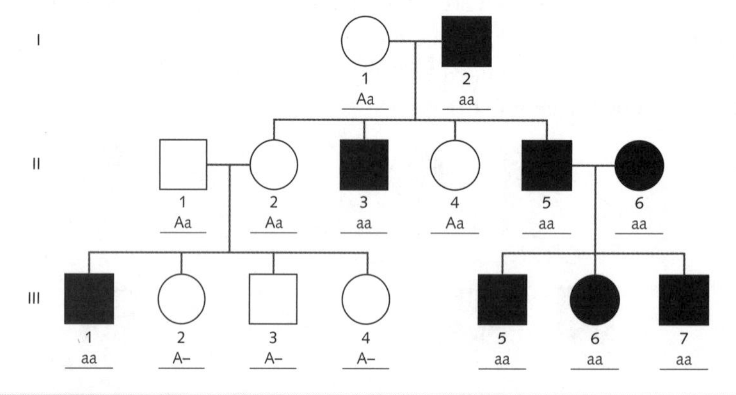

7.5 Predicted genetic outcomes for two autosomal genes p. 187

7.5.1 Genes that assort independently p. 187

1 **a** male: EECC female: eecc

b Male gametes all EC. Female gametes all ec.

c all the F_1 will have the genotype EeCc

d Parents EeCc × EeCc

	EC	Ec	eC	ec
EC	EECC	EECc	EeCC	EeCc
Ec	EECc	EEcc	EeCc	Eecc
eC	EeCC	EeCc	eeCC	eeCc
ec	EeCc	Eecc	eeCc	eecc

Genotypic ratio:

1EECC : 2 EeCC : 2EECc : 4EeCc : 1EEcc : 2Eecc : 1eeCC : 2eeCc : 1eecc

Phenotypic ratio:

9 long ears dark coat : 3 long ears light coat : 3 short ears dark coat : 1 short ears light coat

2 a R – round, r – wrinkled

P – purple flower, p – white flowers

RRPP, RrPP, RRPp, RrPp

b A test cross, cross the plant with a plant homozygous recessive for both traits

c

RRPP × rrpp

	rp
RP	RrPp

All F_1 would be expected to exhibit round seed and purple flowers

RrPP × rrpp

	rp
RP	RrPp
rP	rrPp

50% round seeds, purple flowers
50% wrinkled seeds, purple flowers or a phenotypic ratio of 1:1

RRPp × rrpp

	rp
RP	RrPp
Rp	Rrpp

50% round seeds purple flowers
50% round seed white flowers or a phenotypic ratio of 1:1

RrPp × rrpp

	rp	
RP	RrPp	25% round seeds, purple flowers
Rp	Rrpp	25% round seeds white flowers
rP	rrPp	25% wrinkled seeds purple flowers
rp	rrpp	25% wrinkled seeds white flowers

7.5.2 Linked genes p. 189

3 a $\frac{DR}{dr}$ or $\frac{Dr}{dR}$ b $\frac{DR}{dr} \times \frac{Dr}{dR}$

	DR	dr
Dr	$\frac{Dr}{DR}$	$\frac{Dr}{dr}$
dR	$\frac{dR}{DR}$	$\frac{dR}{dr}$

25% chance a child will be deaf and have retinal degeneration

7.6 Chapter review p. 190

7.6.1 Key terms p. 190

7.6.2 Practice test questions

p. 192

1 C
2 C
3 B
4 A
5 B
6 C
7 C
8 a i Non-patterned is recessive (1).
ii If the patterned trait were recessive, all the offspring would also be patterned as the patterned parent frogs would have to be homozygous recessive and could only pass on alleles for the patterned trait. As some of the offspring are non-patterned, the parents must be heterozygous, and since they show the trait it must be dominant. The non-patterned frogs have inherited an allele for the recessive traits from each of their heterozygous parents. (1)

b P – patterned

p – non-patterned (1)

Pp × Pp (1 mark for Punnet square)

	p	p
P	Pp	Pp
p	pp	pp

50% patterned
50% non-patterned (1 mark)

c As being patterned is the dominant trait, frogs only need one copy of the allele to show the trait. The frogs in cross A were both homozygous dominant for the patterned trait – they could only pass on alleles for the patterned trait, so all offspring were patterned (1). The frogs in cross B were both heterozygous for the dominant trait (Pp), so they could pass on either allele to their offspring producing the two different phenotypes (1).

9 a BBss and bbSS

b Parents: BBss × bbSS

Gametes: all Bs × all bS

F1: all BbSs

All the offspring in the F_1 generation would exhibit the phenotype black and solid coat.

c Parents: BbSs X BbSs

Gametes	¼ BS	¼ Bs	¼ bS	¼ bs
¼ BS	1/16 BBSS	1/16 BBSs	1/16 BbSS	1/16 BbSs
¼ Bs	1/16 BBSs	1/16 BBss	1/16 BbSs	1/16 Bbss
¼ bS	1/16 BbSS	1/16 BbSs	1/16 bbSS	1/16 bbSs
¼ bs	1/16 BbSs	1/16 Bbss	1/16 bbSs	1/16 bbss

Phenotypic ratio: 9 black, solid: 3 black, spotted: 3 brown solid: 1 brown, spotted

Chapter 8 Reproductive strategies

p. 196

Remember

p. 196

1 Mitosis occurs in somatic cells, producing two identical diploid daughter cells.
2 Meiosis occurs in the gonads or sex organs and produces four non-identical haploid daughter cells.
3 Mitosis has one cell division; meiosis has two cell divisions.
Mitosis produces diploid cells; meiosis produces haploid cells.
Mitosis produces two identical daughter cells; meiosis produces four non-identical daughter cells.
4 Diploid cells contain two complete sets of chromosomes, haploid cells contain one set of chromosomes.
5 The process is fertilisation. This process is important as it restores the chromosome number in the next generation. Two haploid gametes are combined to make a diploid zygote by fertilisation.

8.1 Asexual reproduction p. 197

8.1.1 Vegetative propagation p. 197

Answers for this section will vary depending on the question chosen. Below is a sample answer – check your work with your teacher.

1 Your research question should be specific and able to be answered by experimentation; for example, What concentration of fertiliser will produce the fastest growth in a vegetatively propagated carrot?

2 To determine the optimal fertiliser concentration to produce the fastest growth in a vegetatively propagated carrot.

3 a Fertilise concentration
  b Growth of the carrot plant, measured by height
  c Type of carrot, type of fertiliser, environmental temperature, available light, amount of water, volume of fertiliser, time, type of container the carrot is grown in

4 If the carrot is provided with a higher concentration of fertiliser, then it will grow to a taller height.

5 a If the carrots given the highest concentration of fertiliser grow taller than carrots given a lower concentration of fertiliser at the same time
  b The carrots given the highest concentration do not show greater growth in terms of height, compared to carrots given lower concentrations of fertiliser.

6 Sample method:
  1 Collect 12 carrot tops, making sure they are all similar size and health, with no shoots growing.
  2 Collect 12 petri dishes and label three × 0%, three × 10%, three × 20%, three × 30%.
  3 Place 20 mL of water into each of the 0% dishes.
  4 Place 20 mL of 10% concentration fertiliser into each of the 10% dishes.
  5 Place 20 mL of 20% concentration fertiliser into each of the 20% dishes.
  6 Place 20 mL of 30% concentration fertiliser into each of the 30% dishes.
  7 Place one carrot top in each of the dishes.
  8 Place the dishes in the same sunny location, so that they all receive the same amount of light.
  9 Top up each dish with 20 mL of the appropriate solution at the same time in the morning and afternoon of each day.
  10 Every 24 hours, record quantitative data about growth by measuring and recording the length of the longest shoot growing from each of the carrots using a ruler, record measurements in mm and straighten the shoots to measure their length correctly. Record the number of shoots and other observations about the shoot growth.
  11 At the end of the experiment, average the shoot growth for each fertiliser concentration.

Sources of error: inaccuracy in measuring shoot length, plants not receiving the same amount of light, solutions running dry, differential shading of some carrots.

Risks: ingesting fertiliser or getting in eyes

Risk management: wash hands after handling fertiliser or wear gloves, wear safety goggles.

8.1.2 Biological advantages and disadvantages of asexual reproduction p. 199

1 If table salt is added to yeast, then yeast's ability to respire anaerobically will be decreased.

2 a Presence or absence of table salt
  b Amount of CO_2 produced by anaerobic respiration in the yeast
  c Amount of yeast, amount of molasses provided, temperature of the water bath, time between measurements, same type of container.

3 Test tube A1 at 40 minutes

4 Personal error, misreading the scale on the measuring cylinder of misrecording the measurement.

5 Repeat the experiment and see if similar results are recorded.

6

Time	Test tube A mean (g/CO_2)	Test tube B mean (g/CO_2)
0	0.85	0.9
10	1.7	2.1
20	2.9	2.65
30	4.65	5.6
40	10.05	5.75
50	5.15	5.9
60	5.35	6.05
70	5.8	6.25
80	5.85	6.35
90	5.95	0.7
100	8.55	0.6
110	9.85	0
120	13.5	0

7 Sample graph

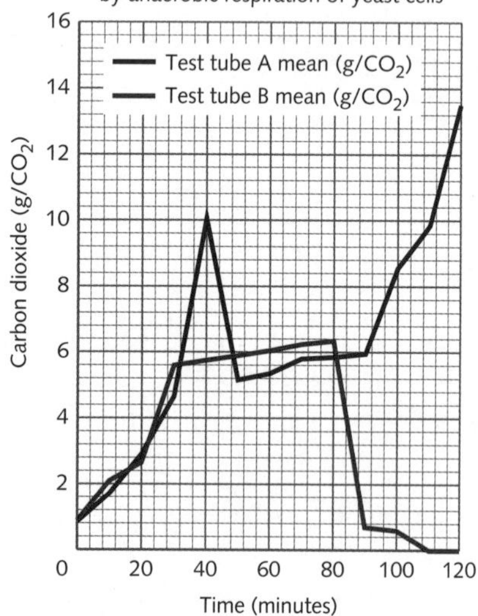

8 The hypothesis is supported, after salt was added to test tube B at the 90-minute mark, production of carbon dioxide stopped in test tube B, dropping down to 0 g/CO_2, 20 minutes after the addition, meaning there was no anaerobic respiration in the yeast. In test tube A, without salt added, the yeast continued to produce CO_2 by anaerobic respiration.

9 Answers may vary. Sample: Yeast cells can reproduce by budding every 90 minutes, the rate of CO_2 for the first 90 minutes shows a linear increase over time, this is because the size of the yeast population is not increasing. After 90 minutes, the yeast population would have doubled due to budding, causing the rate of CO_2 production to increase more rapidly due to their being twice as many yeast cells carrying out anaerobic respiration.

10 As yeast cells reproduce asexually, all new yeast cells will be identical, the population will have no genetic diversity. When the salt was added to the population it affected them all in the same way as they are genetically identical. All the yeast cells were killed by the salt, shown by no CO_2 production 20 minutes after the table salt was added.

11 Answers will vary. Repeat the experiment multiple times, investigate the effect of adding different amounts of salt.

12 Answers will vary. Sample conclusion: The hypothesis, 'if table salt is added to yeast, then yeast's ability to respire anaerobically will be decreased', was supported by the results of this experiment. Yeast stopped producing CO_2 20 minutes after the addition of table salt, yeast cells in similar conditions without the addition of salt continued to produce CO_2. This shows that the presence of salt stopped anaerobic respiration in the yeast cells as CO_2 is a product of this process.

8.2 Sexual reproduction p. 203

8.2.1 Fertilisation p. 203

Answers will vary but may include the following points.

- The body cells of organisms are diploid – they each carry two sets of chromosomes; one set came from the mother and the other set from the father.
- Gametes must only contain one set of chromosomes – be haploid, so that when they join, the resulting zygote will not have two sets of chromosomes again – be diploid.
- A new diploid organism is created by the process of fertilisation – the fusing of a sperm cell and an egg cell to create a zygote.

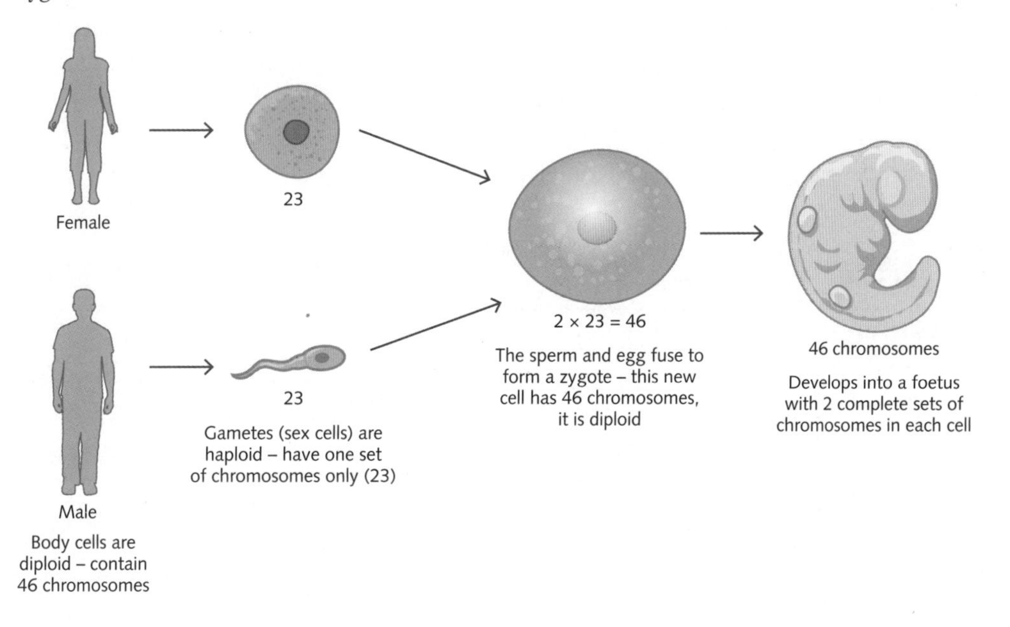

- Due to random assortment and crossing over in the process of meiosis that creates gametes, each gamete will contain a unique combination of alleles from the parent.
- Each fertilisation event involves a different sperm and egg cell, meaning each new zygote will be different depending on which combination of alleles carried by the sperm and egg cell.

8.2.2 Biological advantages and disadvantages of sexual reproduction p. 204

Sample table

	Asexual reproduction	Sexual reproduction
Advantages	• Do not have to spend time and energy finding a mate • Can produce many offspring quickly • Due to being genetically identical the offspring are well suited to the current environment	• All offspring genetically different, creates genetic variety. Increases survival rate of population if the environment changes • Smaller number of offspring, less competition
Disadvantages	• Offspring identical, little genetic diversity puts population at risk if the environment changes • Large number offspring increases competition for resources	• It takes time and energy to find a mate • Fewer offspring, cannot reproduce quickly to take advantage of good conditions • Not all members of population may get to mate • Can be dangerous/lead to death of one of the mating pair

8.3 Cloning p. 205

8.3.1 Cloning p. 205

1 Nuclear transfer. A somatic cell from the mammoth was used to provide a diploid nucleus. An egg cell taken from the mouse had its nucleus removed. The diploid nucleus from the mammoth is then fused with the enucleated mouse egg cell. This creates a mouse egg cell that contains genetic material from the mammoth, and is now diploid.

2 Mitosis, to produce identical diploid cells, that would create an embryo.

3 Answers will vary. Goes against the cultural and religious beliefs of some, see it as playing god and creating life artificially. Welfare of the mice that are used to obtain eggs, this is an invasive procedure. Success rate of the cloning process, welfare of any clones produced by this process. What would the cloned animal be used for; would it be released into the wild or kept in the lab for experimentation. Who owns the clone, how will gained knowledge and information be shared? How will cloning of animals be regulated? What should scientists be able to clone and for what reason?

4 Answers will vary depending on the ethical issue chosen, check your work with your teacher.

8.4 Chapter review p. 208

8.4.1 Key terms p. 208

Term	Definition
Asexual reproduction	Offspring are produced from only one parent
Binary fission	Division of a cell into two, without mitosis; the process by which a prokaryotic cell divides to form two daughter cells
Biodiversity	The full range of different biological entities in a particular area or region; it can be described at various levels, including the range of different species, genetic diversity, or the diversity of ecosystems present in a large area
Bioethics	Ethics relating to biological research
Blastocyst	A hollow ball of cells formed in early embryonic development
Budding	Development of a new organism from an outgrowth of the parent organism
Clone	A genetically identical copy of an organism
Cloning	Process of producing a cell, tissue or organism genetically identical to its parent
Fragmentation	Division of a parent organism into pieces with each piece then giving rise to a complete organism
Grafting	When part of one plant is artificially attached to another plant
Mitosis	A type of cell division that results in two daughter cells each having the same number and kind of chromosomes as the parent nucleus
Monoculture	Agricultural practice of growing a single crop or plant species over a wide area for many consecutive years
Mutation	A change in the nucleotide sequence of the genome of an organism
Parthenogenesis	Form of asexual reproduction where growth and development into a complete organism occurs from an unfertilised egg
Spore	A reproductive body able to withstand harsh environmental conditions
Vegetative propagation	Cloning of plants

8.4.2 Practice test questions p. 209

1 B
2 C
3 C
4 D
5 A
6 C
7 D

8 **a** 15

b Population A (1) – the size of the population stays remains constant, sexual offspring does not produce large numbers of offspring or rapid population growth (1) OR Population B (1) – the introduction of the bacterium at generation 15, resulted in the death of some of the population but some were able to survive due to having genetic diversity because of sexual reproduction (1).

c Population C (1). After the introduction of the bacterium at generation 15, all members of this population died due to a lack of genetic diversity, they were all susceptible to the bacterium (1).

d Population A (1) was not exposed to the bacterium, this is the only population that did not show a decrease in numbers after generation 15 when the bacterium was introduced to the populations (1).

e To show how population numbers would change over the generations when a bacterium was not introduced; to enable a comparison (1) to the effect of the bacterium on the other populations B and C (1).

Chapter 9 Adaptations and biodiversity p. 211

Remember p. 211

1 Abiotic refers to non-living factors and biotic refers to living factors in the environment.
2 An ecosystem is the interaction of abiotic and biotic components.
3 A community is all the living species in a particular area at a particular time.
4 Two of the following:
Mutualism – both organisms benefit
Commensalism – one organism benefits and the other is unaffected
Parasitism – one organism benefits and the other is harmed.
5 How energy is passed between organisms in a community
6 A food web is a collection of food chains.
7 A characteristic that gives an organism an advantage to survive and reproduce in its environment.

9.1 Genetic diversity p. 212

9.1.1 Genetic diversity p. 212

1 Trees, fish, algae
2 Water, rocks, air, soil
3 Fish or tree
4 Freshwater pond/billabong
5 Yes, it has abiotic and biotic factors and interactions between them, it would be self-sustaining.

9.1.2 Genetic diversity within a species or population p. 213

1 Answers may vary, for example: How similar are the genomes of extinct and living lion populations?
2 Indian lions are all genetically similar, meaning that they all share similar phenotypic traits. If there was a change in their environment, there is less chance there will be members of the population with characteristics that would enable them to survive, all the lions would become extinct.
3 Case study – the scientists are looking specifically at the genomes of lions to document the amount of variation in the population.
4 Cave lions and modern lions shared a common ancestor 500 000 years ago. The ancestors of Indian lions diverged from African lions about 70 000 years ago and migrated to India. Indian lion populations lack genetic diversity.
5 The sequence of bases in the genome of 14 extinct lion species and the sequence of the genome of 6 modern lions from India and Africa. Incidence of cranial defects, low sperm count and testosterone levels, as well as smaller manes of lions in Indian populations.

6 The researchers came from University of Copenhagen and Barcelona Institute of Science and Technology, from institutes in two different countries, the article also says that others were also involved.

7 A scientific idea, evidence from the sequencing of the genomes of lions has been used to support ideas.

9.2 Adaptations p. 215

9.2.1 Adaptations of animals: king penguins p. 215

Types of adaptation		
Structural	**Physiological**	**Behavioural**
• Thick layer of blubber • Thick layer of inner feathers • Outside feather layer is waterproof • Webbed feet adapted for swimming • Body shape – small surface area : volume ratio • Streamlined body shape for swimming • Black feathers to absorb heat • Small flippers • Beak for catching prey	• Countercurrent heat exchange • Salt glands to excrete excess salt • Brooding patches in males that are well supplied with blood vessels • Can lower heart rate when diving	• Huddling • Standing on heels of flippers • Monogamous • Male incubates the egg

9.2.2 Adaptations of plants p. 217

Plant adaptation	Type of adaptation	Environmental factor adaptation is suited for	How adaptation assists survival
Sunken stomata in hairy leaves	Structural	Heat/dry air	Traps humid air next to stomata opening and reduces water loss by evaporation
Bulbs exposed to cold temperatures to induce flowering	Physiological	Temperature	Ensures the plant flowers in the right season to enable pollination
Woody fruit	Structural	Arid environment	Conserves water compared to the creation of fleshy fruit, protects seed while it develops
Changes in turgor pressure in the leaf in response to touch causing the leaf to close or fold	Physiological/behavioural	Insects, grazers	Prevents insects or other grazing animals from eating leaves of the plant
Deciduous trees dropping leaves in winter	Physiological	Temperature/available light	Saves energy for the plant, plant does not have to maintain leaf tissue when there is limited sunlight for photosynthesis
Thick waxy cuticles covering leaves	Structural	Dry/hot environment Pathogens Insects/grazers	Prevents water loss from leaves. Prevents entry of pathogens. Deters insects/grazers from eating leaves
Active secretion of salt	Physiological	Salinity of soil or water	Maintain correct osmotic concentrations inside cells and tissues.

9.3 Survival through interdependencies between species p. 218

9.3.1 Interdependencies between species p. 218

	Relationship between two species	Name the relationship between the two species	What happens to each species within this relationship, whether it benefits, is harmed, or neither
1 a	Bee and flower	Mutualism	Both species benefit
b	Lichen on tree	Commensalism	Lichen benefits, tree unharmed
c	Oxpecker and ox	Mutualism	Both species benefit
d	Mosquito and person	Parasitism	Mosquito benefits, person harmed
e	Cheetah and antelope	Predator/Prey	Cheetah benefits, antelope harmed
f	Giraffe and deer	Competition	Both species harmed
g	Mould and bread	Amensalism	Bacteria harmed; mould unaffected

2 Answers may vary, for example:

- Do the roots of the mistletoe grow into the tree's tissues?
- Can a mistletoe survive independently of the tree?
- Does mistletoe growth on the tree cause signs of ill-health in the tree?
- Does mistletoe cause the death of trees they grow on?
- Can a mistletoe carry out photosynthesis?

9.3.2 More complex interdependencies: competition p. 220

1 Answers may vary, for example: To test the competitive exclusion principle by recreating Gause's *Paramecium* experiment with two different paramecium species.

2 Answers may vary, for example: if two different strains of paramecium are competing for the same resources, then only one of the paramecium strains will be successful and survive; the other strain will die.

1 **Results**

Sample

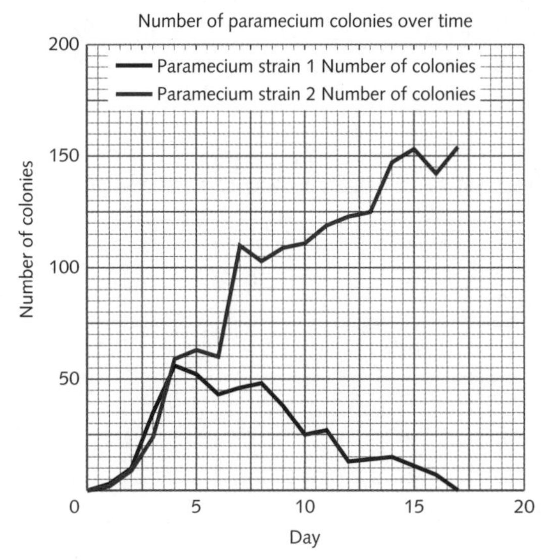

2 The data supports the hypothesis, after both strains reached over 50 colonies, on day 4, the number of colonies for strain 1 started to decrease over time until there were no members of this strain left by day 17, while strain 2 continued to survive and increase the number of its colonies.

3 The data supports the competitive exclusive principle, once the colonies of the two strains reached the level of above 50 colonies each, strain 2 was more successful in competing for resources and was the only strain to survive, preventing strain 1 from surviving in the same area.

4 a No control, to show how both populations would have increased without the population of the other strain There was only one trial of the experiment, only one set of results.

b No control – you cannot be sure that it was the presence of strain 2 that caused the death of all the strain 1 colonies or some other factor.

Only one trial, you cannot be sure that this will always be the observed result, the experiment should be repeated multiple times to check that the results are repeatable.

5

Risk	How to minimise
Paramecia are harmless to humans but there may be pathogenic organisms in water. Possible release of organisms to natural environment	Wear disposable gloves and safety glasses. Wash hands thoroughly. Dispose of agar plates and *Paramecium* colonies correctly – destroy by autoclaving.

9.3.3 More complex interdependencies: predation p. 223

Sample graph

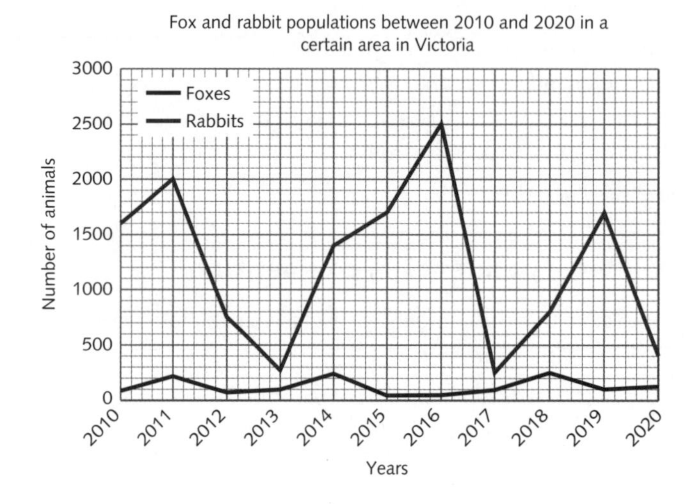

1. There is always a greater number of prey than predators. The number of prey always increases before the number of predators increase. The graph follows the following pattern: the number of rabbits increases, the number of foxes increases because there is more prey, the number of rabbits reduces because there are more foxes, the number of foxes reduces because there is less rabbits.
2. Predator/prey relationship
3. The foxes keep the rabbit population relatively stable, stops the population of rabbits becoming too large, reducing competition for resources within the rabbit population.
4. The fox number may have declined more quickly after 2016.

   The rabbit numbers would not have increased into 2017 and 2018.

   The fox numbers would have not increased as much between 2018 and 2019.

9.4 Keystone species p. 225

9.4.1 Keystone species in a mangrove ecosystem p. 225

1. Mangroves
2. They provide food and shelter for many members of the ecosystem: detritus from the leaves adds nutrients to the mud to helping algae to grow, fiddler crabs mature within the roots and when adult feed on the leaves, shellfish live on the roots of the mangroves. These organisms provide food for larger organisms such as fish and birds. Therefore, the mangroves support the entire ecosystem by helping producers such as algae to grow and providing a habitat for small and immature organisms that are necessary to support the food web of the ecosystem.
3. There would be less nutrients in the mud, leading to less algae growth, crabs would have nowhere to mature, shellfish/molluscs would have nowhere to live, this would lead to a collapse of the food web as higher order consumers would have no energy source and the ecosystem would collapse.

4 Sample food web

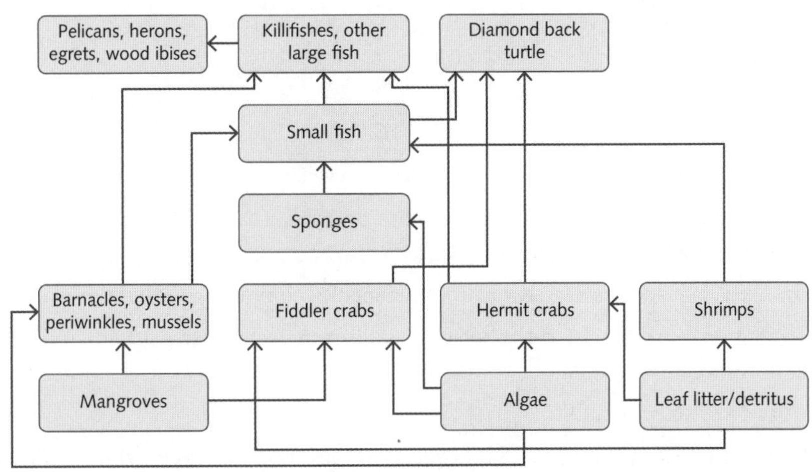

9.5 Adaptations and interdependencies in Australian ecosystems p. 227

1 Hazard reduction burns are started during the day and are hot fires, indigenous fires are cool fires, burn at a lower temperature, and are started in the morning or late afternoon.

2 Bush turkeys and insects – predator/prey. Hawks and small animals – predator/prey. Brolgas and insects – predator/prey. Wallabies/birds and lice –parasitism. Wombats/wallabies/native birds and grass – herbivore/plant. Bilby/bettong/brush tailed possum and feral cats/foxes – predator/prey.

3 The fires flushed out insects and small animals that provided food for the bush turkeys and hawks and dead insects for the brolgas. Fire encourages new grass growth providing more food for herbivores. The ash from the fires is used by animals for an ash bath that helps remove parasites.

4 Answers may vary. For example:

Currently hazard reduction control burning is used annually in Australia to reduce some of the dense grass and scattered trees that cover 23% of mainland Australia to reduce the severity of annual bushfires. In the past indigenous Australians also used controlled burning to manage the ecosystem. The two types of control burning differ in the temperature of the fires, indigenous fires were lit in the early or late in the day and burn with a lower temperature than the hazard reduction burn, that are lit in the middle of the day. The timing of the indigenous fires was based on an understanding of each environment, they reduced the frequency of bushfires, created diverse habitats, and enhanced the survival of native plants and animals. Several native species became extinct in Maralinga when the indigenous fire and hunting practices were stopped. These species returned to the area when the indigenous fire practices were resumed in the area, providing evidence for the effectiveness of their management methods.

9.6 Chapter review p. 229

9.6.1 Key terms p. 229

Correct terms:

1 Biotic
2 Omnivore
3 Parasite, host
4 Prey, physiological
5 Mutual
6 Commensal
7 Diversity
8 Keystone
9 Distribution
10 Abundance

9.6.2 Practice test questions p. 230

1 A
2 B
3 B or C
4 A
5 C
6 C
7 Row 1: Coexistence any correct example, e.g. a starfish and a sea anemone living on the same rock platform

Row 2: Symbiotic relationship where one species benefits, and the other is unharmed, e.g. cattle egret that eats insects that are disturbed when a zebra grazes on grass

Row 3: Competition, e.g. one species of tree and another species competing for light or water in a forest

Row 4: Symbiotic relationship whereby both species benefit from each other, e.g. bees and flowers. Bees get nectar from flowers and flowers get pollinated.

Row 5: Predator/prey, e.g. fox and rabbit

Row 6: Species that has a disproportionately large effect on the communities in which it occurs, regardless of its abundance. Its presence is needed to maintain balance in the ecosystem, e.g. the mangrove tree in a mangrove swamp

Row 7: Symbiotic relationship whereby one species is inhibited or destroyed, while the other remains unaffected, e.g. bread mould and bacteria

8 a Commensalism

b The plant species benefits (1), as its seeds can germinate after passing through the elephant's gut and will grow within a nice pile of dung that will provide nutrients to the growing seedling (1).

Chapter 10 Investigating a biological issue p. 232

Remember p. 232

1 Independent variable – the variable that is changed in the experiment

Dependent variable – the variable that is measured as results, caused by the change in the independent variable

Controlled variables – variables that are kept constant in an experiment

2 An experimental condition set up to compare to the results obtained when changing the independent variable
3 When changing one variable causes the change in the other variable
4 If all variables are controlled except for the independent variable and the results measure what they were supposed to measure
5 Repeatable is when the same experimenter repeats an experiment with similar results; reproducible is when another experimenter repeats the experiment with similar results.

10.1 Beginning an investigation p. 233

10.1.1 Identifying a bioethical issue p. 233

Scenario 1

1 Answers may vary, for example: How to decide which patient is put on the ventilator
2 Answers will vary. The issue concerns what happens when there is a shortage of life saving equipment to treat all patients presenting with illness. This issue affects the survival of the patients, their friends and relatives, the wellbeing of the medical staff that must make the decision. The problem has been created by a pandemic that can be fatal to some individuals if they are not placed on ventilators.

The issue needs to be resolved to provide guidelines that should be followed when determining which patient should get access to limited medical technology. Who should determine if patients are provided with lifesaving technology or not? What are the implications for medical staff for denying access to life-saving technology?

Scenario 2 Sample answers

3 Whether to terminate the pregnancy or proceed and give birth to a disabled child

4 If it is identified that a foetus has a severe genetic disorder, should the mother be pressured to abort the foetus? Some people have religious beliefs that consider abortion wrong. What are the rights of the unborn child in terms of quality of life? Parents should have the right to choose if they terminate the pregnancy or continue. The issue can impact the mental health of the parents, the quality of life of the child and costs to the health care system. Differing opinions in society about what constitutes life and what is more important, protecting the foetus or the future life quality of the child and its parents. What are treatment options and potential costs for this disorder if the child is born. What is the process for providing parents with all necessary information to make an informed choice?

Scenario 3

5 Should euthanasia be made legal in the United Kingdom?

6 Euthanasia is the practice of intentionally ending a life to relieve pain and suffering. Different countries have different euthanasia laws. Many people suffering from painful/untreatable diseases travel to other countries to access euthanasia options. Many religious and cultural group are opposed to the practice of euthanasia. People want to be able to choose if they continue living when they have very poor quality of life due to ill health without having to spend money travelling to another country. If euthanasia is legalised what regulations and processes need to be put in place to protect individuals?

10.1.2 Referencing and acknowledging sources p. 235

1 Answers will vary use below as a guide, check with your teacher.

Source of information	APA citation
Book about genetics	Willett, E., 2006. *Genetics Demystified*. New York: McGraw-Hill.
Book about embryology	Walbot, V. and Holder, N., 1987. *Developmental Biology*. New York: Random House.
Peer reviewed journal article	Mancuso, P., Chen, C., Kaminski, R. *et al*. CRISPR based editing of SIV proviral DNA in ART treated non-human primates. *Nat Commun* 11, 6065 (2020).
Popular science magazine	Michael, M. (2020). North Sea island survived prehistoric tsunami. *New Scientist, 248*(3311), 12–21.
Website on bioethics	Adelaide Centre for Bioethics and Culture. (n.d.). Adelaide Centre for Bioethics and Culture. Retrieved 2020 December 8 from http://www.bioethics.org.au/index.html
Television show on current affairs	Harvey, G. (Producer). (2020). Exposing a wet market: could the next global pandemic start in these cages? [Television series episode in] *60 minutes,* Melbourne, Australia: channel 9.
Film set in another country	Cuaron, A. (Producer), Cuaron, A. (Director). (2018). Roma [Motion picture]. Mexico: Esperanto Filmoj.

10.2 Evaluating evidence p. 236

10.2.1 The distinction between opinion, anecdote and evidence p. 236

1 a Evidence: includes data
b Opinion: this is a claim with no evidence to support it
c Anecdote: this is based on personal observation/experience
d Anecdote: this is based on personal observation/memory
e Opinion: this is a claim made by somebody with no evidence to support it
f Opinion: this is a claim made by somebody with no evidence to support it
g Anecdote: this claim is based on personal observation
h Anecdote: relating personal observation of a friend
i Evidence: information/data from DNA was used to make this claim
j Evidence: this claim is supported by data

10.2.2 The distinction between scientific and non-scientific ideas

p. 237

1 a Scientific idea: supported by experimental data

b Non-scientific: could not be supported by a controlled experiment/reliant of opinions and subjective judgements

c Non-scientific: based on opinion/anecdote and not data from a controlled experiment

d Non-scientific: no objective criteria for what makes a better egg, not supported by data from a controlled experiment, prejudiced

e Scientific: based on data from a controlled experiment

f Scientific: claim supported by data that has been systematically collected

10.2.3 Quality of evidence

p. 237

Part A: Assessing data qualitatively

p. 237

Scenario 1

1 a Primary data

b Could contain personal errors, misreading of measuring equipment, measurement should be taken more than once. Could contain systematic errors from the measurement method used with the metre ruler, not using the same measurement method for all individuals.

Scenario 2

2 a Primary data

b It should be free from errors, appropriate measuring equipment was used and calibrated before measuring each weight. The data seems to follow a normal distribution with no results outside this pattern.

Part B: Organising, analysing, and evaluating data

p. 237

3 & 4

Table 1 Height of bean seedlings grown in different soil pH

	Plant height (cm)									Avg % height change
	Week 1			Week 2			Week 3			
Soil pH	Plant 1	Plant 2	Avg	Plant 1	Plant 2	Avg	Plant 1	Plant 2	Avg	
6.2	3.4	3.2	3.3	6.9	4.7	5.8	13.5	12.1	12.8	287
6.4	2.8	3.8	3.3	7.2	6.8	7.0	16.2	17.2	16.7	406
6.6	4.1	2.1	3.1	5.1	16.0	10.55	17.8	14.8	16.3	425
6.8	3.8	3.6	3.7	4.4	5.4	4.9	9.3	10.1	9.7	162
7.0	4.2	3.6	3.9	6.0	6.2	6.1	10.2	11.2	10.7	174
7.2	1.9	2.9	2.4	4.2	3.9	4.05	6.7	7.4	7.05	193

5 Plant 2 week2 pH 6.6, personal error in using measuring equipment or recording data

6 Repeat each height measurement more than once, calculate % height increase, ensure they record the starting height, use more accurate measuring equipment, measure in mm, repeat the experiment multiple times and average the results at each pH.

7 See table above

8 See table above

9 Sample graph

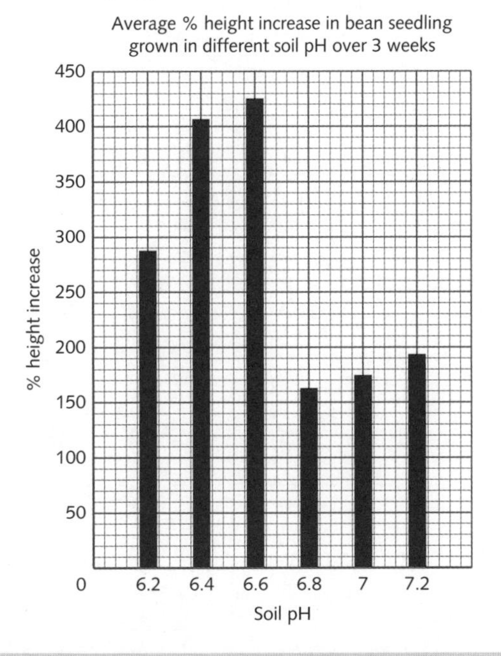

10.3 Evaluating a bioethical issue p. 241

1 Answers may vary, for example: Are participants provided with all information about possible outcomes before consenting to participate. Do the benefits from using challenge trials outweigh the potential risks to participants. Is this a large enough sample size to guarantee the vaccine is safe and reliable. Is the vaccine safe to use on the population outside the 20–29 age group? Is the vaccine safe to give to individuals with compromised health? Is it ethical to deliberately expose the control group with no vaccination to a potentially fatal virus? Impact on the mental health of the participants, anxiety of not knowing if they are in the vaccine group or not, if they will get ill or not? Potential side effects of the untested vaccine on humans. How will all participants be safely quarantined from vulnerable members of the population?

2, 3 & 4 Answers will vary depending on bioethical issue chosen, check with your teacher.

10.4 Preparing a report p. 241

Answers will vary. Check with your teacher. This is a suggested marking guide.

Criteria	4	3	2	1	0
Use of clear, coherent expression. Maximum number of words is 500; minimum number of words is 100	Excellent expression of information in a coherent and concise manner. Close to maximum word limit	Well expressed information in a coherent and mostly concise manner. Close to maximum word limit	Fair expression of information, but lacking coherence and/or conciseness. Closer to minimum word limit	Poorly expressed information or lack of conciseness and coherency. Too far below minimum word limit	Not shown
Use of correct scientific terminology	Precise and appropriate use of scientific terminology, demonstrating an excellent understanding of concepts	Appropriate use of scientific terminology, demonstrating a good understanding of concepts	Some appropriate terminology correctly used	Little or incorrect use of scientific terminology	Not shown
Social, economic, legal or political factors	Detailed explanation of multiple social, economic, legal, or political factors relevant to the issue	Explanation of multiple social, economic, legal, or political factors relevant to the issue	Explanation of more than one social, economic, legal or political factors relevant to the issue	One relevant social, economic, legal or political factors, little or no explanation given	Not shown

(continued)

Criteria	4	3	2	1	0
Provide arguments for your point of view	Opinions regarding the issue are presented impartially with relevant evidence is used to clearly support arguments	Opinions regarding the issue are presented impartially with some relevant evidence used to support arguments	Little evidence is used to support arguments for given point of view	A point of view is detailed with no evidence to support arguments	Not shown
References	Multiple reliable references have been used and recorded using correct APA referencing	Numerous references used and correctly listed. Some sources may be questionable, such as the Wikipedia and blogs	More than one reference used and partially listed. General interest sites heavily used	One reference used and correctly listed or partially listed, general use websites used	Not shown

10.5 Chapter review p. 245

10.5.1 Key terms p. 245

Order of correct terms: justice, public opinion, integrity, respect, beneficence, quoting, paraphrasing, anecdote, plagiarism, opinion, summary, dogma

10.5.2 Practice test questions p. 246

1 (1 mark for one ethical and one social point)

	Some possible issues /implications
Ethical	What happens to male/unsuitable embryos, is it right to destroy embryos based on gender? Conflict with religious cultural beliefs- playing God, interfering in natural process Should healthy embryos be chosen/discarded only on the basis of gender – gender discrimination? The rights of the embryo are not being considered
Social	Lead to an imbalance of genders in some cultures where one gender is preferred May lead to one gender being valued above the other gender May lead to pressure on families to make use of this technology

2 Each implication must be different – cannot use the same implication twice (1 mark for each point)

	Economic implication	Biological implication
Bt cotton	Higher cost for farmers who must buy new seeds each year Farmers may save money by not having to pay for pesticide	Pickers have developed a skin condition from Bt cotton Less pesticide used, reducing harmful chemicals in environment
Golden rice	Increased health of population may lead to more involvement in work force Farmers can save money by collecting seeds to replant crop the next year	Higher levels of vitamin A in rice, improving health of population Shown to be safe for humans to eat

9780170452632

Notes